Moreton Morrell Site

This book is to be returned on or before the
last date stamped

An Introduction to Plant Structure and Development

Plant Anatomy for the Twenty-First Century

An Introduction to Plant Structure and Development incorporates basic knowledge of plant anatomy with contemporary information and ideas about the development of structure and form. This textbook has been designed for undergraduates with a background in introductory botany or biology and basic knowledge of plant systematics and evolution.

Descriptions of plant structure and development are integrated with discussions of the results from some of the most significant recent research on plant development. Topics include the integrative significance of plasmodesmata and the concept of the symplast, the concept of multicellularity, the role of the cytoskeleton in development, signal transduction, and the genetic control of development. Brief sections on evolution and function are also included. The book is profusely illustrated with line drawings and photographs, closely integrated with the text, which will enhance students' understanding, as will the comprehensive glossary. Extensive bibliographies provide the basis for in-depth study in major areas by students and researchers.

CHARLES B. BECK is Professor Emeritus of Botany at the University of Michigan.

An Introduction to Plant Structure and Development

Plant Anatomy for the Twenty-First Century

Charles B. Beck
University of Michigan

CAMBRIDGE
UNIVERSITY PRESS

CAMBRIDGE UNIVERSITY PRESS
Cambridge, New York, Melbourne, Madrid, Cape Town, Singapore, São Paulo

Cambridge University Press
The Edinburgh Building, Cambridge, CB2 2RU, UK

Published in the United States of America by Cambridge University Press, New York

www.cambridge.org
Information on this title: www.cambridge.org/9780521837408

First published 2005

Printed in the United Kingdom at the University Press, Cambridge

A catalogue record for this publication is available from the British Library

Library of Congress Cataloguing in Publication data
Beck, Charles B.
An introduction to plant structure and development : plant anatomy for the 21st
century / Charles B. Beck.
    p.   cm.
Includes bibliographical references and index.
ISBN 0 521 83740 5 (hardback)
1. Plant anatomy.   2. Plants – Development.   I. Title.
QK641.B38   2005
571.3′2 – dc22

ISBN-13 978-0-521-83740-8 hardback
ISBN-10 0-521-83740-5 hardback

To
My wife, Janice,
and our daughters, Ann and Sara
for their love, encouragement,
and enduring support,

and

to my students,

David Benzing
Robert Chau
Crispin Devadas
Margaret Knaus
G. Kadambari Kumari
Rudolf Schmid
William Stein
Garland Upchurch
Richard White
David Wight

who are a continuing inspiration and
from whom I have learned much

It is important that students bring
a certain ragamuffin barefoot
irreverence to their studies; they
are not here to worship what is
known but to question it.

Jacob Bronowski
The Ascent of Man (1975)

Contents

Preface	*page* xiii
Acknowledgements	xv

Chapter 1	Problems of adaptation to a terrestrial environment	1
Perspective	1
Structural adaptations	3
Preview of subsequent chapters	5
References	6
Further reading	7

Chapter 2	An overview of plant structure and development	8
Perspective: origin of multicellularity	8
Some aspects of the shoot system of the vascular plant	10
Apical meristems	13
Primary tissue regions of the stem and root	16
Vascular bundle types	23
Secondary growth	25
Cells of the xylem	28
Cells of the phloem	33
References	35
Further reading	36

Chapter 3	The protoplast of the eukaryotic cell	38
Perspective	38
Morphology of the protoplast	39
Movement of organelles in the protoplast	50
Ergastic substances	53
References	54
Further reading	56

Chapter 4	Structure and development of the cell wall	57
Perspective	57
Structure and composition of the cell wall	58
Growth of the cell wall	64
Cell wall development	67
Plasmodesmata	71
References	75
Further reading	79

Chapter 5 | Meristems of the shoot and their role in plant growth and development 81

Perspective 81
Apical meristems 81
Formation of leaf primordia 88
Transitional tissue regions 90
Intercalary meristems 93
The primary peripheral thickening meristem of
  monocotyledons 94
Cell growth and development 96
The effect of hormones on cell growth and development 96
Genetics and cell growth 97
The cytoskeleton and cell growth 98
Cell shaping by microtubules 99
References 100
Further reading 103

Chapter 6 | Morphology and development of the primary vascular system of the stem 105

Perspective 105
Cellular composition and patterns of development of
  primary xylem 106
Cellular composition and patterns of development of
  primary phloem 109
Differentiation of primary vascular tissues 110
The role of auxin in the development of the primary
  vascular system 116
References 118
Further reading 119

Chapter 7 | Sympodial systems and patterns of nodal anatomy 120

Perspective: leaf traces 120
Nodal structure of pteridophytes 120
Sympodial systems of seed plants 122
Leaf trace lacunae 131
The cauline vs. foliar nature of vascular bundles
  in the eustele 132
Phyllotaxy 133
References 136
Further reading 137

Chapter 8 | The epidermis 138

Perspective 138
Epidermis of the shoot 138
Epidermis of the root 146

Stomata 147
The mechanism of movement in guard cells 149
Development of stomata 149
References 152
Further reading 152

Chapter 9 | The origin of secondary tissue systems and the effect of their formation on the primary body in seed plants 154

Perspective: role of the vascular cambium 154
The effect of secondary growth on the primary body 157
The effect of secondary growth on leaf and branch traces 159
References 162
Further reading 162

Chapter 10 | The vascular cambium: structure and function 163

Perspective 163
Structure of the vascular cambium 163
General overview of cambial activity 166
Plant hormones and cambial activity 170
Submicroscopic structure of cambial initials 171
The onset of dormancy and the reactivation of dormant cambium 172
Cytokinesis in fusiform initials 174
The problem of differential growth of cambial cells and immature cambial derivatives 175
References 176
Further reading 179

Chapter 11 | Secondary xylem 180

Perspective 180
Overview of the structure of secondary xylem 180
Secondary xylem of gymnosperms 183
Resin ducts 189
Secondary xylem of dicotyledons 190
Differentiation of tracheary elements 195
Patterns of distribution of xylary elements and rays 202
Tyloses 207
Mechanism of water transport 209
References 210
Further reading 213

Chapter 12 | The phloem 215

Perspective: evolution of the phloem 215
Gross structure and development of the phloem 216

The nature and development of the cell wall
   of sieve elements   221
Role of the cytoskeleton in wall development   224
The nature and development of the protoplast
   of sieve elements   225
Nature and function of P-protein   227
Distinctive features of the phloem of gymnosperms   229
The nature and function of companion cells and
   Strasburger cells   230
The mechanism of transport in the phloem   234
References   235
Further reading   238

Chapter 13 | Periderm, rhytidome, and the nature
             of bark   240

Perspective   240
Periderm: structure and development   240
Formation of rhytidome   243
Lenticels   245
The outer protective layer of monocotyledons   246
References   247
Further reading   247

Chapter 14 | Unusual features of structure and
             development in stems and roots   248

Perspective   248
Primary peripheral thickening meristem   248
Secondary growth in monocotyledons   249
Anomalous stem and root structure   250
References   256
Further reading   256

Chapter 15 | Secretion in plants   257

Perspective   257
Substances secreted by plants   257
Mechanisms of secretion   258
Internal secretory structures   259
External secretory structures   263
References   269
Further reading   271

Chapter 16 | The root   272

Perspective: evolution of the root   272
Gross morphology   274
Contractile roots and other highly specialized root systems   275
Apical meristems   276

The quiescent center and its role in development 280
Primary tissues and tissue regions 281
Lateral transport of water and minerals in the young root 289
Development of primary tissues 291
Auxin and tissue patterning 295
Lateral root development 295
Adventitious roots 297
Secondary growth 299
The root cap: its function and role in gravitropism 301
Mycorrhizae 302
Nitrogen fixation in root nodules 305
Root–stem transition 306
References 308
Further reading 313

Chapter 17 | The leaf — 316

Perspective: evolution of the leaf 316
Basic leaf structure 317
Leaf development 326
Role of the cytoskeleton in leaf development 331
Variations in leaf form, structure, and arrangement 333
Structure in relation to function 336
Photosynthesis and assimilate loading 336
Leaf structure of C_3 and C_4 plants 338
Supporting structures in leaves 339
Transfusion tissue in conifers 340
Leaf abscission 341
References 343
Further reading 347

Chapter 18 | Reproduction and the origin of the sporophyte — 350

Perspective: the plant life cycle 350
Reproduction in gymnosperms 351
Reproduction in angiosperms 355
Development of the seed in angiosperms 362
Fruit development and the role of fruits in seed dispersal 367
Seed germination and development of the seedling 369
Floral morphogenesis 370
Pollen–pistil interactions 374
Self-incompatibility 377
The role of the cytoskeleton in pollen tube growth 378
References 379
Further reading 383

Glossary 387
Index 421

Preface

Since my introduction to plant anatomy by William Strickland at the University of Richmond and my interaction with Arthur Eames and Harlan Banks at Cornell University during graduate study, I have been entranced by the elegant beauty of plant structure. At the University of Michigan I taught both paleobotany and plant anatomy for many years, and served as committee chair for graduate students, some of whom studied fossil plants and others of whom worked on the structure and development of extant taxa. During the past several decades during which the introduction of new techniques of study at the subcellular and molecular levels has resulted in a resurgence of research throughout the world, my interest in the development of plant structure has grown steadily.

Many books on plant structure, some highly technical, have appeared since the publication of the seminal textbooks of Katherine Esau during the 1950s and 1960s, but no single book that, in my opinion, incorporates both the basic knowledge of plant anatomy and contemporary information and ideas about the development of structure and form that could be used as an effective introductory textbook. Consequently, I have tried to meet the challenge of preparing such a book. In each chapter I have presented what I consider to be the fundamental knowledge essential for an understanding of basic plant structure and development and have integrated with this the results of some of the most significant recent research on plant development. Whereas emphasis throughout the book is on structure and development, I have also included sections on evolution and function where it seemed essential and appropriate to do so. The application of cellular and molecular biological approaches and techniques in the study of plant development has revolutionized the field. Understanding of the integrative significance of plasmodesmata and the concept of the symplast have led to an appreciation and widespread acceptance of the organismal theory of plant multicellularity which in turn has influenced research on plant development. Exciting and significant areas of research such as the role of the cytoskeleton in development, signal transduction, genetic control of development, among others have greatly advanced our understanding. I have not treated the very important subject of the genetic control of development in any depth because it requires a much deeper knowledge of genetics than the undergraduate for which this book is written is likely to have attained. I have, however, included references to important genetic studies in the bibliographies of several chapters. Other subjects may not be as fully covered as some teachers and researchers would desire, but they are very likely to find pertinent references to literature on those subjects in the extensive bibliographies to which they can direct their students who have the necessary backgrounds.

Diverging from the approach in many textbooks, I have included in this book tentative conclusions that are essentially still hypotheses, and discussions of research that is controversial, often providing opposing viewpoints. I believe that, in addition to providing well-established information on a subject, a textbook should also provide the student with an understanding of the nature of ongoing scientific research.

In order to make this book more readable for the undergraduate, I have omitted most literature citations in sections of the text in which the basic, widely accepted knowledge in the field is presented, but have included some references of historical importance in the references at the end of each chapter. On the other hand, when presenting new information, ideas, and conclusions that are not yet widely accepted, I have cited in the text and included in the references the sources of this information. Thus, students as well as researchers who wish to consult the original papers may find the reference sections useful.

My objective has been to prepare a new plant anatomy textbook for a new century, incorporating the best research in the most active and significant areas with the widely accepted common knowledge that provides the foundation of the field. Only you the readers can decide if I have succeeded.

Charles B. Beck
Ann Arbor, 2004

Acknowledgements

One's knowledge comes from many sources. Not least are the research and writings of many predecessors in the field. Men and women such as Nägeli, De Bary, Strasburger, Haberlandt, Van Tieghem, Solereder, Jeffrey, Eames, Bailey, Metcalfe and Chalk, Esau and countless others have provided the foundation upon which current-day researchers are building. To these, whom sometimes we forget, we owe a debt of gratitude. I acknowledge a profound debt to my college and university teachers, William Strickland and Robert Smart who introduced me to plant structure in the first place, Arthur Eames and Harlan Banks who widened my horizons and reinforced in my understanding the fundamentals of plant anatomy, and to John Walton who encouraged me to take risks and taught me how to write. I acknowledge, as well, the significant contributions to my knowledge of the many researchers who are currently active in the field.

Direct assistance during the preparation of this book has come from many sources. I feel particularly indebted to colleagues who have critically read chapters in manuscript and made important suggestions for change and improvement. These are Professor William Stein of the State University of New York, Binghamton who read several chapters, Professor Shirley Tucker of the University of California at Santa Barbara, Professor Nancy Dengler of the University of Toronto, and Professor Darleen DeMason of the University of California, Riverside. Other colleagues have provided information on special topics. Professor Peter Ray of Stanford University provided information on the functional significance of the optical qualities of epidermal cells in leaves, Professor Judy Jernstedt of the University of California at Davis provided information on contractile roots, Professor Larry Nooden of the University of Michigan was a source of important information on several aspects of plant physiology, Professor Robert Fogel of the University of Michigan provided information on mycorrhizae, and Professor Edward Voss and Dr. Christiane Anderson of the University of Michigan were valuable sources of information on plant taxonomy. To all of these I express my sincere appreciation.

Professor Philip Gingerich, Director of the Museum of Paleontology at the University of Michigan, made available to me the resources and services of the Museum. Preparation of this book would not have been possible without this assistance, and to Phil I express my sincere gratitude. The illustrations are nearly as important as the text in a plant anatomy book. In this book all original line drawings were finished by Bonnie Miljour, artist *par excellence* of the Museum of Paleontology. Ms. Miljour also grouped and placed all illustrations in electronic files. The importance to this project of her great expertise cannot be overemphasized. Thank you, Bonnie, for the beauty of your work and for your very important contribution to this book.

Two members of the Museums office staff, Cindy Stauch and Meegan Novara, were also of inestimable assistance in many ways. I express my sincere appreciation to them.

The original photographs were taken primarily by two University of Michigan photographers, Louis Martonyi, now deceased, who was photographer for the Department of Biology during the 1980s, and David Bay, current photographer for the Department of Ecology and Evolutionary Biology. Thank you, David, for your excellent work. A few photographs were taken by the author in the facilities of the Microscopy and Image-analysis Laboratory of the University of Michigan Medical School. This was made possible by the kindness of the Laboratory Manager, Chris Edwards and with the technical assistance of Shelley Almburg, to both of whom I express my appreciation. I express my sincere gratitude to colleagues who provided photographs: Professor Pedro J. Casero of Universidad de Extremadura, Badajoz, Spain; Professor P. Dayanandan of Madras Christian College, India; Dr. Elisabeth de Faÿ of Université Henri Poincaré, Nancy, France; Professor Nancy Dengler of the University of Toronto, Canada; Dr. Katrin Ehlers of the Justus-Liebig-Universität, Giessen, Germany; Dr. Irene Lichtscheidl of Universität Wien, Austria; Dr. E. Panteris of the University of Athens, Greece; and Dr. Koichi Uehara of Chiba University, Japan. Professor P. Maheshwari of the University of New Delhi sent me many excellent slides during his lifetime, many of which have been photographed for use in this book. I have also used many illustrations from published sources, and I express my gratitude to the individuals, commercial publishers, university presses, and professional societies that have granted permission for the use of their copyrighted materials.

Although every effort has been made to secure necessary permissions to reproduce copyrighted material in this work, it has proved impossible in two cases to trace the copyright holders. The copyright holder of the original illustration from Lehninger (1961), which I have used as my Fig. 3.7a, is Dr. A. E. Vatter. The copyright holder of the original illustrations from Eames and MacDaniels (1925), which I have used as my Figs. 13.3, 13.4, 14.5, 16.15c, and 16.21 is David Eames. Appropriate acknowledgements will be included in any reprinting or in any subsequent edition of this book if the copyright holders are located.

In order to understand copyright law, which varies somewhat throughout the world, I called on my friend, Professor John Reed of the University of Michigan Law School, for advice. He directed me to Professor Molley Van Houweling, a specialist in copyright law, who gave me valuable information. I am grateful to these colleagues.

Without the resources of the University of Michigan Library this book could not have been written, and the excellent assistance of the reference librarians in the Shapiro Science Library is acknowledged with gratitude.

To the editors of Cambridge University Press I express my gratitude, and acknowledge their important roles in the preparation and production of this book. In particular, I wish to thank Ward Cooper, commissioning editor, life sciences; Clare Georgy, assistant editor, life sciences; Sue Tuck and Joseph Bottrill, production editors; and Anna Hodson, copyeditor.

Finally, I must acknowledge friends and family who through their interest and support have made a contribution to this project greater than they can imagine. Every morning for many years past and during the several years of this project I have joined friends for coffee. We call the group the Coffee Klatch. Members have included Robert Lowry, cytogeneticist and microscopist, Erich Steiner, plant geneticist, Norman Kemp, animal morphologist, Ralph Loomis, teacher of English literature, Harry Douthit, microbiologist, James Cather, developmental biologist, Michael Wynne, phycologist, Barbara Brown, university bus driver, and me. Conversation has ranged over a broad spectrum of interests and activities, but almost never on "the book." Interaction with this wonderful group of university colleagues has provided me with a daily means of relaxation and a time to forget about cells, tissues, microtubules, and actin microfilaments. On the other hand, I have felt the subtle but genuine support for me and this project by members of the group. So I express my sincere appreciation to my friends of the Coffee Klatch.

One person, however, stands out above all others in importance. My wife, Janice, has supported me with remarkable patience and understanding during work on this book. She has added to her busy schedule many activities for which I would ordinarily have taken responsibility and has been a constant source of support and encouragement. Thank you, Sweetheart, for being the wonderful person you are, and for your absolutely essential contributions to this project.

Problems of adaptation to a terrestrial environment

Perspective

Land plants, plants that complete their life cycle entirely in a terrestrial environment, are represented largely by bryophytes and vascular plants. In all taxa except seed plants, however, at least a thin film of water is required for fertilization; and even in two primitive groups of seed plants, the cycads and *Ginkgo*, fertilization is by free-swimming spermatozoids released into a liquid medium in the archegonial chamber. A few angiosperms, although terrestrial in origin, have reverted to an aquatic existence.

Vascular plants are by far the dominant groups on the Earth comprising over 255 000 species in contrast to about 22 000 species of bryophytes and approximately 20 000 species of algae. The first vascular plants appear in the fossil record in the late Silurian, about 420 million years ago, but their green algal ancestors are thought to have appeared nearly 400 million years earlier! Shared features comprise the major evidence that vascular plants, possibly also bryophytes, evolved from green algae: both synthesize chlorophylls a and b, both store true starch in plastids; both have motile cells with whiplash flagella, and both (but only a few green algae) are characterized by phragmoplast and cell plate formation following mitosis. A green alga, with these and other significant characteristics, that may provide a model of an algal ancestor of vascular plants is *Coleochaete*, a member of the Charophyceae. Features in addition to those listed above that lead to this conclusion are the development in *Coleochaete* of a zygote in which cell division begins while embedded in the gametophyte thallus, the presence of sporopollenin in the wall of the zygote, and the presence of lignin in the gametophyte. It is widely believed that the Embryophyta (bryophytes and vascular plants) and the Charophyceae evolved from a common aquatic ancestor. Detailed presentations of the evidence for this viewpoint and the nature of the presumed common ancestor can be found in major works by Graham (1993) and Niklas (1997, 2000).

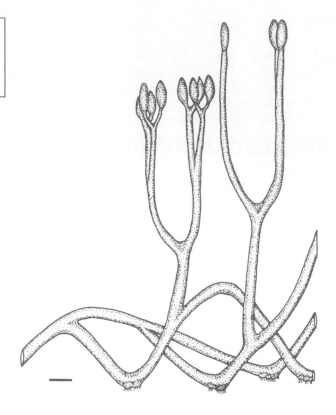

Figure 1.1 Reconstruction of *Aglaophyton major*. Bar = 10 nm. From Edwards (1986). Used by permission of The Linnaean Society of London.

The first, indisputable vascular plants were characterized by a conducting system containing xylem and phloem, a waxy cuticle, epidermal stomata, and a reproductive system that produced **trilete spores** (spores with a triradiate scar resulting from their development in spherical tetrads) and probably containing sporopollenin in the walls. Such plants appear first in the late Silurian, but *Aglaophyton major* (see Edwards, 1986) from the Lower Devonian, which has morphologic and structural features of both some bryophytes and primitive vascular plants, provides perhaps the best available model of a vascular plant precursor. *Aglaophyton* was a small plant, probably no more than 180 mm (about 7 inches) high, composed of dichotomous, upright axes that branched from rhizomes on the surface of the substrate (Fig. 1.1). The epidermis of all axes was covered by a cuticle and contained stomata. Some upright axes were terminated in pairs of sporangia, containing small spores of one size only. Edwards suggested that the plant probably formed extensive mats, consisting largely of vegetative axes, but produced fertile axes, bearing sporangia, "at irregular intervals." The rhizomes were probably vegetative axes that formed clusters of **rhizoids** (absorbing structures) where some axes arched over and contacted the substrate. One of the most interesting structural features of the axes of *Aglaophyton* was the central conducting strand. Although appearing superficially as a vascular strand consisting of xylem and phloem, and described that way by earlier workers, Edwards was unable to detect characteristic structural features of tracheary elements (that is, cells with secondary wall material deposited in the form of rings, helices, or a reticulum) or of sieve elements. Instead, he found three regions of cells, an inner column of thin-walled cells surrounded by thick-walled cells, the walls of both of which were dark in color. These were enclosed by an outer zone of thin-walled cells with light-colored walls. He concluded that the two inner regions of cells with dark cell walls were probably analogous to tracheids but most similar to the **hydroids** (water-conducting cells) of some mosses and that the outermost cells with light-colored walls were analogous to sieve elements and very similar to the **leptoids** (photosynthate-conducting cells) of mosses. *Aglaophyton* was, therefore, a non-vascular plant sporophyte in which the sporophyte was the dominant phase in a system of pteridophytic (free-sporing) reproduction. In gross morphology and branching pattern, and the presence of an epidermis covered by a cuticle and containing stomata, it was very similar to primitive vascular plants that lived during Upper Silurian and Lower Devonian times. In its water- and photosynthate-conducting cells closely resembling, respectively, hydroids and leptoids, as well as in its small size and free-sporing reproduction, it closely resembled mosses. It is reasonable, therefore, to hypothesize that vascular plants evolved from this or plants of similar morphology and anatomy. (For detailed information on the earliest vascular plants, see Taylor and Taylor (1993) and Stewart and Rothwell (1993).)

Structural adaptations

During the past 350–400 million years many structural and physiological changes occurred as vascular plants evolved. Evolution on land posed many problems for plants such as *Aglaophyton* and its descendants not shared with their marine algal ancestors. In an aquatic environment, conditions are equable, and problems of water loss, support, absorption of water and minerals, and transport of water and minerals, photosynthate, and hormones, are either minimal or non-existent. This is true also, in large part, for very small plants such as most mosses. For example, the absence of efficient water-conducting cells in mosses apparently does not pose a problem for them since it is well known that in many taxa water and minerals are absorbed through the external surfaces of the sporophytes and gametophytes. This is not unlike the situation in aquatic plants in which water and minerals are absorbed by all parts of the plant directly through the epidermis, which lacks a cuticle. Consequently, there is no need for a highly efficient system of transport of water and minerals. Likewise, with few exceptions, the transport of hormones and photosynthate is also not a problem since these substances are produced in all cells. On land, however, solar radiation, wind, and temperature extremes result in a much harsher environment. As *Aglaophyton* and its descendants evolved on land, structural features evolved as adaptations to both this harsher environment and to their increase in size.

Adaptations reducing water loss were the evolution of a three-dimensional, rod-like plant body which decreased the surface/volume ratio, and an epidermis covered by a waxy cuticle largely impermeable to the passage of water. Although the evolution of a rod-like form was advantageous in restricting the surface area from which water could be lost, an optimal surface area in relation to volume was required through which transpiration as well as gaseous exchange could occur. The evolution of stomata in the epidermis allowed the exchange of O_2 and CO_2, essential in respiration and photosynthesis, and by their ability to control the size of pores through which water vapor diffused, stomata also contributed to a restriction of water loss from the plant. Adequate surface area was also required, however, through which the plant could receive signals from the environment – signals such as light, temperature, or the presence of other organisms such as pathogens or symbionts as well as chemical signals from the atmosphere or from other organisms. We now know that chemicals produced by plants living today are also released through the surface and may elicit responses from other organisms such as moths and hummingbirds that function as pollinators. The response of plants to environmental signals, referred to as **signal transduction**, is a new and active area of research in plant biology.

Protection of spores and gametes, so very important in a terrestrial environment, was accomplished through the evolution of sporangia and gametangia enclosed in sterile jackets of cells. The spores themselves became encased in walls containing **sporopollenin,** a substance which restricts water loss and is highly resistant to decay.

Absorption of water and minerals from the soil was facilitated by the evolution of rhizoids and roots, the latter often containing symbiotic fungi forming mycorrhizae which, as we shall see in detail later, enhanced their absorptive function. Roots, in particular, also served to anchor the plant in its substrate and to prevent its displacement by wind and flowing water.

The effective transport of water and minerals as well as hormones, photosynthate, and other substances became increasingly important with increase in size of the descendants of *Aglaophyton* or other vascular plant precursors. This was accomplished by the evolution of complex vascular tissues containing tracheids and vessel members in the xylem and sieve cells and sieve tube members in the phloem, conducting cells especially adapted structurally for the transport of these materials. Associated with the evolution of cellular transport systems, specialized mechanisms evolved which facilitated the efficient transport of water and minerals from the roots to and out of the leaves of tall trees. Concurrently, mechanisms for the translocation of photosynthate and other assimilates throughout the plant evolved.

The problem of support of the plant body also became increasingly severe with increase in size and was solved by structural adaptations. In plants, or parts of plants, consisting largely of living tissues, support was provided by their enclosure by an epidermis as well as by turgor pressure within the cells. Ultimately, some of the functions of the epidermis were taken over by **periderm** (a major component of bark) consisting largely of non-living cells, the walls of which are impregnated with **suberin** which restricted the passage of water through them. Support was also provided by the production of tissues consisting largely of non-living, longitudinally elongate cells with thick, lignified, cellulosic walls. The major supporting tissue in large plants is the xylem, consisting in pteridophytes and their ancestors as well as in gymnosperms primarily of tracheids, and in angiosperms of fibers and vessel members. **Lignin** in the cell walls increased the tensile strength of elongate cells comprising the xylem, thus endowing vascular plants with both strength and flexibility, so very important in conditions of high wind velocity.

The above-ground parts of the plant bodies of primitive vascular plants consisted primarily of radially symmetrical branching axes, all of which were photosynthetic. With the evolution of larger vascular plants consisting of stems and branches covered with bark, an adaptation that facilitated the process of photosynthesis was necessary. This was accomplished by the evolution of leaves which increased the surface/volume ratio of photosynthetic tissues in the plant. Structural adaptations in the leaves, such as the orientation of thin-walled elongate cells at right angles to the upper surface which channeled light at relatively high intensity into the leaves, and the complex system of intercellular channels which provided extensive wet surface area for the absorption of CO_2 facilitated efficient photosynthesis. For further information on adaptations by early plants to a terrestrial environment, and the evolution of plant body plans, please see Niklas (2000).

Preview of subsequent chapters

As we proceed through this book we shall encounter progressively detailed information on the structure and development as well as some aspects of evolution and function of the descendants of primitive plants such as *Aglaophyton*. We shall consider many members of the Embryophyta, including the Lycophyta (lycopods and their relatives), Sphenophyta (sphenophytes), and Pterophyta (ferns), but the major emphasis will be on the seed plants (gymnosperms and angiosperms). In order to provide an orientation to all who have had little or no training in plant anatomy, and to introduce some important concepts, the following chapter will be an overview of plant structure and development. If you have had a good course in introductory botany or biology, you may wish to proceed to later chapters. Chapters 3 and 4 present, respectively, basic information on the cell protoplast and the cell wall. The cell protoplast is usually covered in some detail in introductory courses, but the cell wall is often neglected. Consequently this book provides a fairly comprehensive discussion of its structure and development.

Chapter 5 presents very important information on apical meristems of the shoot, apical regions from which other cells and tissues in the shoot system are derived, and which provide to vascular plants their distinctive characteristic of indeterminate growth. Appendages such as leaves and lateral branches also are ultimately derived from the apical meristems. In Chapters 6, 7, and 8 we consider the structure and development of the various tissues and tissue systems that result from the activity of apical meristems. These chapters include, in sequence, discussions of the primary vascular tissues (xylem and phloem) that are embedded in the parenchyma of the pith and cortex; the architecture of the primary vascular system and its relationship to the arrangement of leaves; and the epidermis, the single layer of tissue that bounds all of these other primary tissues and tissue systems, and which forms an outer protective and supportive layer of the plant prior to the development of secondary tissues.

The second part of the book, Chapters 9 through 12, consists of discussions of the vascular cambium, a lateral meristem, the activity of which results in the formation of secondary vascular tissues, and the effects of their formation on the tissues and tissue systems produced by the apical meristems early in the development of the plant body. It also includes detailed presentations of the structure, development, and to a lesser extent evolution and function of the secondary xylem and the secondary phloem.

Chapters 13 through 18 deal with secretory structures and functions; anomalous stem and root structure; the outer protective tissues and tissue regions of stems that produce secondary tissues (the periderm and rhytidome) that comprise the bark; the structure, development, and function of leaves as well as a brief discussion of their evolution; a presentation of the structure, development, and function of roots, with some comments on their evolution; and finally, a chapter on reproduction which includes some basic life cycles, discussions of the structure and morphology of flowers, the structure and development of fruits and seeds, and some aspects of the ecology of reproduction in angiosperms.

The author hopes that you will enjoy this book and that by the end of your course in plant anatomy you will be as enthusiastic about this exciting field as he is.

REFERENCES

Edwards, D. S. 1986. *Aglaophyton major*, a non-vascular land-plant from the Devonian Rhynie Chert. *Bot. J. Linn. Soc.* **93**: 173–204.

Graham, L. E. 1993. *Origin of Land Plants.* New York: John Wiley and Sons.

Niklas, K. J. 1997. *The Evolutionary Biology of Plants.* Chicago, IL: University of Chicago Press.

2000. The evolution of plant body plans: a biomechanical perspective. *Ann. Bot.* **85**: 411–438.

Stewart, W. N. and G. W. Rothwell. 1993. *Palaeobotany and the Evolution of Plants.* Cambridge, UK: Cambridge University Press, Chapters 7 and 9.

Taylor, T. N. and E. L. Taylor. 1993. *The Biology and Evolution of Fossil Plants.* Englewood Cliffs, NJ: Prentice-Hall, Chapter 6.

FURTHER READING

Banks, H. P. 1975. The oldest vascular plants: a note of caution. *Rev. Palaeobot. Palynol.* **20**: 13–25.

Delwiche, C. F., Graham, L. E., and N. Thomson. 1989. Lignin-like compounds and sporopollenin in *Coleochaete*, an algal model for land plant ancestry. *Science* **245**: 399–401.

Edwards, D. 1970. Fertile Rhyniophytina from the Lower Devonian of Britain. *Palaeontology* **13**: 451–461.

Edwards, D. and J. Feehan. 1980. Records of *Cooksonia*-type sporangia from the Wenlock strata in Ireland. *Nature* **287**: 41–42.

Graham, L. E. 1984. *Coleochaete* and the origin of land plants. *Am. J. Bot.* **71**: 603–608.

1985. The origin of the life cycle of land plants. *Am. Scientist* **73**: 178–186.

Reuzeau, C. and R. F. Pont-Lezica. 1995. Comparing plant and animal extracellular matrix–cytoskeleton connections: are they alike? *Protoplasma* **186**: 113–121.

Shute, C. H. and D. Edwards. 1989. A new rhyniopsid with novel sporangium organization from the Lower Devonian of South Wales. *Bot. J. Linn. Soc.* **100**: 111–137.

Taylor, T. N. 1988. The origin of land plants: some answers, more questions. *Taxon* **37**: 805–833.

Chapter 2

An overview of plant structure and development

Perspective: origin of multicellularity

Since early in the study of plants botanists have been interested in the structure, function, development, and evolution of cells, tissues, and organs. Because some green plants are very small and unicellular, but others are large and multicellular, the origin of multicellularity in plants also has been of great interest to botanists. Among the green algae from which higher plants are thought to have evolved, some colonial taxa such as *Pandorina*, *Volvox*, and relatives consist of aggregations of motile cells that individually appear identical to apparently related unicellular forms (Fig. 2.1). Consequently, it was concluded early in the history of botany, and widely accepted, that multicellular plants evolved by the aggregation of unicellular organisms. This viewpoint led to the establishment of the **cell theory of multicellularity** in plants which proposes that cells are the building blocks of multicellular plants (Fig. 2.2). As early as 1867, however, Hoffmeister proposed that cells are simply subdivisions within an organism. This viewpoint, supported and expanded upon in 1906 by Lester Sharp at Cornell University, has been elucidated and clarified more recently by Hagemann (1982), Kaplan (1992), and Wojtaszek (2000) among others. These workers conclude on the basis of abundant evidence that a unicellular alga and a large vascular plant are organisms that differ primarily in size and in the degree to which they have been subdivided by cells (Fig. 2.2). This **organismal theory of multicellularity** has gained many adherents within the past several decades (see Kaplan, 1992), and is of great importance because of ways in which it has influenced the thinking of botanists about the processes of development.

A primary and convincing basis for the organismal theory is the nature and result of cell division in plants. Following mitosis and cell plate formation, the protoplasts of the two resulting daughter cells maintain continuity through highly specialized cytoplasmic strands called **plasmodesmata** (Fig. 2.3) (see Chapter 4 for a detailed discussion of plasmodesmata). Thus, although the plant is blocked off in regions called cells, the plasmodesmata provide for an interconnected

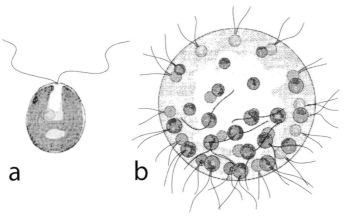

Figure 2.1 Unicellular and colonial body plans among green algae. (a) *Chlamydomonas* sp. (b) *Pandorina morum*. From Niklas (2000). Used by permission of Oxford University Press.

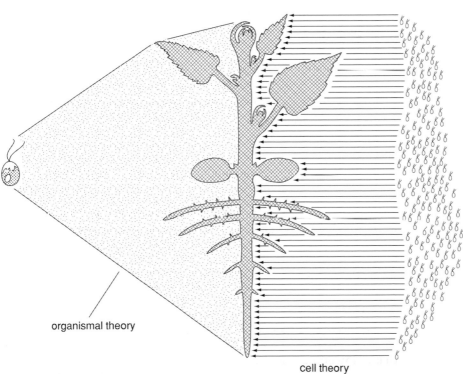

organismal theory

cell theory

Figure 2.2 Diagram showing the relationship between a unicellular and a multicellular organism according to the organismal theory of multicellularity and the cell theory. From Kaplan (1992). Used by permission of the University of Chicago Press. © 1992 The University of Chicago. All rights reserved.

system of protoplasts called the **symplast**. The plasmodesmata function as passageways for communication between living cells, that is, for the transmission between cells of molecules of varying size including even large molecules such as proteins and nucleic acids (see, e.g., Lucas *et al.*, 1993; Kragler *et al.*, 1998; Ehlers and Kollmann, 2001). It has become clear in recent years that this communication

Figure 2.3 Diagrams of primary pit fields traversed by plasmodesmata in primary cell walls. The plasmodesmata connect the protoplasts of adjacent parenchyma cells.

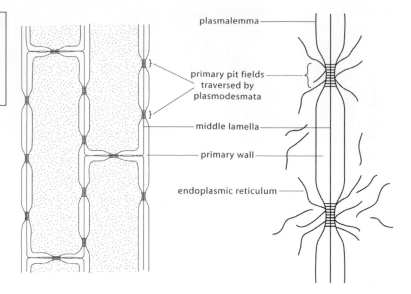

between cells has a profound influence in plant development (see, e.g., Verbeke, 1992). These and other workers believe that plasmodesmata may exert "controlling influence" on cell differentiation, tissue formation, organogenesis, and specialized physiological functions.

For more detailed discussions of evidence in support of the organismal theory of multicellularity in plants and the significance of this theory in understanding plant development, see Kaplan and Hagemann (1991), Niklas and Kaplan (1991), Kaplan (1992), and Wojtaszek (2000).

Let us now look at the vascular plant body in general terms and obtain an overview of its structure and development. In subsequent chapters we shall consider in more detail many aspects of plant structure and development as well as of function and evolution.

Some aspects of the shoot system of the vascular plant

The vascular plant consists of an aerial shoot system and, typically, a subterranean root system (Fig. 2.4). The **shoot system** consists of a main axis that bears lateral branches. Leaves may be borne on both the main axis and lateral branches in plants that complete their life cycle in one growing season, but in those that live for several to many years, leaves are usually found only on the parts of lateral branches that have developed in the past year or the last several years. For example, in deciduous plants, (e.g., many woody angiosperms) leaves develop only on the most distal segments of the laterals, i.e., the parts produced during the most recent growing season (Fig. 2.5) and will fall from the plant at the end of the same growing season. In most conifers and other evergreens, however, leaves may stay on the plant

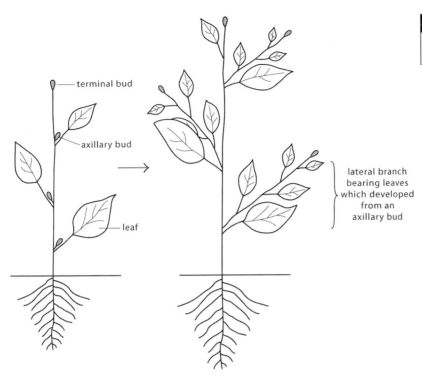

Figure 2.4 Diagrams of two stages in the development of an annual vascular plant.

terminal bud

axillary bud

leaf

lateral branch bearing leaves which developed from an axillary bud

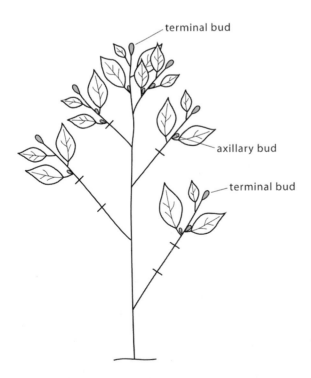

terminal bud

axillary bud

terminal bud

Figure 2.5 Diagram of the shoot of a deciduous, woody perennial with leaves and axillary buds, on terminal twigs only, produced during the current growing season. Each bar (at right angles to a lateral branch) indicates the previous site of a terminal bud.

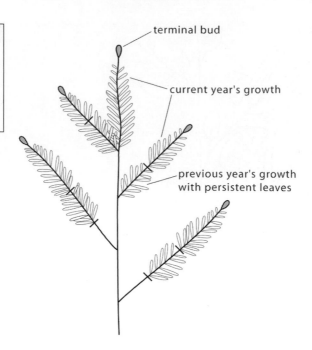

Figure 2.6 Diagrams of an "evergreen" conifer shoot with leaves on the current and previous year's growth. The bars at right angles to the lateral branches indicate previous positions of terminal buds.

terminal bud

current year's growth

previous year's growth with persistent leaves

for several years, but the youngest, i.e., the most recently developed leaves, are borne on twigs produced in the current or most recent past growing season (Fig. 2.6).

A unique feature of vascular plants is the presence of buds which occur at the tips of the main and lateral branches and, in gymnosperms and angiosperms, commonly (but not always in gymnosperms) in the **axils of leaves** (the angle between leaves and the stem to which they are attached) (Fig. 2.5). A **bud** consists of an apical meristem enclothed by protective bud scales (modified leaves) (Fig. 2.7). A **meristem** is a localized region of cells that is characterized by active cell division, the ultimate result of which is the addition of new cells, tissues, and organs (such as leaves) to the plant body. It is, thus, the structural feature that imparts to plants their distinctive serial mode of development – so different from that of animals.

In contrast to animals whose development can be characterized as determinate, plant development, by virtue of the presence of apical meristems, is **indeterminate**. That is, plants have the ability to add new cells and tissues to the plant body during each growing season as long as the plant lives. This makes possible the enormous size of very old trees such as the redwoods of California or the very large deciduous trees of virgin hardwood forests of northeastern USA. The development of some parts of plants such as leaves and components of flowers, however, is **determinate** in that their form, and to some extent, size are genetically predetermined. Once these plant parts have completed their development, they do not grow further no matter how long they remain on the plant as functional entities.

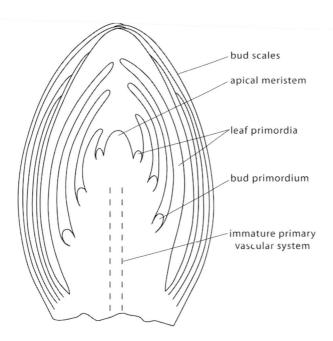

bud scales

apical meristem

leaf primordia

bud primordium

immature primary
vascular system

Figure 2.7 Diagram of a terminal bud as viewed in a median longitudinal section. Space is left between foliar components for clarity.

Apical meristems

We shall now look at the gross internal structure of buds and, in particular, consider the activity of apical meristems and some aspects of the development of cells and tissues resulting from this activity. The apical meristem of the shoot is a dome-shaped structure that comprises the apical-most region of the main and lateral axes. In the root, the apical meristem is covered by the root cap. In this chapter we shall consider only the apical meristem of the shoot, deferring discussion of the apical meristem of the root to Chapter 16. In the shoot, small protuberances, the **leaf primordia**, from which leaves will develop, form around the base on the periphery of the apical meristem (Fig. 2.8). **Bud primordia** develop in the axils of older leaf primordia. Because of its permanently apical position, cells, tissues, and structures such as leaf and bud primordia close to the apical meristem are younger, and consequently, less mature than those farther away. It is apparent, therefore, that the direction in which development proceeds is **acropetal**, that is, from the more mature, proximal part of the shoot toward the less mature, distal region near the apical meristem. In other words, cells, tissues, and lateral appendages (e.g., leaves) become progressively more mature in the direction of the apical meristem which is actively producing new cells that are added on to those produced earlier. If this concept of acropetal development is difficult for you, consider this analogy. In building a wall a bricklayer starts at the base, adding layer after layer of bricks until the wall is completed. The mortar between the first two layers of bricks at the base of the wall sets (i.e., hardens or "differentiates") first.

Figure 2.8 Diagram of a shoot apex as viewed in a median longitudinal section.

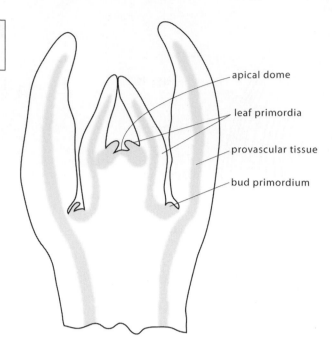

apical dome

leaf primordia

provascular tissue

bud primordium

That between subsequent layers sets later and later until finally that between the last two layers at the top of the walls has set. In other words, the mortar in the wall can be thought of as having differentiated sequentially from bottom to top, that is, acropetally.

Cells produced by the apical meristem initially resemble those of the apical meristem in size and form. Plant tissues are composed of many different types of cells, and the cells produced by apical meristems must grow and **differentiate** during development, ultimately attaining the specific characteristics of particular, mature, functional cells (Fig. 2.9). If we consider the development through time of a single cell derivative of an apical meristem, we shall see that from the time it is first produced it begins to grow, to some degree in diameter, but primarily in length, with its long axis paralleling the long axis of the stem, root, or branch in which it resides. In other words, the derivative grows longitudinally. (Although this is a common pattern of cell growth in plants, some cells grow almost equally in diameter and length.) It is generally agreed that the motive force for cell growth is turgor pressure within the cell. In order for the protoplast to grow, the cell wall must also grow, concurrently, by increasing in surface area. Growth of the wall, which includes synthesis of the chemical and structural components of which it is composed, is an area of great interest and active research, and will be considered in detail later. Changes that occur in the cell protoplast during growth and differentiation include the fusion of small vacuoles to form larger vacuoles, and increase in the volume of cytoplasm within which develop various cell organelles such as mitochondria, plastids, Golgi bodies, endoplasmic reticulum, microtubules, and microfilaments. Eventually, the total complement of originally small vacuoles will have fused into a

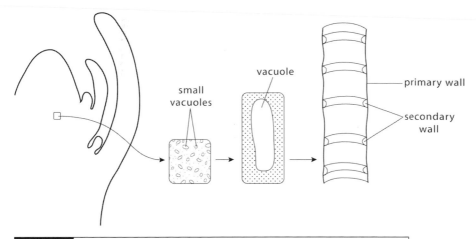

Figure 2.9 Diagrammatic representation of growth and differentiation of a type of water-conducting cell produced by the apical meristem. Note that growth consists primarily of increase in length. Very small vacuoles fuse to form a large central vacuole resulting in the peripheral location of the cytoplasm. Prior to death of the protoplast, secondary wall is deposited in the form of rings.

single central vacuole forcing the cytoplasm with its organelles and nucleus into a peripheral position (Fig. 2.9). Other aspects of dramatic changes in cells during differentiation are exemplified by conducting cells in the phloem and xylem. In the phloem the organelles and even the nucleus of a developing conducting cell become modified or may even degenerate prior to achievement of functional maturity by the cell. Concurrently, the end walls become highly perforate, which facilitates the transport of photosynthate, hormones, and other chemical substances from cell to cell. The protoplasts in conducting cells of the xylem actually die, but prior to death the cell walls become highly modified. The walls increase in thickness, often differentially, resulting in structural features that strengthen the cells, and the walls may become pitted. They may also become heavily impregnated with lignin, a compound that increases their tensile strength. The absence of a protoplast and the modifications of the cell walls facilitate the transport of water and dissolved mineral nutrients. In later chapters we shall consider in detail the structural changes of conducting cells in the phloem and xylem and the relationship of these to the processes of transport.

A major activity of apical meristems is the production of **primary tissues** in stems and roots, resulting in an increase in the length and, to a lesser extent, an increase in thickness (diameter) of these axes. Among the primary tissues produced are the primary phloem and primary xylem which, in seed plants, comprise vascular bundles, and the parenchyma, collenchyma, and sclerenchyma of which the pith and cortex are composed (Fig. 2.10). Meristematic activity in shoot systems also results in the production of leaf primordia that define **nodes**, sites of attachments of leaves, and **internodes**,

Figure 2.10 (a) Primary tissue regions of a dicotyledon stem shown in transverse section. (b) Primary tissue regions of a seed plant root. Note the ribbed primary xylem column, and the separate bundles of primary phloem.

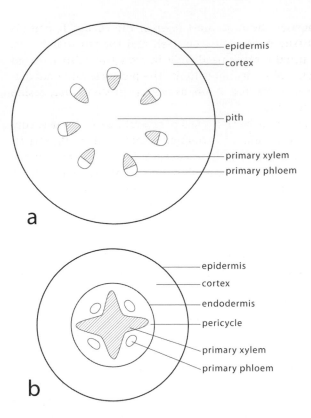

segments of the stem between nodes. Close to the apical meristem, internodes are very short, but increase in length with distance from it. In most vascular plants internodal elongation results primarily from the growth of cells produced earlier by the apical meristem and subjacent tissues. As meristematic activity continues, the internodal tissues differentiate largely acropetally. In some plants, e.g., *Equisetum* (horsetail) and grasses, however, differentiation in internodes is primarily **basipetal** (i.e., tissues differentiate progressively toward the base). We shall consider development in these plants in some detail in Chapter 5.

Primary tissue regions of the stem and root

As a result of the activity of apical meristems, and subsequent differentiation of cells and tissues, axes of the shoot and root systems are composed of several distinct **tissue regions** (Fig. 2.10). In the stem and the root there may be a central pith surrounded by vascular tissue (xylem and phloem) variously arranged. The vascular tissue may be continuous, or comprised of vascular bundles that may or may not be interconnected. The central region in some stems and most roots is a solid column of vascular tissue and, thus, there is no pith (Fig. 2.10b). To the exterior of the region of vascular tissue are the cortex and epidermis. In some stems a single-layered region called the endodermis comprises the innermost layer of the cortex. It bounds the pericycle, a tissue region between itself and the vascular tissue. The pericycle is usually recognized only in axes that also have an endodermis which defines its outer limit. The pericycle and endodermis are common in roots (Fig. 2.10b) as well as in stems that reside in aquatic environments.

Each tissue region may be composed of several **simple tissues**, containing a single type of cell, or a **complex tissue**, comprising several types of cells. As we now know, tissues derived from the apical meristems are **primary tissues**, whereas those derived from lateral meristems are **secondary tissues**. In order to better understand the nature of these tissues we must be able to visualize their component cells in three dimensions. Different types of cells have evolved as adaptations to the terrestrial environment in which most vascular plants live and have become specialized for the performance of one or more functions. Some cells function largely in the synthesis of important compounds or in other metabolic processes such as respiration, digestion, etc. Others function solely or largely in transport (or conduction) of substances whereas some function mainly in providing mechanical support. Thus there are many different types of cells with different morphologies that we must understand. Since all of these types often occur together in complex tissues this poses a difficult, but not insoluble problem. There are two ways in which we can comprehend the morphology of a particular cell in three dimensions. For example, we can macerate a piece of tissue by immersing it in an acid solution that dissolves the intercellular cementing material which holds cells together in tissues. Thus separated, we can observe the individual cells. We may, however, wish to comprehend the three-dimensional form of cells in intact tissues such as sections. To do this we must learn to think in three dimensions on the basis of sections of cells in two dimensions. If we observe the cell in question in both transverse and longitudinal views, we can then combine these, mentally, to obtain a three-dimensional conception of the cell (Fig. 2.11a). To understand the detailed morphology of a cell type we will usually have to observe the cell in transverse view as well as in two longitudinal views: radial and tangential (Fig. 2.11b). Transverse, radial, and tangential views of cells can be observed in transverse, radial, and tangential sections (Fig. 2.12). A **transverse section** is a thin sheet of tissue cut at right angles to the long axis of a stem or root, a **radial section** is one cut along a radius of the circle formed by the exterior boundary of the axis, whereas a **tangential section** is one cut on a tangent perpendicular to the radius (Fig. 2.12). One can refer to the walls of cells that lie in the same plane as radial or tangential sections as radial or tangential walls. Likewise, end walls of cells may be called transverse walls when they occur in a plane parallel to a transverse section.

Let us now consider the tissues of the pith and cortex. Three simple tissues, parenchyma, collenchyma and sclerenchyma, singly, in combinations of two, or all three, may comprise either or both of

Figure 2.11 (a) With knowledge of the shape of a transverse wall and a longitudinal wall of a cell, one can envision its three-dimensional form. (b) If a wall of a cell is parallel to a transverse plane it is called a transverse wall; if parallel to a radial plane it is called a radial wall; if parallel to a tangential plane, it is called a tangential wall.

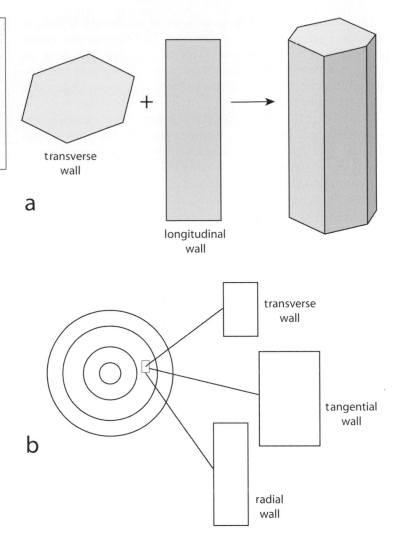

these tissue regions. **Parenchyma** is thought of as the ground tissue of an axis since it occurs in greatest abundance and is the tissue in which the vascular tissues are embedded. Parenchyma cells (Fig. 2.13a, b) may be variably isodiametric in both the pith and the cortex, but are more commonly longitudinally elongate in the cortex. They have relatively thin walls that almost always consist of both primary and secondary wall layers. Wall layers are continuous except in the regions of **simple pits**, circular to irregularly shaped regions lacking secondary wall material. These simple pits usually occur opposite each other in contiguous cells forming **pit-pairs** (Fig. 2.14). Simple pit-pairs in parenchyma and other cells with living protoplasts are regions through which multiple plasmodesmata form interconnections with the protoplasts of adjacent cells. Plasmodesmata are not restricted to pit-pairs, however, and may traverse the wall in other regions as well. These interconnections make possible communication between cells of the symplast. The transport of water and inorganic

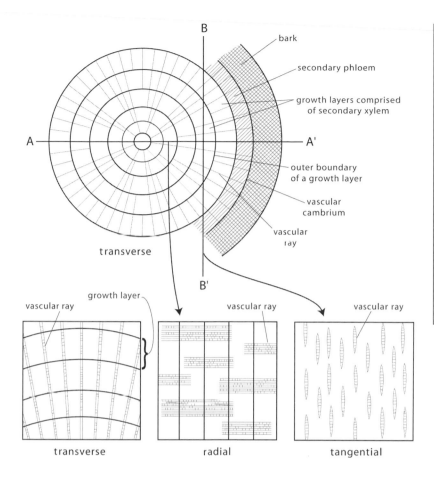

transverse

B

bark

secondary phloem

growth layers comprised
of secondary xylem

outer boundary
of a growth layer

vascular
cambrium

vascular
ray

B'

vascular ray   growth layer

vascular ray   vascular ray

transverse   radial   tangential

Figure 2.12 Planes of section through secondary xylem. Any section cut parallel to the plane of the page is a transverse section. Note that vascular rays cross growth layers at right angles. Any section cut parallel to a radius of the circle formed by any growth layer is a radial section as, for example, a section cut parallel to plane A–A'. In a radial section, rays appear in side view, crossing growth layers at right angles. Any section cut parallel to a tangent to any imaginary circle parallel to the boundary of a growth layer is a tangential section as, for example, a section cut parallel to plane B–B'. In a tangential section vascular rays appear in end view.

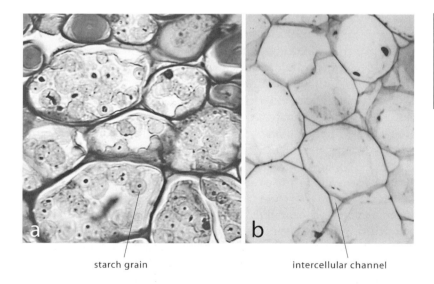

starch grain   intercellular channel

Figure 2.13 (a) Starch grains in parenchyma cells of the cortex. (b) Parenchyma tissue in the pith containing conspicuous intercellular channels. Magnification × 408.

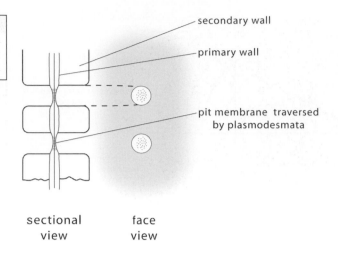

Figure 2.14 Simple pit-pairs in sectional and face views with plasmodesmata connecting the protoplasts of contiguous cells.

secondary wall

primary wall

pit membrane traversed by plasmodesmata

sectional view

face view

solvents can occur solely through the cell walls which, collectively, comprise the **apoplast**. Long-distance transport, however, occurs primarily through the **lumina** (the cavities within the cell walls) of non-living cells such as tracheids and vessel members. Some workers include the lumina of these cells as components of the apoplast.

Parenchyma tissues are regions of metabolism in the plant including the processes of synthesis of various compounds (hormones, enzymes, pigments, essential oils, toxic substances, etc.). One of the processes of synthesis most important to both the plant and to animals including humans, is photosynthesis which produces glucose, the basic food substance for all living beings. Another equally important metabolic process is respiration which provides the source of energy utilized by the plant in carrying out its various activities. Other important processes are the conversion of glucose to starch, the form in which it is stored in parenchyma tissues, and the reverse process of digestion of starch which makes glucose available for use by the plant. Another highly important compound composed of glucose molecules is **cellulose** which is the primary structural material of cell walls.

A system of **intercellular channels** (Figs. 2.13b, 2.15), often referred to as gas spaces, characterizes the parenchyma tissue of plants. The intercellular channels, commonly connected to the outside atmosphere through stomata (or lenticels in older stems), comprise an extensive system in the pith and cortex in stems, roots, and some rhizomes as well as in leaves and some fruits. They are especially well developed in roots and rhizomes in aquatic plants or plants that live in wet soils. The gas-filled spaces vary from the narrow channels between cells to the large, irregular spaces in the mesophyll of leaves and in the aerenchyma of aquatic plants (Fig. 2.16) (Raven, 1996; Prat *et al.*, 1997). Intercellular channels are significant in providing aeration, i.e., facilitating an interchange of O_2 and CO_2 in this tissue which is the site of both photosynthesis (which utilizes CO_2 and produces O_2) and respiration (which utilizes O_2 and produces CO_2). In

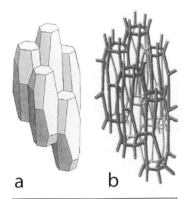

a   b

Figure 2.15 Diagram of the system of intercellular channels (b) between parenchyma cells (a).

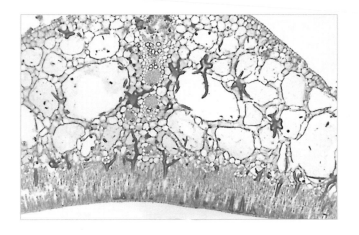

Figure 2.16 Large gas-filled spaces in the leaf of *Nymphaea* (water lily), an aquatic dicotyledon. Note, also, the large astrosclereids. Magnification × 65.

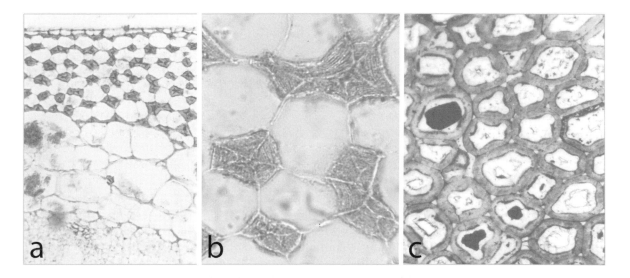

a b c

aquatic plants the system of gas-filled intercellular channels is also important in providing buoyancy.

Collenchyma, like parenchyma, is a tissue in which the various metabolic processes mentioned above also occur. Unlike parenchyma, however, collenchyma forms only a small part of the pith or cortex. In fact, it is only rarely found in the pith, but occurs commonly in the cortex, especially in the outer cortex of herbaceous stems (Fig. 2.17). It is also an important component of leaves and some flower parts. In the stem it often occurs as a continuous layer or as longitudinal ribs in the outermost cortex. In leaves and flower parts it often accompanies major vascular bundles. Wherever it occurs it provides a supporting function.

Collenchyma cells are characterized by relatively thick, but unevenly thickened and unlignified primary walls. The walls are usually highly hydrated, and often conspicuously lamellate (Fig. 2.17b). In the most common type of collenchyma the greatest wall thickening occurs in the corners of the longitudinally elongate cells which

Figure 2.17 Collenchyma. (a) Angular collenchyma in the outer cortex just beneath the epidermis with thickened walls in the corners of the cells. Magnification × 18. (b) Enlargement of a part of (a) showing the lamellate nature of the thickened regions of the cells walls. Magnification × 1150. (c) Lamellar collenchyma with regions of thickened walls between cell corners. Magnification × 220.

Figure 2.18 Sclereids of diverse forms. (a, b) Brachysclereids (stone cells) from the fruit of *Pyrus* (pear). (c, d) Sclereids from the stem cortex of *Hoya* (wax plant), in sectional (c) and surface (d) views. (e, f) Macrosclereids from the endocarp of *Malus* (apple). (g) Columnar sclereid with branched ends from the palisade mesophyll of *Hakea*. (h, i) Sclereids of variable form from the petiole of *Camellia*. (j) An astrosclereid from the stem cortex of *Trochodendron*. (k) An extensive layer of macrosclereids from the epidermis of a clove scale of *Allium sativum* (garlic). From Esau (1977). Used by permission of John Wiley and Sons, Inc.

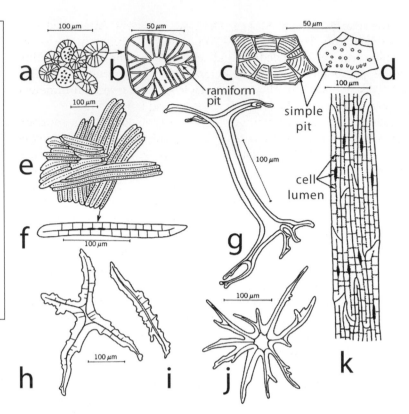

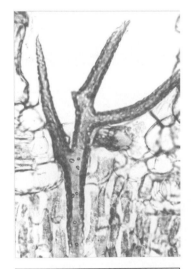

Figure 2.19 A large sclereid from the leaf of *Nymphaea*. Note the thin cell wall containing numerous simple pits. Sclereids of this type often retain a living protoplast. Magnification × 370.

are polygonal in transverse shape. Collenchyma of this type is called **angular collenchyma** (Fig. 2.17a, b) if there are no, or only very small, intercellular channels in the tissue, or **lacunar collenchyma** if the tissue contains conspicuous intercellular channels. In a third type of collenchyma, however, the regions of greatest wall thickening occur between the corners. This type is called **lamellar collenchyma** (Fig. 2.17c).

Sclerenchyma is the major supporting tissue in primary tissue regions and consists of either fibers or sclereids. **Sclereids** are more or less isodiametric, but of highly variable form (Fig. 2.18), and have relatively thin to very thick lignified walls containing simple pits which are canal-like and often branched (ramiform) in very thick walls. Many sclereids, especially those with relatively thin walls, have living protoplasts (Fig. 2.19). Sclereids with different shapes are given different names (Fig. 2.18). **Brachysclereids**, a common type of ovoid to somewhat irregular shape, and often called "stone cells," are common in the flesh of fruits occurring singly or in clusters, as in pear. In fruits with stony endocarps, e.g., peaches, almonds, cherries, etc., they may be the sole cell type comprising the stony layer. They also occur singly or clustered in the pith and/or cortex of some stems. **Macrosclereids** are columnar or rod-shaped and are common components of the epidermis of seeds and the endocarp of fruits. **Osteosclereids** are bone-shaped and are common in leaves and seed coats. **Astrosclereids**

are irregularly star-shaped with elongate, relatively thin-walled pro-cesses extending from a central region. They are common in the leaves of many tropical dicotyledons and in the stems and leaves of some aquatic plants (Fig. 2.16). **Trichosclereids** are very slender, hair-like, and sparsely branched. They may reach lengths of several millimeters and are common in the stems and roots of aquatic plants.

Fibers that occur in primary tissues of roots and stems are com-monly very elongate cells with relatively thick, lignified walls con-taining simple pits (Fig. 2.20). They may comprise separate bundles or bands, often in the peripheral regions of stems, sometimes in the inner cortex (Fig. 2.21a) or, very commonly, may be associated with vascular bundles as bundle sheaths or bundle caps (Fig. 2.21b, c). Fibers serve a largely supporting function in the plant. They tend to be flexible and have great tensile strength and, thus, allow bend-ing without breaking of plant axes. Fibers in primary tissue regions, especially those obtained from some monocots (Fig. 2.21c), are utilized commercially in the production of twine, rope, doormats, burlap, etc. Somewhat similar fibers occur in the secondary phloem, and fibers that are shorter, narrower and ofter thinner-walled are very common in the secondary xylem. In this book we shall restrict the term scle-renchyma fibers to those that occur in primary tissues.

We shall defer a discussion of the pericycle and endodermis until a later chapter on the root (Chapter 15). The very important single-layered tissue region, the epidermis, which comprises the outer boundary of stems, roots, and other plant organs in regions in which periderm has not formed will receive detailed treatment in Chapter 11, devoted solely to it. All other important tissues and tis-sue regions will also be discussed in detail in later chapters.

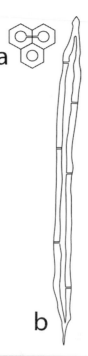

Figure 2.20 (a) Sclerenchyma fibers in transverse view, showing a simple pit-pair. (b) Sclerenchyma fiber in longitudinal view.

Vascular bundle types

Primary xylem and primary phloem, are complex tissues, compris-ing several distinctive cell types. In later chapters we shall consider in detail the composition, development, and functions of these tis-sues, but shall now consider the morphology of the different types of vascular bundles in which they occur in seed plants.

Vascular bundles vary in morphology largely on the basis of the topographic arrangement of their constituent primary xylem and pri-mary phloem. The vascular tissues are, typically, though not always, enclosed by one or more layers of either parenchyma or sclerenchyma cells which comprise the bundle sheath (Fig. 2.21b). Many vascular bundles are also characterized by a bundle cap of fibers (Fig. 2.21b, c).

Four bundle types are widely recognized. The most common type is the **collateral bundle** in which the primary xylem is inner-most and primary phloem comprises the outer part (Fig. 2.22a). The collateral bundle is of frequent occurrence in both gymnosperms and angiosperms. In a **bicollateral bundle** the primary xylem is bounded both to the inside and outside by strands of primary phloem

Figure 2.21 (a) Sclerenchyma fibers in the inner cortex of *Pelargonium*. (b) Fibers forming the bundle caps of vascular bundles of *Triticum* (wheat) and extending to the epidermis. (c) Very thick-walled fibers comprising the bundle cap, and enclosing the conducting cells, of a vascular bundle of *Agave*. Magnification × 255.

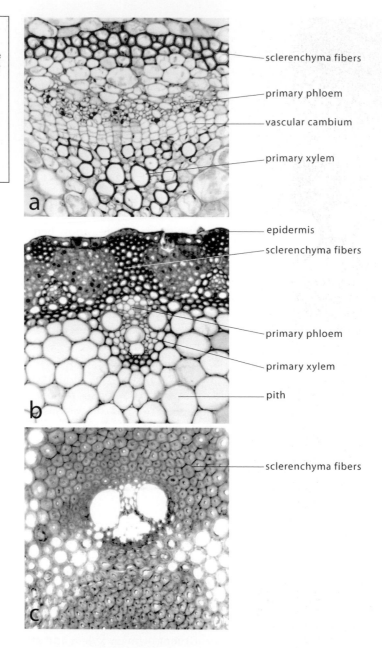

(Fig. 2.22b). These two strands of phloem are referred to, respectively, as internal and external phloem. Bicollateral bundles occur only in some angiosperms. There are two types of **concentric bundles** in which one vascular tissue completely encloses the other. In **amphivasal bundles** (Fig. 2.22c), of frequent occurrence in some families of monocotyledons and a few dicotyledons, the primary xylem surrounds the primary phloem. In **amphicribral bundles** (Fig. 2.22d), common in some ferns, but also found in a few aquatic angiosperms, the primary phloem surrounds the primary xylem.

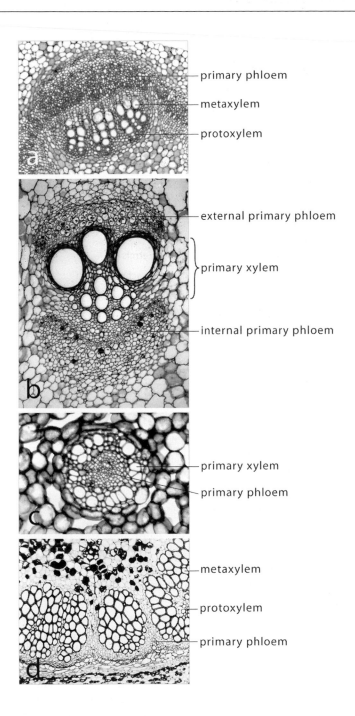

— primary phloem

— metaxylem

— protoxylem

a

— external primary phloem

} primary xylem

— internal primary phloem

b

— primary xylem

— primary phloem

c

— metaxylem

— protoxylem

— primary phloem

d

Figure 2.22 Types of primary vascular bundles. (a) A collateral bundle of *Cassia*. Magnification × 120. (b) A bicollateral bundle of *Cucurbita* (squash). Magnification × 70. (c) An amphivasal bundle of *Acorus calamus*. Magnification × 157. (d) An amphicribral bundle in the rhizome of the fern *Osmunda*. Magnification × 47.

Secondary growth

Activity of the apical meristems results in the development of the primary tissues and tissue regions and the lateral appendages of the shoot and root, all of which comprise what is often termed the **primary plant body**. As growth continues, in many plants, additional tissues are added laterally to stems and roots which add substantially

Figure 2.23 Sites of differentiation of the vascular cambium and the phellogen (cork cambium) lateral meristems in (a) a dicotyledon stem and (b) a root, both shown in transverse section.

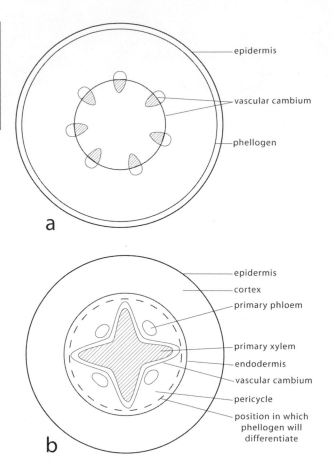

to their thickness. These tissues, called **secondary tissues**, are produced by lateral meristems, the vascular cambium and the phellogen or cork cambium. In seed plant stems, the **vascular cambium** differentiates from immature tissue between the primary xylem and phloem in the vascular bundles as well as from tissue between the bundles, forming a continuous circle as viewed in transverse section (Fig. 2.23a). In roots it differentiates between the central primary xylem and bundles of primary phloem as well as around the ends of the ribs of primary xylem (Fig. 2.23b). This meristem, one cell layer in thickness, which produces several layers of immature derivatives called the cambial zone, extends from near the stem and root tips to the base of the plant. In three dimensions, the vascular cambium in stems is essentially conical. In roots it is of irregular form initially, becoming more regularly conical as secondary xylem is produced. During periods of growth, cells of the vascular cambium divide in a manner that results in the production of new cells to both the interior and to the exterior of the meristem (Fig. 2.24). This cambial activity results in layers of secondary tissue called **growth layers** and leads to an increase in diameter of the stem or root (see also Fig. 2.12). Although these layers are often called cylinders, they are, in reality,

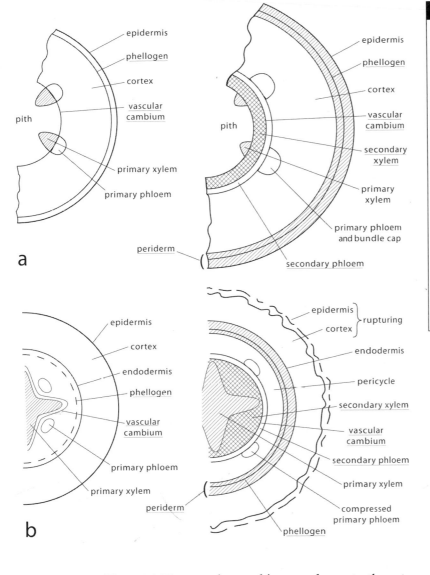

a

b

Figure 2.24 Diagrams showing the lateral meristems (vascular cambium and phellogen) and the secondary tissues which are formed by their cell divisional activity. (a) Production of secondary xylem and phloem, and periderm in a dicotyledon stem. (b) Production of secondary vascular tissues and periderm in a root. Note the position of secondary vascular tissues and periderm in relation to primary tissue regions. Because, in the root, the phellogen develops to the inside of the endodermis, the epidermis, cortex, and endodermis eventually disintegrate and slough off. The labels indicating the meristems and the secondary tissues they produce are underlined.

elongate cones (Fig. 2.25). The vascular cambium produces, to the exterior, secondary phloem in which photosynthate and hormones are transported, primarily from the leaves downward into the roots and, to the interior, secondary xylem in which water and inorganic solvents are transported primarily from the roots upward into the stem. In temperate climatic zones the layers of secondary xylem are usually called annual rings since each layer usually represents the result of cambial activity during one year. The boundaries of these layers are recognizable because the cells produced near the end of a growing season are typically much smaller than those produced earlier. The phellogen, also a single-layered meristem, located in the outer cortex, often differentiates in the layer just inside the epidermis. It produces **phellem** to the exterior and **phelloderm** to the interior, tissues which comprise the **periderm**, a major component of the bark (Fig. 2.24).

Figure 2.25 Stem tip of a woody dicotyledon shown diagrammatically in longitudinal section, illustrating the secondary tissues produced by the vascular cambium and the phellogen. Primary xylem, primary phloem, and lateral appendages are omitted to simplify the diagram.

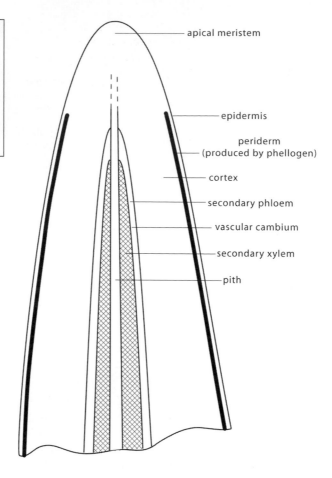

apical meristem

epidermis

periderm (produced by phellogen)

cortex

secondary phloem

vascular cambium

secondary xylem

pith

The secondary xylem consists largely of **tracheary elements** (tracheids and/or vessel members) and associated fibers and parenchyma cells, all of which are elongate cells that parallel the long axis of stems and roots. These cells comprise the **axial system. Rays**, thin ribbons of, largely, parenchyma cells traverse the secondary xylem at right angles to the elongate cells, that is, along radii of the circles formed by the growth layers of secondary tissues (Fig. 2.22a–d). They comprise the **radial system**, extending from the secondary xylem across the vascular cambium and through the secondary phloem. The axial system in the secondary phloem consists of elongate conducting cells and associated companion cells, fibers, and phloem parenchyma cells.

Cells of the xylem

Both primary and secondary xylem may contain tracheids, vessels (consisting of vessel members), fibers, and parenchyma cells. The xylem of the earliest vascular plants in the fossil record as well as the most primitive living vascular plants contain only tracheids.

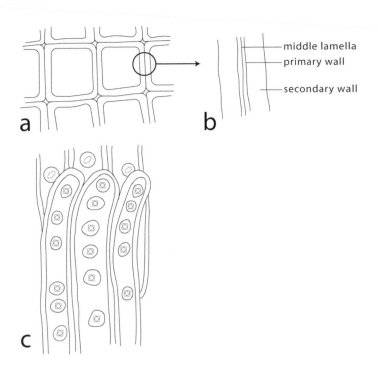

a

b

— middle lamella
— primary wall
— secondary wall

c

Figure 2.26 Conifer tracheids.
(a) Transverse view.
(b) Enlargement of a section of the cell walls of contiguous tracheids showing wall layers.
(c) Longitudinal section illustrating bordered pits on radial walls and overlapping ends.

Tracheids serve two important functions: support, and transport of water and minerals. Vessel members and wood fibers appear in the xylem of plants that evolved later in geologic time. The tracheid is, therefore, considered the basic conducting cell in the xylem and the type from which vessel members and fibers evolved. During its evolution the vessel member became highly specialized for transport and the wood fiber for support (for more detail, see Bailey (1953) and references therein).

The tracheid (Fig. 2.26), at functional maturity, is a non-living, longitudinally elongate cell usually with tapered ends, although the ends may be rather blunt in some taxa (Fig. 2.26c), especially some conifers. In sectional view the tracheid may be circular or polygonal, the latter being characteristic of tracheids when compactly arranged. Tracheids in secondary xylem are characterized, typically, by several to many wall facets, and in conifers they often appear rectangular in sectional view (Fig. 2.26a). The walls of tracheids, as in most other mature cells in vascular plants, are comprised of long strands of cellulose, called **microfibrils**, embedded in an amorphous matrix of lignin and other substances. Cell walls are usually composed of two layers, an outer **primary wall layer** and an inner, thicker **secondary wall layer** (Fig. 2.26b). The thin primary wall is produced following mitosis and cell plate formation, and the secondary wall is added on to the inner surface of the primary wall as development of the cell proceeds. The secondary wall layer may be largely discontinuous, with secondary wall material deposited on the primary wall in helical, reticulate, or scalariform patterns, or largely continuous, but containing pits, well-defined areas lacking secondary wall (see below). More detail,

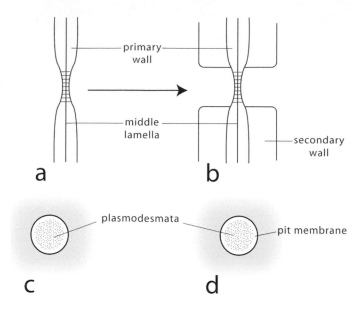

Figure 2.27 (a, c) Primary pit fields in (a) sectional and (c) face views, showing plasmodesmata. (b, d) Simple pit-pairs in (b) sectional and (d) face views. Note that the simple pit-pair develops in relation to the primary pit field by the production of secondary walls in contiguous cells.

and aspects of development of these variations in the morphology of tracheid walls will be presented in Chapters 5 and 9.

In the secondary xylem the tracheid ends overlap those of other tracheids (Fig. 2.26c). In a living tree the tracheid lumina are filled with water which, during transpiration, moves in a helical path from one tracheid end into that of the adjacent, overlapping tracheid. Transport is facilitated by the lack of a protoplast in the tracheids, by the permeability of their cellulosic cell walls, and by the presence of pits which occur, in abundance, in pairs in the walls of the contiguous, overlapping tracheid ends.

Before describing vessel members and wood fibers, let us consider in more detail the nature of pits. **Pits** (Figs. 2.27, 2.28) are canals in the secondary wall layer, but these canals do not extend through the primary wall. They develop because of the lack of deposition, or the deposition of very little, secondary wall material at the sites of pit formation during wall development. Pits commonly develop opposite sites of **primary pit fields** (Fig. 2.27a), thin regions in the primary wall that are traversed by plasmodesmata, specialized strands of endoplasmic reticulum that connect and provide avenues for molecular transport between the protoplasts of adjacent cells. The failure of secondary wall to be synthesized at these sites of pit formation may be controlled by informational molecules that are transported by the plasmodesmata (see Chapter 3 on the cell wall).

Typically pits occur in pairs (Figs. 2.27c, 2.28a, d, f) with each pit of a pit-pair occurring opposite the other in a contiguous cell wall of adjacent cells. The primary walls of the two cells and the middle lamella lie between the two pits of a pair and form the pit-pair membrane, usually called simply the **pit membrane**. There are two types of pits, simple and bordered. A **simple pit** is a canal in a secondary wall with straight sides, which in face view may be

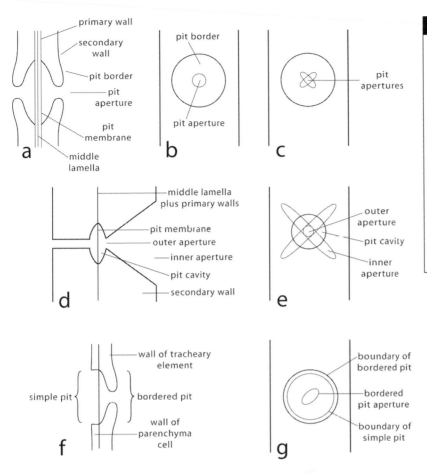

Figure 2.28 Bordered and half-bordered pit-pairs. (a) Sectional view of a bordered pit-pair. (b) Face view of a bordered pit-pair with circular apertures. (c) Face view of a bordered pit-pair with crossed, elliptical apertures. (d, e) Sectional and face views of a bordered pit-pair in thick secondary walls. The pit canal, in the form of a flattened cone, has inner and outer apertures. The inner, crossed apertures extend beyond the boundary of the pit cavity. (f, g) A half-bordered pit-pair in sectional and face views.

circular, oval, or somewhat irregular (Fig. 2.27b, d). A **face view** is obtained when one looks directly into the pit from the inside or the outside of the cell. In **sectional view**, obtained when the wall in which the pit-pair occurs is cut perpendicular to the plane in which the wall lies, the pit-pair may be seen as two opposing canals separated by the pit membrane (Fig. 2.27b). A **bordered pit** differs from a simple pit primarily in the presence of a **pit border**, composed of secondary wall, that overhangs and partially encloses the **pit cavity** thus formed (Fig. 2.28a, d). The opening into the pit cavity (the **pit aperture**) varies in size and shape with the degree of overhang of the border. When a pit is viewed in face view, the aperture may appear circular (Fig. 2.28b), elliptical, or lenticular. When elliptical or lenticular, the apertures of the two pits of the pit-pair will be crossed (Fig. 2.28c), the angle of each aperture reflecting the predominant angle of the cellulose microfibrils in the respective secondary cell walls. In pit-pairs in very thick walls, the pit canal is often shaped like a greatly flattened cone with two apertures, an outer, circular aperture adjacent to the pit cavity and an inner elliptical aperture bordering the cell lumen (Fig. 2.28d). In such pits the crossed, inner apertures of the pit-pair often extend beyond the boundary of the pit cavity (Fig. 2.28e). If a

Figure 2.29 Diagrams of vessel members. (a) A vessel member with an oblique end wall and a compound perforation plate, representative of a relatively primitive angiosperm. (b, c) Vessel members with simple perforation plates, representative of more evolutionarily advanced angiosperms. Note the elliptical to circular bordered pits in the lateral walls. (d) A series of superposed vessel members comprising a vessel.

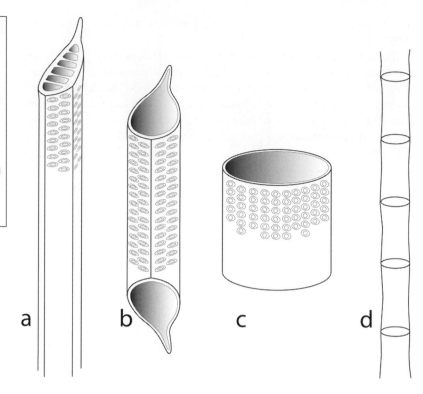

a b c d

tracheid or vessel member is adjacent to a parenchyma cell a **half-bordered pit-pair** (Fig. 2.28f, g) may develop consisting of a simple pit in the wall of the living parenchyma cell and a bordered pit in the wall of the non-living, water-conducting cell. From this observation you might conclude that simple pits occur commonly in living cells and that bordered pits occur in cells that, when functionally mature, are non-living. You should note, however, that whereas simple pits are not exclusive to living cells, bordered pits occur only in cells that when functionally mature are non-living.

The pit membrane may be highly specialized and, in the bordered pits of water-conducting cells, is highly porous which facilitates the transport of water and solutes from one cell to an adjacent one. In some conifers, pit membranes are characterized by a central, thickened region called the torus which functions as a valve that, under certain conditions, closes the aperture, preventing the entry of air into the xylem. (We shall consider the structure and function of these pit membranes in detail in Chapter 9 on the secondary xylem.)

Vessel members (Fig. 2.29a–c), characteristic of the xylem of most angiosperms, differ from tracheids in several fundamental ways. On average they are shorter and of greater diameter, and have perforate end walls. In the xylem they are superposed, end to end, forming long tubes, the vessels (Fig. 2.29d). Transport of water and inorganic solutes may thus follow a longitudinal rather than the helical course characteristic of plants that contain only tracheids. Transport efficiency may be enhanced thereby and by the perforate structure of the end walls.

Since vessel members evolved from tracheids, it is not surprising that those in the most primitive angiosperms are similar to tracheids, being relatively long, of small diameter, angular in section, and with oblique end walls containing numerous, usually elliptical, perforations (Fig. 2.29a). In more advanced angiosperms, vessel members are shorter, of greater diameter, circular in section, and with transverse end walls with a single (simple) perforation (Fig. 2.29c). Variation in these characteristics is continuous between these extremes. Pits on the lateral walls of vessel members are generally small and vary from elliptical- to circular-bordered.

Wood fibers, unlike tracheids and vessel members, often retain their protoplasts at functional maturity – some for a relatively short time, others apparently for many years, possibly for the life of the plant. They are usually longer than the longest vessel members in the same wood, of relatively small diameter, are circular to polygonal in section, have relatively thick secondary walls (although in fibers in the wood of different species the wall thickness may be highly variable), and have pointed ends. Pits on the lateral walls are bordered, but are commonly greatly reduced in size, and may appear to be simple. Fibers provide the major support in secondary wood of angiosperms and, in volume occupied, comprise the major component of many woods.

In some woods, especially those of relatively primitive species, the supporting cells are intermediate in morphology between tracheids and the fibers of woods in more advanced species. Such cells are called **fiber-tracheids**. Some fiber-tracheids possess living protoplasts and are characterized by septa of secondary wall material that divide the cells into compartments that apparently function much like **wood parenchyma** as sites of storage of starch and other metabolites. Parenchyma in the secondary xylem is distributed in axial columns and in the vascular rays, ribbons of cells that run radially through the tissue (see Chapter 10 on the secondary xylem).

Cells of the phloem

There are two types of conducting cells in the phloem, sieve cells and sieve tube members. Both are living cells, but both at functional maturity have highly modified protoplasts lacking vacuolar membranes, an adaptation to their function in conduction. **Sieve cells** (Fig. 2.30a), characteristic of primitive vascular plants and gymnosperms, are longitudinally elongate, and have tapered to blunt ends that overlap the ends of other sieve cells. Specialized perforate regions called **sieve areas** (Fig. 2.30a, c) occur in the lateral walls, and are the sites of transport of photosynthate and other organic substances (enzymes, hormones, etc.) from one cell to another. Sieve areas are abundant

Figure 2.30 Diagrams of sieve elements. (a) A sieve cell. Although sieve areas occur over the entire length of the cell they are shown here only at one end. (b) A sieve tube member with associated companion cells. Note the compound sieve plate and the lateral sieve areas. (c) A sieve area in face view with callose cylinders enclosing the sieve pores. (d) The sieve area in sectional view.

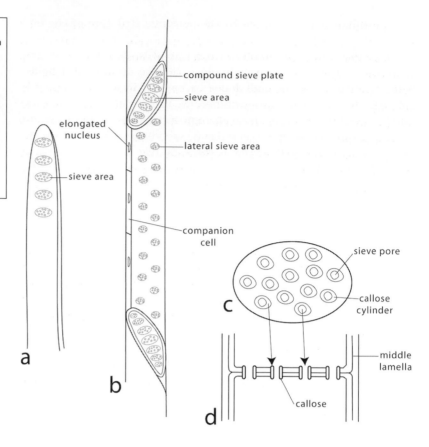

in the walls of the overlapping ends of sieve cells. **Sieve tube members** (Fig. 2.30b), which occur in angiosperms, have sieve areas on their end walls, and are arranged in superposed columns called **sieve tubes**. Although small, indistinct sieve areas may occur in the lateral walls of sieve tube members, well-developed sieve areas are usually restricted to the end walls and provide the predominant pathway for the passage of photosynthate and other organic solutes from one cell to the next. Sieve tube members in primitive angiosperms are commonly long and have oblique end walls containing **compound sieve plates** consisting of several to many sieve areas (Fig. 2.30b) whereas those in the most advanced angiosperms are relatively short and have transverse ends and **simple sieve plates** consisting of single sieve areas. The sieve areas are specialized regions which consist of the walls and middle lamella of contiguous cells (Fig. 2.30d). These regions contain numerous **sieve pores**, often lined with **callose** (Fig. 2.30c, d), through which the protoplasts of contiguous sieve tube members are interconnected. As we shall see in the chapter on the phloem, the callose cylinders may not be present in the functioning phloem, developing in response to the trauma of cutting the plant for sectioning.

In intimate contact with sieve tube members are **companion cells** (Fig. 2.30b), which in some taxa are **transfer cells**. Companion cells and transfer cells facilitate the transfer of photosynthate from sites of its production in leaves into the sieve tube members. Transfer cells have distinctive secondary walls with extensive ingrowths that increase the inner wall area and, thus, the area of the plasmalemma which lines the wall. Increase in the area of the plasmalemma increases the efficiency of active transfer of photosynthate into the cell protoplast. We shall look at companion cells and transfer cells in more detail in Chapter 10. In conifers, specialized marginal ray cells, called **Strasburger cells**, serve the same function as companion cells in angiosperms.

Fibers are also characteristic of the phloem. In primary vascular bundles, strands of phloem fibers develop along the outer edge forming what appear in transverse section to be bundle caps. **Phloem fibers** are especially abundant in secondary phloem, occurring in dense tissues between and often enclosing strands of conducting cells. Phloem fibers, which have important supporting and protective functions, are very long with sharply tapered ends and often very thick secondary walls containing simple pits. **Axial parenchyma cells**, which typically occur in longitudinal columns, as well as **ray parenchyma cells** are present in the phloem and probably function with those in the secondary xylem as a system in which photosynthate and hormones are transported both longitudinally and laterally in secondary vascular tissues.

REFERENCES

Bailey, I. W. 1953. Evolution of the tracheary tissue of land plants. *Am. J. Bot.* **40**: 4–8.

Carlquist, S. 1961. *Comparative Plant Anatomy*. New York: Holt, Rinehart and Winston.

Ehlers, K. and R. Kollmann. 2001. Primary and secondary plasmodesmata: structure, origin, and functioning. *Protoplasma* **216**: 1–30.

Esau, K. 1977. *Anatomy of Seed Plants*, 2nd edn. New York: John Wiley and Sons.

Hagemann, W. 1982. Vergleichende Morphologie und Anatomie-Organismus und Zelle: ist eine Synthese möglich? *Ber. Deutsch. Bot. Ges.* **95**: 45–56.

Kaplan, D. R. 1992. The relationship of cells to organisms in plants: problem and implications of an organismal perspective. *Int. J. Plant Sci.* **153**: S28–S37.

Kaplan, D. R. and W. Hagemann. 1991. The relationship of cell and organism in vascular plants. *BioScience* **41**: 693–703.

Kragler, F., Lucas W. J., and J. Monzer. 1998. Plasmodesmata: dynamics, domains and patterning. *Ann. Bot.* **81**: 1–10.

Lucas, W. J., Ding, B., and C. van der Schoot. 1993. Plasmodesmata and the supracellular nature of plants. *New Phytol.* **125**: 435–476.

Niklas, K. and D. R. Kaplan. 1991. Biomechanics and the adaptive significance of multicellularity in plants. In *Proc. 4th Int. Congr. Syst. Evol. Biol. Dioscorides*, Portland, OR, pp. 489–502.

Prat, R., Andre, J. P., Mutaftschiev, S., and A. M. Catesson. 1997. Three-dimensional study of the intercellular gas space in *Vigna radiata* hypocotyl. *Protoplasma* **196**: 69–77.

Raven, J. A. 1996. Into the voids: the distribution, function, development and maintenance of gas spaces in plants. *Ann. Bot.* **78**: 137–142.

Verbeke, J. A. 1992. Developmental principles of cell and tissue differentiation: cell–cell communication and induction. *Int. J. Plant Sci.* **153**: S86–S89.

Wojtaszek, P. 2000. Genes and plant cell walls: a difficult relationship. *Bot. Rev.* **75**: 437–475.

FURTHER READING

Bailey, I. W. 1936. The problem of differentiating and classifying tracheids, fiber-tracheids, and libriform wood fibers. *Trop. Woods* **45**: 18–23.

Baluska, F., Volkmann, D., and P. W. Barlow. 2004. Eukaryotic cells and their cell bodies: cell theory revised. *Ann. Bot.* **94**: 9–32.

Beer, M. and G. Setterfield. 1958. Fine structure in thickened primary walls of collenchyma cells of celery petioles. *Am. J. Bot.* **45**: 571–580.

Eames, A. J. and L. H. MacDaniels. 1947. *An Introduction to Plant Anatomy*, 2nd edn. New York: McGraw-Hill.

Esau, K. 1936. Ontogeny and structure of collenchyma and of vascular tissues in celery petioles. *Hilgardia* **10**: 431–476.

1965. *Plant Anatomy*, 2nd edn. New York: John Wiley and Sons.

Evans, P. S. 1965. Intercalary growth in the aerial shoot of *Eleocharis acuta* R.Br. Prodr. I. Structure of the growing zone. *Ann. Bot.* **29**: 205–217.

Fahn, A. 1974. *Plant Anatomy*, 2nd edn. Oxford, UK: Pergamon Press.

Fahn, A. and B. Leshem. 1963. Wood fibers with living protoplasts. *New Phytol.* **62**: 91–98.

Foster, A. S. 1955. Structure and ontogeny of terminal sclereids in *Boronia serrulata*. *Am. J. Bot.* **42**: 551–560.

Foster, A. S. and E. M. Gifford, Jr. 1974. *Comparative Morphology of Vascular Plants*, 2nd edn. San Francisco, CA: W. H. Freeman.

Gaudet, J. 1960. Ontogeny of the foliar sclereids in *Nymphaea odorata*. *Am. J. Bot.* **47**: 525–532.

Haberlandt, G. 1914. *Physiological Plant Anatomy*. London: Macmillan.

Hayward, H. E. 1938. *The Structure of Economic Plants*. London: Macmillan.

Jane, F. W. 1970. *The Structure of Wood*, 2nd edn (revd K. Wilson and D. J. B. White). London: A. and C. Black.

Linsbauer, K. (ed.) 1922–43. *Handbuch der Pflanzenanatomie*. Berlin: Gebrüder Borntraeger.

Metcalfe, C. R. 1960. *Anatomy of the Monocotyledons*, vol. 1, *Gramineae*. Oxford, UK: Clarendon Press.

1971. *Anatomy of the Monocotyledons*, vol. 5, *Cyperaceae*. Oxford, UK: Clarendon Press.

Metcalfe, C. R. and L. Chalk. 1950. *Anatomy of the Dicotyledons*, vols. 1 and 2. Oxford, UK: Clarendon Press.

Robards, A. W. 1967. The xylem fibres of *Salix fragilis* L. *J. Roy. Microscop. Soc.* **87**: 329–352.

Romberger, J. A., Hejnowicz, Z., and J. F. Hill. 1993. *Plant Structure: Function and Development*. Berlin: Springer-Verlag.

Solereder, H. (English transl. L. A. Boodle and F. E. Fritsch) 1908. *Systematic Anatomy of the Dicotyledons*, vols. 1 and 2. Oxford, UK: Clarendon Press.

Tomlinson, P. B. 1959. Structure and distribution of sclereids in the leaves of palms. *New Phytol.* **58**: 253–266.

1961. *Anatomy of the Monocotyledons*, vol. 2, *Palmae*. Oxford, UK: Clarendon Press.

1969. *Anatomy of the Monocotyledons*, vol. 3, *Commelinales–Zingiberales*. Oxford, UK: Clarendon Press.

Wardlaw, C. W. 1965. *Organization and Evolution in Plants*. London: Longman, Green.

Chapter 3

The protoplast of the eukaryotic cell

Perspective

The eukaryotic cell is composed, with a few exceptions, of both a living protoplast, the site of cellular metabolism, and an enclosing cellulosic wall of one or more layers (Fig. 3.1). While not alive as a structural unit, the wall is commonly traversed by living components, **plasmodesmata**, which connect adjacent protoplasts and, thus facilitate communication between, and the integration of, cells within a tissue. All plant cells possess a protoplast during development, and in many it persists throughout the life of the plant. Some cells, however, do not achieve their ultimate functional state until the protoplast dies as, for example, a specialized water-conducting cell such as a vessel member.

The protoplasts of all plant cells are basically similar, but may differ in relation to the function of the mature cells. For example, the protoplast of a parenchyma cell in the outer cortex or in a leaf will contain many chloroplasts since a major function of these cells is photosynthesis. In contrast, a cell of the pith (a storage region) in the center of the stem may lack chloroplasts but will contain unpigmented plastids in which starch is synthesized (amyloplasts). The protoplast of an immature vessel member, however, destined to die, may contain no plastids at all, or plastids of a highly modified type.

Each cell protoplast is characterized by the potential for the development of an entire organism (see Steward *et al.*, 1964). This total potentiality is, however, rarely achieved under normal conditions. It is, indeed, primarily the evolution of mechanisms that control (restrict) the expression of this potentiality that accounts for the differences in the morphology and function of cells.

The unity of individual cell protoplasts notwithstanding, cells are influenced during development by the environment resulting from their association in tissues. They may function in concert with other cells during the differentiation of tissues and the development of organs. This coordination of development is facilitated by

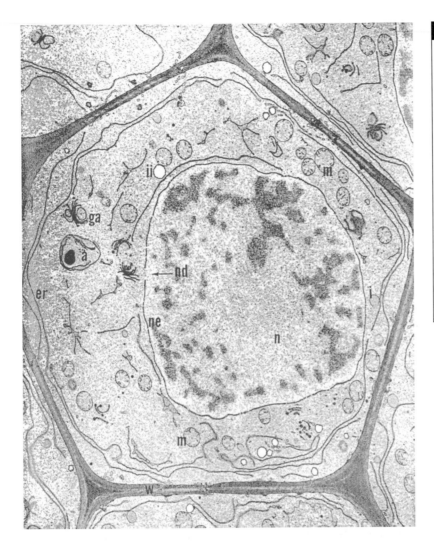

Figure 3.1 Transmission electron micrograph of a meristematic cell from the root cap of *Zea mays* (maize). The protoplast, dominated by the nucleus (n), contains mitochondria (m), Golgi bodies (ga), and endoplasmic reticulum (er). The protoplast is enclosed by a cellulosic primary wall (W). Note that in this meristematic cell only a few small vacuoles (white) are visible. The dark structures in the nucleus are chromosomes. nd, nuclear membrane discontinuity; ne, nuclear envelope. Magnification × 5973. From Whaley *et al.* (1960). Used by permission of the Botanical Society of America.

the plasmodesmata which connect the protoplasts of adjacent cells and which we shall consider in detail in the following chapter on the cell wall.

Morphology of the protoplast

The protoplast consists of a relatively liquid, colloidal phase, the cytoplasm, in which reside a number of morphologically distinct membrane systems such as the endoplasmic reticulum and Golgi bodies, and membrane-bound organelles among which are included the nucleus, mitochondria, and plastids (Figs. 3.1, 3.2). In addition, microtubules and actin microfilaments also reside within the cytoplasm. The protoplast, itself, is enclosed by a membrane.

Figure 3.2 Enlargement of part of a cell protoplast of a meristematic cell similar to that shown in Fig. 3.1. The double membrane comprising the nuclear envelope (ne) is continuous with endoplasmic reticulum (er). Note the nuclear pores (np) and the plasmalemma (pl) bounding the protoplast and lining the primary cell wall (W). ga, Golgi body; m, mitochondrion. Magnification × 13 248. From Whaley *et al.* (1960). Used by permission of the Botanical Society of America.

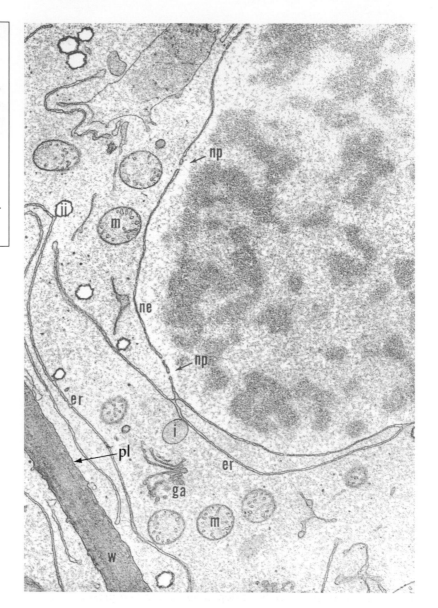

The cytoplasm

The cytoplasm is the relatively liquid matrix of the protoplast, often referred to as the **hyaloplasm**, in which the organelles and membrane systems are suspended. Its viscosity may vary in different regions of the cell, and may fluctuate during different stages of development. It is composed, primarily, of protein macromolecules in colloidal suspension.

The plasmalemma

Enclosing the protoplast is a bounding membrane, the plasmalemma (Fig. 3.2), or plasma membrane, sometimes also called the ectoplast. As observed in section, the plasmalemma is three-layered consisting of two electron-dense (dark) layers separated by an electron-transparent (clear) layer. Such a membrane is referred to as a **unit membrane**. Although the outer and inner layers commonly appear of similar thickness and density, it has been observed that in some plants the inner layer may be denser than the outer, suggesting that the layers vary in chemical composition. Several models of membrane structure have been proposed. Most are characterized by globular proteins that comprise both the inner and outer layers between which is a lipid bilayer. It is important to emphasize that membrane structure is not static but, rather, dynamic, possibly constantly changing depending on the particular function the membrane serves. Unit membranes function not only as structural units but also as foundations for enzymes as well as in the control of the passage of various substances into and out of the protoplast and the various membrane-bound organelles of the protoplast. Membrane function may also vary during different stages of development. For example, during cell growth when new cell wall is being formed, clusters of globules called **rosettes** form in the plasmalemma. These are clusters of enzymes which mediate the synthesis of cellulose microfibrils (see Chapter 4 for detail).

The plasmalemma (Fig. 3.2) usually lies in contact with or closely parallel to the cell wall. In certain specialized absorptive and secretory cells (**transfer cells**) which accumulate and transfer substances into other cells of the plant body or the exterior, the wall is characterized by the development of numerous ingrowths. Consequently, the plasmalemma becomes highly invaginated, resulting in an increase in surface area which facilitates transfer of materials. In some other types of secretory cells, the plasmalemma may be highly convoluted even though the wall is of uniform thickness. This may result from the continued incorporation into the plasmalemma of membrane from Golgi vesicles after the cells have ceased to increase in size. It is well established that the plasmalemma functions in the control of the entry into and exit from the protoplast of various substances. Furthermore, the establishment of turgor pressure within the cell, a condition essential to growth as well as to maintenance of normal function in living cells, is dependent upon the presence of this selectively permeable surface membrane.

The nucleus and ribosomes

As the control center of the cell, the **nucleus** is the dominant organelle, often in size, always in influence (Figs. 3.1, 3.2). Within its chromosomes is the DNA (deoxyribonucleic acid), the repository of information encoded in the genes throughout the evolution of the group. The nucleus is also the major site of RNA (ribonucleic acid) synthesis in the cell. The information of DNA is transferred to RNA during its synthesis. RNA (of which there are several kinds) is transferred to the exterior of the nucleus where it acts as an intermediary in protein synthesis. The actual sites of protein synthesis are the **ribosomes** (Fig. 3.3a, b), small spherical bodies about 15–20 nm in diameter which

Figure 3.3 (a) Rough endoplasmic reticulum (ER) in a mesophyll cell of *Nicotiana tabacum* (tobacco). Note the numerous ribosomes on the surface of the ER. (b) Polyribosomes on the surface of an ER cisterna in a leaf of tobacco. To the left is a highly fenestrated Golgi body seen in surface view. Compare this with that in Fig. 3.5a. From Esau (1977). Used by permission of John Wiley and Sons, Inc.

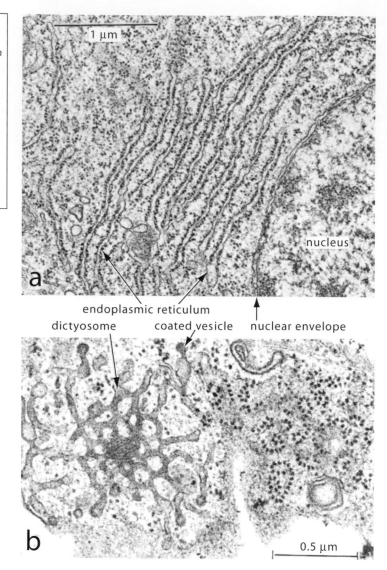

consist of protein and RNA. During protein synthesis the ribosomes aggregate along tubules of endoplasmic reticulum forming **polyribosomes** (Fig. 3.3b). Ribosomes may be dispersed within the cytoplasm or bound to the surface of the internal membrane system (Fig. 3.3a). They also occur in mitochondria and plastids.

The internal membrane system

One of the most interesting morphological features of the protoplast is the intricate membrane system which consists of the nuclear envelope and the endoplasmic reticulum. The **nuclear envelope** consists of two closely parallel membranes perforated by nuclear pores (Fig. 3.2). The pores are apertures through which molecules of RNA may leave the nucleus and enter the cytoplasm.

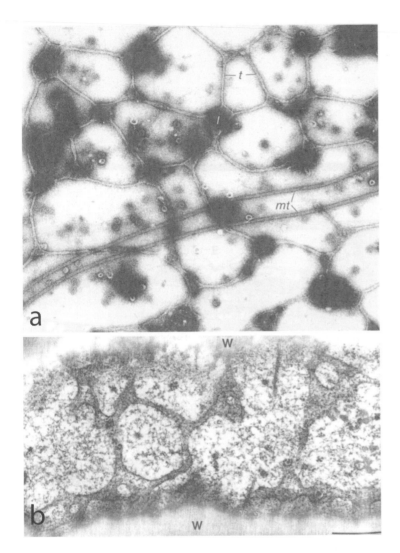

a

b

Figure 3.4 (a) Smooth, tubular endoplasmic reticulum from the protoplast of a guard cell of *Vicia faba* (vetch). The ER has a predominantly reticulate form, but contains some largely obscured lamellar regions. This peripheral (cortical) ER was stationary, and adhered to the plasmalemma but was not fused to it. Note the associated microtubules (mt). t, tubular ER; l, lamellar ER. Magnification × 22 000. (b) ER in the peripheral cytoplasm of cells of *Drosera*. Note the flattened, fenestrated regions. W, cell wall. Bar = 0.5 μm. (a) From a thesis by Wiedenhoeft (1985) reproduced in Hepler *et al.* (1990). Used by permission of The Company of Biologists, Ltd. (b) From Lichtscheidl *et al.* (1990). Used by permission of Springer-Verlag Wien.

In continuity with the nuclear envelope is the extensive system of membranes called the **endoplasmic reticulum** (Fig. 3.2) (Porter and Machado, 1960; Porter, 1961) and commonly indicated simply as ER. This reticulate system consists of tubules, vesicles, and cisternae (regions enclosed by parallel membranes). Recent studies have demonstrated that, in general, the ER in mature cells forms a well-defined, polygonal network closely associated with the plasmalemma (Fig. 3.4a, b) (see, e.g., Hepler *et al.*, 1990). This peripheral ER is stationary and may contact the plasmalemma, but whether it actually fuses with the plasmalemma is unclear. Although connected to the peripheral ER, the interior part of the system is more labile and, because of cytoplasmic streaming, its reticulate morphology may be obscured.

The endoplasmic reticulum is characterized as smooth or rough on the basis of the presence or absence on its surface of ribosomes. Because of its extensive surface area and vesiculate nature, it is generally thought to provide a system through, or along which, various

Figure 3.5 Golgi bodies (ga) in a meristematic cell of *Zea mays*, on the left in sectional view showing the stack of membrane cisternae, and on the right in surface view. Note the vesicles produced at the edges of the cisternae. Compare these images with that of the Golgi body in Fig. 3.3b. m, mitochondrion; er, endoplasmic reticulum; ne, nuclear envelope; gv, Golgi vesicles. Magnification × 32 860. From Whaley *et al.* (1960). Used by permission of the Botanical Society of America.

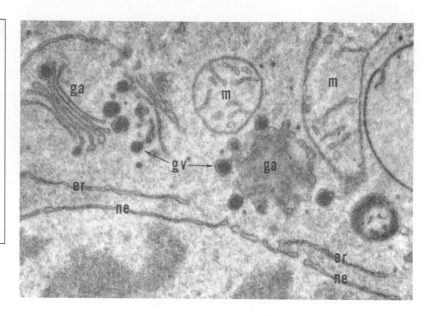

cellular substances, especially precursor compounds of large molecules, are transported within the cell. Smooth ER is common in cells in which metabolic activity is reduced whereas rough ER is characteristic of cells with high rates of metabolism such as actively dividing cambial cells. It is probable that the products of protein synthesis in the ribosomes on the surface of rough ER are deposited within the ER cisternae and then transported to various regions of the cell. Pectic compounds and simple carbohydrates essential to formation of the middle lamella and the cell wall may also be transported through the endoplasmic reticulum and transferred in ER or Golgi vesicles to the immediate sites of biosynthesis. The peripheral network of endoplasmic reticulum, in association with a system of actin microfilaments, also functions as a pathway along which move organelles such as Golgi bodies and mitochondria (Boevink *et al.*, 1998; Nebenführ *et al.*, 1999; Hawes and Satiat-Jeunemaitre, 2001).

Golgi bodies

Golgi bodies (Figs. 3.3b, 3.5) (also called dictyosomes, Golgi stacks, Golgi apparatus, or simply Golgi) consist of stacks of cisternae, essentially circular in surface view and appearing as flattened sacs in sectional view (Mollenhauer and Morre, 1966). Interconnected tubules from which small vesicles originate comprise the border of each cisterna (Figs. 3.3b, 3.5) and may be highly fenestrated. Golgi cisternae range in diameter from about 0.5 to 1.0 μm (Robards, 1970). The Golgi vesicles, apparently produced in succession from the margins of the cisternae, contain a variety of compounds, some probably absorbed from the surrounding cytoplasm, some synthesized within the Golgi body, others transferred through the endoplasmic reticulum for deposition at sites of synthesis. Golgi bodies play an important role in the development of both the middle lamella and the new cell wall.

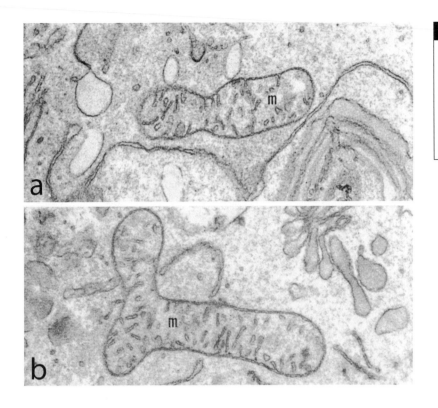

Figure 3.6 (a, b) Mitochondria. Note the two bounding unit membranes and the cristae which are invaginations of the inner membrane; also the variation in form. Magnification × 22 000. From Whaley *et al.* (1960). Used by permission of the Botanical Society of America.

Precursor compounds of the components of which these structures are comprised are transferred in Golgi vesicles to the exterior of the plasmalemma (Northcote and Pickett-Heaps, 1966). Golgi bodies have been observed to aggregate in regions of cell wall synthesis. The method of their movement within the protoplast has been a mystery until recently (see pp. 50–52), and earlier workers assumed that they were formed anew in the regions in which they aggregated.

Mitochondria

Mitochondria (Figs. 3.2, 3.6) are sites of respiration and synthesis of adenosine triphosphate (ATP) which supplies the energy required for the numerous metabolic activities of the cell (see Lehninger, 1964). Mitochondria, usually less than 1.0 µm in diameter and no more than 5.0 µm long, are commonly rod-shaped, but may be spherical, dumb-bell-shaped or sometimes branched; and they may vary from one form to another. Each mitochondrion is composed of two unit membranes, an outer bounding one and an inner one that is invaginated in varying degrees (Fig. 3.6a, b). The invaginations, called **cristae**, often oriented at approximately right angles to the surface, extend in some cases nearly across the organelle. It is difficult to determine whether these invaginations are tubules or flattened vesicles although the latter structure is probably the more common in higher plants. The enzymes involved in the complex series of reactions which culminate in the production of ATP are thought to be localized in the membranes, especially the inner membrane.

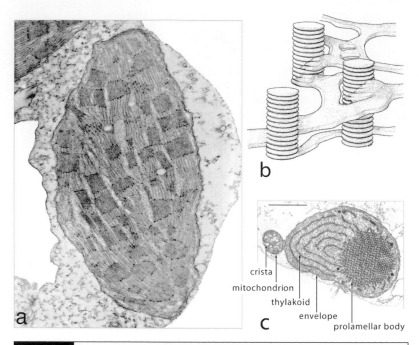

crista
mitochondrion
thylakoid
envelope
prolamellar body

Figure 3.7 (a) Chloroplast from a mesophyll cell of *Zea mays* showing the extensive and intricate membrane system within a matrix termed the stroma. Compact stacks of specialized membranes called thylakoids comprise grana. Since chlorophyll is restricted to the grana, they are the sites of photosynthesis in the chloroplast. Note also the numerous ribosomes in the stroma. Magnification × 16 636. (b) Diagram illustrating the relationship between the grana and the stroma lamellae. (c) Etiolated chloroplast containing a paracrystalline, prolamellar body in a mesophyll cell of *Saccharum officinarum* (sugar cane). (a) From Lehninger (1961). (b) From Weier *et al.* (1967). Used by permission of Brookhaven National Laboratory. (c) From Esau (1977). Used by permission of John Wiley and Sons, Inc.

Apparently, small densely staining granules called **osmiophilic granules** that occur to the interior of the two membranes are a constant feature of mitochondria. These granules are thought to be the sites of deposition of calcium phosphate during oxidative phosphorylation (Peachy, 1964). Mitochondria both divide and fuse, and are highly mobile within the cell, moving to and aggregating near sites requiring sources of energy.

Plastids

Whereas most mitochondria are too small to be easily observed with the light microscope, plastids are much larger and one of the distinctive and conspicuous features of plant cell protoplasts (Fig. 3.7). They may be unpigmented and, thus, called **leucoplasts**, or pigmented and called **chromoplasts**. Chromoplasts are commonly grouped in two categories, one characterized by orange and yellow carotenoid pigments only, the other by a predominance of chlorophylls. The latter type, the **chloroplast**, also contains the orange and yellow pigments which, during most of a growing season, are masked by chlorophyll. The cessation of chlorophyll synthesis in late summer and early fall, and the resultant visibility of the carotenoid pigments, is largely responsible for the appearance of orange and yellow autumnal coloration in the leaves of trees and shrubs. The deep purplish-red leaf color as well as that of flower petals results from increased synthesis of anthocyanins, water-soluble pigments, in the cell vacuoles.

Some chloroplasts, such as those of green tomatoes, or immature red peppers, change during development into chromoplasts of the orange type. Whether this is the result primarily of a cessation of chlorophyll synthesis, or an increase in the synthesis of carotenoid pigments, or a combination of these, is not clear. Some carotenoid chromoplasts have an angular, crystalline appearance, apparently the result of crystallization of the carotene within them (Ledbetter and Porter, 1970). In such cases it is not certain whether they retain an enclosing plastid membrane.

It was not until the advent of the electron microscope that plastids were observed to be of membranous construction. The chloroplast (Fig. 3.7), extensively studied because of its role in photosynthesis, is by far the best understood of plastids (see Gunning, 1965a, b; Laetsch and Price, 1969). It is bounded by a double membrane which encloses an intricate membrane system supported within a matrix termed the **stroma**. The membrane system (Fig. 3.7a, b) consists of parallel **stroma lamellae** which connect compact stacks of specialized membrane, each component of which is called a **thylakoid**. Stacks of thylakoids comprise **grana**. In the chloroplasts of vascular plants, chlorophyll is restricted to the grana. Chromoplasts that lack chlorophyll also lack a complex internal membrane system. When plants are grown in the dark, the development of typical chloroplasts is inhibited, chlorophyll is not synthesized in the plastids that develop, and they are, thus, called **etioplasts**. They are characterized by **prolamellar bodies** (Fig. 3.7c) composed of tubular membranes which comprise a paracrystalline lattice. When exposed to light thylakoids develop from the components of the prolamellar body. It should not be surprising, therefore, that prior to seed germination, cells of the suspensor and embryo in some taxa are characterized by leucoplasts that contain prolamellar bodies (see, e.g., Johansson and Walles, 1993).

Chloroplasts, like mitochondria, contain ribosomes and, thus, have the ability to manufacture ATP. Both contain DNA and RNA, and are capable of protein synthesis. Both, furthermore, can replicate by simple fission. For these and other reasons it has been suggested that these organelles might have evolved from bacterial symbionts (see, e.g., Cohen, 1970; Raven, 1970).

Spherosomes and microbodies

Spherosomes are small bodies (about 1.0 μm in diameter) bounded by a single membrane and contain enzymes that synthesize oils and fats. They are thought to be derived from endoplasmic reticulum (Robards,

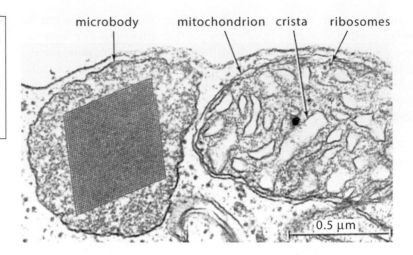

microbody **mitochondrion** **crista** **ribosomes**

0.5 µm

Figure 3.8 A microbody from a mesophyll cell of *Nicotiana tabacum* containing a large crystal. The microbody contrasts with plastids in having a single bounding unit membrane. From Esau (1977). Used by permission of John Wiley and Sons, Inc.

1970). **Microbodies** (Fig. 3.8) like spherosomes are bounded by a single membrane. They have a granular appearance and sometimes contain a conspicuous crystal. Some microbodies play an important role in photorespiration whereas others contain enzymes required for the conversion of fats to carbohydrates during seed germination.

Microtubules

Plant microtubules are slender tubes of indeterminate length, usually straight, with a sectional diameter of about 25 nm. These common constituents of eukaryotic cells were discovered in 1963 by Ledbetter and Porter who demonstrated that they are composed of spherical protein subunits (dimers of alpha and beta tubulin), forming a circle of 13 when observed in transverse section. Microtubules grow at each end by polymerization of tubulin dimers, and may become up to 1000 times as long as they are thick. Microtubules are commonly observed in the peripheral regions of cell protoplasts (Fig. 3.9a, b). They also comprise the spindle fibers in dividing cells. The microtubules of the nuclear spindle are, however, somewhat smaller in diameter than others having a diameter of about 20 nm (Robards, 1970). In regions of cell wall growth and cellulose microfibril synthesis, microtubules, just below the plasmalemma, are routinely observed in an orientation parallel to that of the newly synthesized microfibrils (Fig. 3.9b). This correlation suggests that they may exert some control over the orientation of the microfibrils. (For more detail, see Chapter 4).

Actin microfilaments

These long, thin, protein filaments, about 8 nm in diameter, are frequently found in association with tubules of endoplasmic reticulum in the peripheral region of the protoplast. Interaction of actin with the myosin of cell organelles results in movement of the organelles along actin microfilaments or bundles of microfilaments (Williamson, 1993). In cells of the Characean alga *Nitella*, just to the inside of the stationary, peripheral endoplasmic reticulum where cytoplasmic

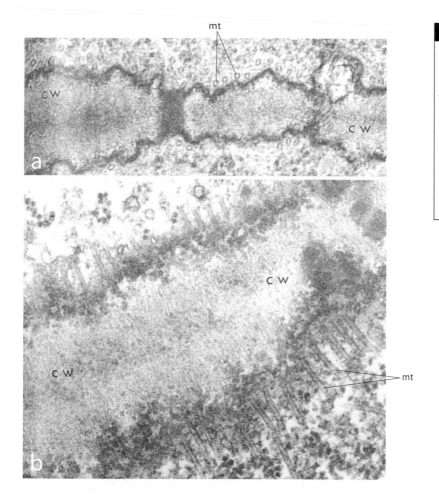

Figure 3.9 (a) Microtubules (mt), in sectional view, just inside the plasmalemma (and cell wall, CW) of contiguous cells of *Juniperus*. Magnification × 64 150. (b) Microtubules in an oblique section showing their parallel relationship to cellulose microfibrils in the cell wall. Magnification × 74 600. From Ledbetter and Porter (1963). Used by copyright permission of the Rockefeller University Press.

streaming occurs, the ER itself has been shown to move along bundles of actin microfilaments (actin cables) (Fig. 3.10a, b), exerting shear force that seems to result in cytoplasmic streaming throughout the interior of the cells (Kachar and Reese, 1988).

Vacuoles

A vacuole is a region within the cell bound by a differentially permeable membrane called the **tonoplast**. It contains water and dissolved substances such as pigments, salts, sugars, and compounds such as calcium oxalate which often crystallizes. Vacuoles may also contain tannins, organic acids, and amino acids. An immature cell usually contains many small vacuoles which fuse to form a single large central vacuole as the cell matures. As a consequence, the cytoplasm with its cell organelles becomes restricted to the peripheral region of the cell. One of the most significant functions of the vacuole is its role in the water balance of the cell. Active absorption of ions in excess of the concentration to the exterior results in increase in turgor pressure in the cell and its expansion and growth. (For a discussion of cell growth, please see Chapter 5.) The vacuole also has a hydrolytic function. In

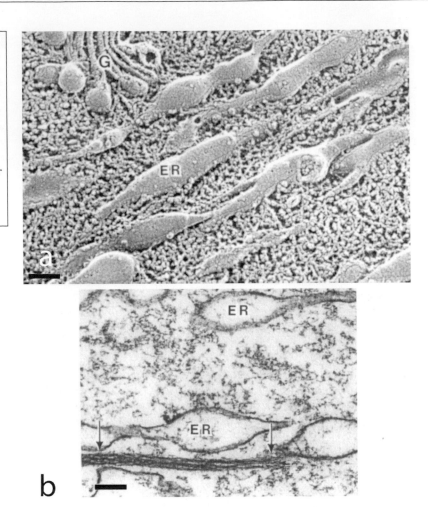

Figure 3.10 (a) Freeze-etch views of tubules of endoplasmic reticulum in the alga *Nitella*. The reticulate form of the ER has been distorted by cytoplasmic streaming. G, Golgi body. Bar = 0.17 μm. (b) The ER is underlain by cables of actin microfilaments (arrows) along which it moves. Bar = 0.17 μm. (a, b) From Kachar and Reese (1998). Used by copyright permission of the Rockefeller University Press.

some taxa, for example, in the cotyledons of some embryos, protein grains develop within small vacuoles. Upon digestion, the vacuoles fuse to form a larger vacuole. An especially interesting aspects of vacuoles is the incorporation within them of organelles such as mitochondria, ribosomes, and plastids. This occurs when the tonoplast becomes invaginated (i.e., extended inwardly) and organelles in the adjacent cytoplasm become enclosed by a vesicle composed of vacuolar membrane which pinches off within the vacuole. After lysis of the organelles the enclosing membrane disappears. The molecular components of the organelles can then be recycled within the cell. This is a function comparable to that of the lysosomes of animal cells.

Movement of organelles in the protoplast

It has been known for many years that organelles such as Golgi, mitochondria, and plastids move within the protoplast in what has been described as cytoplasmic streaming. Golgi have been observed to

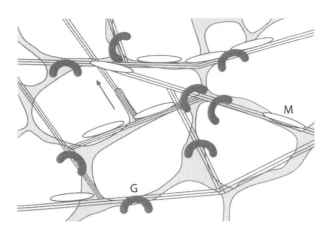

Figure 3.11 Diagram of a model of the organization of the peripheral (cortical) cytoplasm of a cell. The reticulate ER network (light shading) overlies a cytoskeleton of actin cables over which move Golgi bodies (G) and mitochondria (M). ER tubules can grow along actin cables and fuse with other tubules. The microtubule cytoskeleton has been omitted from the diagram. From Hawes and Satiat-Jeunemaitre (2001). Used by permission of the American Society of Plant Biologists.

aggregate in regions of synthesis, such as sites of new cell wall formation, and mitochondria in regions requiring energy such as regions of active transport through the plasmalemma or at the ends of elongating cells. It is clear, therefore, that such movement is not simply random as might be implied from the term cytoplasmic streaming. Recent research has demonstrated that organelle movements are related to tubules of endoplasmic reticulum (or in some cases to microtubules), closely associated actin microfibrils, and myosin "motors" (Williamson, 1993). A study by Boevink *et al.* (1998) shows conclusively that Golgi move along the ER tubules of the polygonal network just beneath the plasmalemma, and that a system of actin microfilaments precisely parallels the ER network. These results are consonant with reported observations of ER–actin complexes in several taxa by other researchers (e.g., Lichtscheidl *et al.*, 1990; Nebenführ *et al.*, 1999; and others). It is now widely accepted that the motility of organelles such as Golgi, mitochondria, plastids, etc. results from an interaction of myosin motors with the stationary actin microfilaments, utilizing energy from ATP (Fig. 3.11). It is known that Golgi accumulate various precursor compounds which are transported in Golgi vesicles to sites of synthesis as, for example, a region of cell wall in which wall thickening is occurring. The presumed function of Golgi as they move along the actin network is the absorption of these compounds from the stationary peripheral endoplasmic reticulum (Boevink *et al.*, 1998).

It seems unlikely that the release and transport of Golgi vesicles to the plasmalemma or vacuoles, or from the ER to Golgi, are random. In fact, Golgi have been observed to exhibit rapid stop and go movements. An interesting hypothesis of Nebenführ *et al.* (1999)

Figure 3.12 A diagrammatic representation of an hypothesis of regulated stop and go movement of Golgi bodies and its relationship to the transport of products through the secretory pathway. (a) A Golgi body with myosin motors attached moves along actin filaments. PM, plasma membrane. (b) At an ER exit site a local signal is produced that inhibits movement of the Golgi by releasing it from the actin tracks and allowing uptake from ER to Golgi transport vesicles. (c) At cell wall expansion sites, or sites of secondary wall thickening, a signal is produced that stops the Golgi movement and leads to the release of Golgi secretory vesicles. For more detail, please see the text. From Nebenführ *et al.* (1999). Used by permission of the American Society of Plant Biologists.

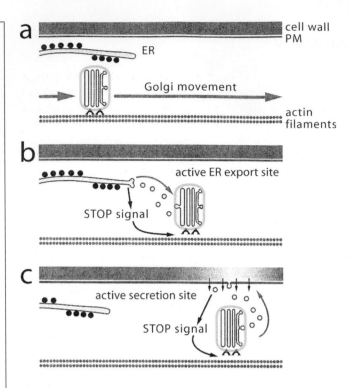

(Fig. 3.12) describes a possible regulated stop and go system as follows: Golgi bodies with attached myosin motors move along actin microfilaments. At an ER exit site, a local "stop" signal disconnects the Golgi body from the actin track which allows the transfer of ER vesicles to the Golgi body. At a site of cell wall thickening, for example, another local stop signal results in uncoupling the Golgi from the actin track and the transfer of vesicles containing precursor compounds of cellulose or cell wall matrix to the plasmalemma where the contents are expelled by exocytosis into the sites of cellulose and /or matrix synthesis. A candidate for the stop signal is calcium (Nebenführ *et al.*, 1999) which, in high concentrations, is known to block cytoplasmic streaming (Shimmen and Yokota, 1994). (For more detail, see the cited papers by Nebenführ *et al.* and Shimmen and Yokota.)

Movement of structures within pollen tubes has been shown to result from the association of actin microfilaments and microtubules. In *Nicotiana alata* (tobacco), for example, Lancelle *et al.* (1987) demonstrated actin microfilaments oriented in parallel with and cross-bridged to microtubules in the peripheral region of pollen tubes. It is generally believed that myosin is chemically bound to the tube nucleus and the generative cell in pollen tubes and that interaction between myosin and actin results in the movement of these structures (Williamson, 1993). The mechanism of nuclear movement is, however, complex and unclear at present. For an in-depth discussion of nuclear movement and extensive reference to the literature, see Williamson (1993).

Ergastic substances

Certain products of cell metabolism are appropriately referred to as **ergastic substances**, that is, substances resulting from the "work" of the cell. These substances are categorized as either **storage products** or **waste products**. Among the storage products are starch which accumulates in bodies of characteristic morphology termed **starch grains**, protein which forms **aleurone grains**, and oils often contained in plastids termed **elaioplasts** or in **spherosomes**, oil droplets sometimes enclosed in single membranes, but often lacking membranes.

Starch grains begin their development in either chloroplasts or leucoplasts. A leucoplast in which starch accumulates is termed an **amyloplast**. Starch grains vary in size from 12 to 100 μm in greatest dimension. They are variable in form as well, but are often ovoid, occurring singly or in aggregates. Each grain is composed of eccentric or concentric layers of starch around a **hilum**, the site around which starch synthesis is initiated. In the light microscope the hilum often appears dark and highly refractive. Although it has sometimes been considered a structure, its true nature is unclear. Since there is no evidence of the hilum in electron microscope photographs, it is probable that its appearance under the light microscope is an optical artifact. The layering of starch grains, often visible with the light microscope, is also not apparent in electron micrographs. The visible layering might be caused by imbibition of water and differential swelling of alternating layers of different composition (Esau, 1965). Starch, like cellulose, is composed of chains of glucose residues in orderly array. Consequently it is optically anisotropic and doubly refractive when viewed with polarized light.

Storage protein may occur in several forms, but it is best known as aleurone grains which occur in seeds. Fats and oils are of common occurrence in plant cells and take the form of solid, spheroidal particles or liquid droplets. Fats and oils may form either within the cytoplasm, where they are known as spherosomes, or within leucoplasts termed elaioplasts. Certain aromatic and highly volatile essential oils provide the distinctive odors associated with plants such as mint (*Mentha*), cedar (*Juniperus*), nutmeg (*Myristica*), etc. (see Chapter 15 on secretory structures).

Waste metabolites of the protoplast fall into two major categories, calcium oxalate crystals and tanniniferous substances. **Calcium oxalate crystals** (see Arnott and Pautard, 1970) often occur as very small, prismatic crystals in the cytoplasm. Since such crystals are sometimes so small that their edges and angles could not be resolved by nineteenth-century microscopes, botanists of that period often referred to them as crystal dust. Prismatic crystals may, however, also occur as single, large crystals which nearly fill a cell. Such cells which apparently function primarily as repositories for crystals are called **crystal idioblasts**. Aggregations of needle-like crystals, termed **raphides**, and compact, spherical aggregates of angular crystals, called **druses**, also occur in crystal idioblasts. Idioblasts containing solitary prismatic crystals are of common occurrence in the phloem of certain plants, especially in the Pomoideae. Raphide and druse idioblasts occur in many diverse groups, but the former are especially common in the monocotyledons, the latter possibly more common in dicotyledons. Crystal idioblasts may occur singly or in longitudinal rows. Although calcium oxylate crystals generally have been considered waste products, Arnott and Pautard (1970) suggest that some crystalline deposits may be a form of stored calcium. The crystal or crystal cluster apparently develops in the cell vacuole, and it is known that, in many cases, the protoplast dies following crystal formation. There is, however, good evidence that it may remain alive in some plants and retain a relatively normal appearance (Price, 1970).

Solitary protein crystals occur in some microbodies, organelles bounded by a single membrane. Crystalline calcium carbonate is rare in plants, and is best known in **cystoliths** that occur in specialized epidermal cells of the leaves of *Ficus elastica* (rubber plant). **Silica** is a common constituent of epidermal cell wall of grasses and also occurs as **silica bodies**, masses of silica that fill some epidermal cells.

Tannins and **tanniniferous substances** are phenolic by-products and are usually considered to be waste products. Although generally toxic to the protoplast in certain concentrations, they may be converted to non-toxic substances and stored or transferred to other regions of the plant. Some, such as cinnamic acid, are converted to lignans, and subsequently to lignin, an important component of the cell wall matrix (see Chapter 4). Tannin often appears microscopically as fine to coarsely granular material, but sometimes may occur in large, compact masses. It can occur in almost any tissue, and is characteristic of all major groups of plants. It is translocated into both the bark and the central secondary xylem of trees where its accumulation apparently contributes importantly to the death of these regions. It is very conspicuous in tissues of the cortex and pith of pteridophytes where it occurs in nontoxic form.

REFERENCES

Arnott, H. J. and F. G. E. Pautard. 1970. Calcification in plants. In H. Schraer, ed., *Biological Calcification: Cellular and Molecular Aspects*. New York: Appleton-Century-Crofts, pp. 175–446.

Boevink, P., Oparka, K., Santa Cruz, S., *et al.* 1998. Stacks on tracks: the plant Golgi apparatus traffics on an actin/ER network. *Plant J.* **15**: 441–447.

Cohen, S. S. 1970. Are/were mitochondria and chloroplasts microorganisms? *Am. Scientist* **51**: 281–289.

Esau, K. 1965. *Plant Anatomy*, 2nd edn. New York: John Wiley and Sons.
   1977. *Anatomy of Seed Plants*, 2nd edn. New York: John Wiley and Sons.

Gunning, B. E. S. 1965a. The fine structure of chloroplast stroma following aldehyde osmium-tetroxide fixation. *J. Cell Biol.* **24**: 79–93.

1965b. The greening process in plastids. I. The structure of the prolamellar body. *Protoplasma* **60**: 111–130.

Hawes, C. R. and B. Satiat-Jeunemaitre. 2001. Trekking along the cytoskeleton. *Plant Physiol.* **125**: 119–122.

Hepler, P. K., Palevitz, B. A., Lancelle, S. A., McCauley, M. M., and I. Lichtscheidl. 1990. Cortical endoplasmic reticulum in plants. *J. Cell Sci.* **96**: 355–373.

Johansson, M. and B. Walles. 1993. Functional anatomy of the ovule in broad bean, *Vicia faba* L. II. Ultrastructural development up to early embryogenesis. *Int. J. Plant Sci.* **154**: 535–549.

Kachar, B. and T. S. Reese. 1988. The mechanism of cytoplasmic streaming in Characean algal cells: sliding of endoplasmic reticulum along actin filaments. *J. Cell Biol.* **106**: 1545–1552.

Laetsch, W. M. and I. Price. 1969. Development of the dimorphic chloroplasts of sugar cane. *Am. J. Bot.* **56**: 77–87.

Lancelle, S. A., Cresti, M., and P. K. Hepler. 1987. Ultrastructure of the cytoskeleton in freeze-substituted pollen tubes of *Nicotiana alata*. *Protoplasma* **140**: 141–150.

Ledbetter, M. C. and K. R. Porter. 1970. *Introduction to the Fine Structure of Plant Cells*. Heidelberg, Germany: Springer-Verlag.

Lehninger, A. L. 1961. How cells transform energy. *Sci. American* **205**: 62–73. 1964. *The Mitochondrion*. New York: Benjamin.

Lichtscheidl, I. K., Lancelle, S. A., and P. K. Hepler. 1990. Actin–endoplasmic reticulum complexes in *Drosera*: their structural relationship with the plasmalemma, nucleus, and organelles in cells prepared by high pressure freezing. *Protoplasma* **155**: 116–126.

Mollenhauer, H. H. and D. J. Morré. 1966. Golgi apparatus and plant secretion. *Annu. Rev. Plant Physiol.* **17**: 27–46.

Nebenführ, A., Gallagher, L. A., Dunahay, T. G., *et al.* 1999. Stop-and-go movements of plant Golgi stacks are mediated by the acto-myosin system. *Plant Physiol.* **121**: 1127–1141.

Northcote, D. H. and J. D. Pickett-Heaps. 1966. A function of the Golgi apparatus in polysaccharide synthesis and transport in the root-cap cells of wheat. *Biochem. J.* **98**: 159–167.

Peachy, L. D. 1964. Electron microscope observations on accumulation of divalent cations in intramitochondrial granules. *J. Cell Biol.* **20**: 95–111.

Porter, K. R. 1961. The endoplasmic reticulum: some current interpretations of its forms and functions. In T. W. Goodwin and O. Lindberg, eds., *Biological Structure and Function*. New York: Academic Press, pp. 127–155.

Porter, K. R. and R. D. Machado. 1960. Studies on the endoplasmic reticulum. IV. Its form and distribution during mitosis in cells of onion root tip. *J. Biophys. Biochem. Cytol.* **7**: 167–180.

Price, J. L. 1970. Ultrastructure of druse crystal idioblasts in leaves of *Cercidium floridum*. *Am. J. Bot.* **57**: 1004–1009.

Raven, P. H. 1970. A multiple origin for plastids and mitochondria. *Science* **169**: 641–646.

Robards, A. W. 1970. *Electron Microscopy and Plant Ultrastructure*, London: McGraw-Hill.

Shimmen, T. and E. Yokota, 1994. Physiological and biochemical aspects of cytoplasmic streaming. *Int. Rev. Cytol.* **155**: 97–139.

Steward, F. C., Mapes, M. O., Kent, A. E., and R. D. Holsten. 1964. Growth and development of cultured plant cells. *Science* **143**: 20–27.

Weier, T. E., Stocling, C. R., and L. K. Shumway. 1967. The photosynthetic apparatus in chloroplasts of higher plants. *Brookhaven Symp. Biol.* **19**: 353–374.

Whaley, W. G., Mollenhauer, H. H., and J. H. Leech. 1960. The ultrastructure of the meristematic cell. *Am. J. Bot.* **47**: 319–399.

Wiedenhoeft, R. E. 1985. *Comparative aspects of plant and animal coated vesicles.* M.S. thesis, University of Georgia, Athens, GA.

Williamson, R. E. 1993. Organelle movements. *Annu. Rev. Plant Physiol. Plant Mol. Biol.* **44**: 181–202.

FURTHER READING

Allen, N. S. and D. T. Brown. 1988. Dynamics of the endoplasmic reticulum in living onion epidermal cells in relation to microtubules, microfilaments, and intracellular particle movement. *Cell Motil. Cytoskeleton* **10**: 153–163.

Buvat, R. 1963. Electron microscopy of plant protoplasm. *Int. Rev. Cytol.* **14**: 41–155.

Clowes, F. A. L. and B. E. Juniper. 1968. *Plant Cells.* Oxford, UK: Blackwell.

Frey-Wyssling, A. and K. Mühlethaler. 1965. *Ultrastructural Plant Cytology.* New York: Elsevier.

Ledbetter, M. C. and K. R. Porter. 1963. A "microtubule" in plant cell fine structure. *J. Cell Biol.* **19**: 239–250.

1964. Morphology of microtubules in plant cells. *Science* **144**: 872–874.

Loewy, A. G. and P. Siekevitz. 1969. *Cell Structure and Function*, 2nd edn. New York: Holt, Rinehart and Winston.

Mühlethaler, K. 1967. Ultrastructure and formation of plant cell walls. *Annu. Rev. Plant Physiol.* **18**: 1–23.

Sheetz, M. P. and J. A. Spudich. 1983. Movement of myosin-coated fluorescent beads on actin cables *in vitro. Nature* **303**: 31–34.

Vanderkooi, G. and D.C. Green. 1971. New insights into biological membrane structure. *BioScience* **21**: 409–415.

Chapter 4

Structure and development of the cell wall

Perspective

Nearly all plant cells are characterized by an enclosing, cellulosic wall. Those that are not, such as gametes, are either very short-lived or are protected by enclosure within a sheath or tissue of walled cells. In addition to its vital role in communication between cells, the wall serves both supporting and protective functions. Cell walls were first observed, in cork, by Robert Hooke in 1663 and considered to be "dead" structures. Furthermore, the cell wall, produced by and to the exterior of the protoplast, has been considered by some biologists to be an extracellular structure. Most botanists, however, have persisted in considering the wall to be the outer part of the cell, a view based largely on the integration of cytokinesis and cell wall formation. Strong justification for this viewpoint has been provided by research during the past several decades which has shown that the wall is a dynamic structure that receives biochemical information from the protoplast and sends information to it. Recent studies suggest that the wall is an integral component of a cell wall–plasma membrane–cytoskeleton continuum which provides a pathway for molecular and mechanical signals between cells in a tissue, or between cells and the external environment (Wyatt and Carpita, 1993; Reuzeau and Pont-Lezica, 1995; see also Wojtaszek, 2000). Major components of this continuum are plasmodesmata, highly specialized regions of endoplasmic reticulum which traverse the walls and connect the protoplasts of adjacent cells, microtubules, thought to play important roles in determining the orientation of cellulose microfibrils in the cell wall (Baskin, 2001), and actin microfilaments which have been implicated in cytoplasmic streaming and in the transport of vesicles containing precursor compounds to the sites of wall synthesis (Chaffey *et al.*, 2000).

The realization that the protoplasts of living cells of plants are interconnected by plasmodesmata has led to the concept of the **symplast**, a system that encompasses the totality of living protoplasts in a plant. The concept of the symplast can be traced back to Eduard

Tangl who in 1897 observed fine strands connecting the protoplasts of adjacent cells which, he noted, "unite them to an entity of higher order" (Oparka and Roberts, 2001). Similarly, the interconnected system of plant cell walls comprises the **apoplast** through which water and small molecules can be transported within the plant as well as to and from the external environment. The presence of the symplast is one of several major bases for the organismal theory of multicellularity in plants which views cells as compartments of the organism rather than as the building blocks as suggested by the cell theory (Kaplan and Hagemann, 1991; Kaplan, 1992; see also Wojtaszek, 2000; for more detail, see Chapter 2).

Structure and composition of the cell wall

As we have learned in Chapter 2, most cell walls are layered, consisting of both a **primary wall**, produced by the protoplast following mitosis and formation of the middle lamella, and a **secondary wall**, deposited subsequently on the inner surface of the primary wall. The secondary wall is commonly composed of three distinct layers, S1, S2, and S3 (Figs. 4.1a, b, 4.2). The microfibrils in the S1 layer, the first and outermost layer to be formed, are usually arranged in a shallow helix and, thus, the angle with the long axis of the cell is great. The S2 layer, typically the thickest layer of the secondary wall, is characterized by a steep helix of microfibrils, and the angle with the long axis of the cell is small. The S3 layer, the innermost layer of the secondary wall, is characterized by a shallow helix of microfibrils, thus the angle with the long axis of the cell, like that in the S1 layer, is great. Each of these layers is often composed of several to many thin lamellae (Fig. 4.2) (see Preston, 1974 and references therein for early studies) in which parallel microfibrils are oriented at slightly and progressively different angles, thus forming a helicoid pattern (Roland, 1981; Roland et al., 1987; see also Neville and Levy, 1985). It has been shown, further, that the change in direction of microfibrils between S1, S2, and S3 layers in secondary walls also results from a helicoidal rotation of microfibrils between these layers (Roland, 1981; Roland et al., 1987).

The microfibrillar structure of primary and secondary walls differs in part because the primary walls are synthesized during expansion growth (primarily elongation) of a cell whereas secondary walls develop after most cell growth has ceased. In primary walls, microfibrils appear to be randomly arranged (Figs. 4.1a, 4.3a), but with a generally transverse orientation in the inner part of the wall and a generally longitudinal orientation in the outer part. In contrast, microfibrils in the secondary wall are arranged in parallel within wall lamellae, but at different angles in different lamellae (Figs. 4.2, 4.3b). The structure of certain macroscopic features, for example the shape and orientation of pit borders and apertures, is related to the orientation of cellulose microfibrils in the middle (S2) layer of the secondary wall.

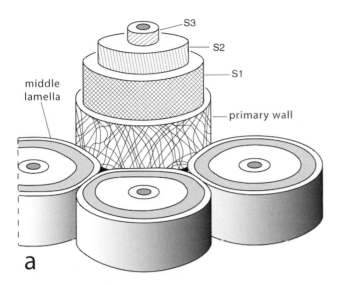

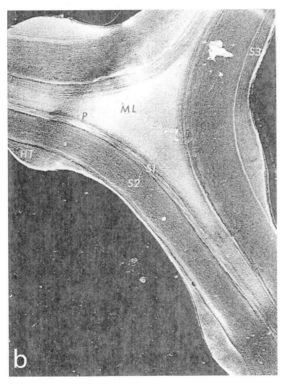

Figure 4.1 (a) Diagram illustrating the orientation of cellulose microfibrils in the primary cell wall, and the S1, S2, and S3 secondary wall layers and their relative thicknesses. (b) Transmission electron micrograph of a transverse section of parts of the walls of three tracheids in the secondary xylem of *Pseudotsuga* (Douglas fir) showing the middle lamella (ML), primary walls (P), and the three secondary wall layers (S1, S2, S3). Note also the sectional views of helical thickenings (HT) that form part of the S3 layer. Magnification × 6424. (b) From Côté (1967). Used by permission of the University of Washington Press.

The cellulose microfibrils, the basic units of wall structure, are embedded in a matrix of carbohydrates and other compounds (see below). If, however, we exclude, for the time being, consideration of the matrix, primary and secondary wall will be seen to contain cellulose microfibrils (Fig. 4.3a, b), often bound together into **macrofibrils** (Fig. 4.4). A **cellulose microfibril** consists of a core of glucose residues comprising cellulose molecules enclosed in hemicelluloses and pectins. As viewed in transverse section, microfibrils are oval to

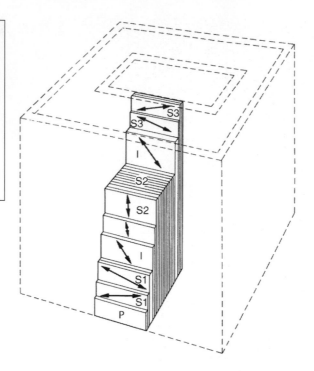

Figure 4.2 Diagrammatic representation of the wall layers of a tracheid. Arrows indicate the general orientation of the cellulose microfibrils in the several layers. Note the lamellae in the S1, S2, and S3 layers. The intermediate layers (I) and the S3 layer are absent in some species. From Preston (1974). Used by permission of Springer-Verlag New York, Inc.

circular in section, and vary in diameter from about 5 to 35 nm. In most plants microfibrils average about 10 nm in diameter, but in some algae they are much larger, averaging about 25 nm in diameter. Microfibrils are composed of groups of cellulose molecules arranged in crystalline lattices called **micelles** (Fig. 4.4) (Robards, 1970) which, together, form what is often referred to as the **crystallite**. The concept of the micelle as a distinct and separable entity, consisting of strands of cellulose molecules, was established in the nineteenth century by the German botanist Carl von Nägeli. The Swiss researcher Frey-Wyssling (1954) considered micelles to be subunits of strands of cellulose molecules that comprised what he called "elementary fibrils." Preston (1974) concluded that "the elementary fibril is an illusion based in part upon a confusion between microfibril size and crystallite size and in part upon a misinterpretation of electron microscope images," a viewpoint that has been widely accepted. Today, therefore, the basic cell wall unit is considered to be the microfibril, the central crystalline cellulose core of which contains micellar strands (Fig. 4.4). Whether or not the strands of cellulose of the crystallite are "distinct and separable entities" is still controversial.

Cellulose microfibrils are separated by **interfibrillar capillaries** (Fig. 4.4) containing a **matrix** composed of various mixtures of some, or all, of the following compounds: hemicelluloses, lignins, pectic compounds, suberin, cutin, simple sugars, proteins containing the amino acids, proline and hydroxyproline, and water. The molecules of the hemicelluloses (e.g., xylan, mannan, and glucan), as well as the pectin residues (galacturonic acid), are linked to the cellulose

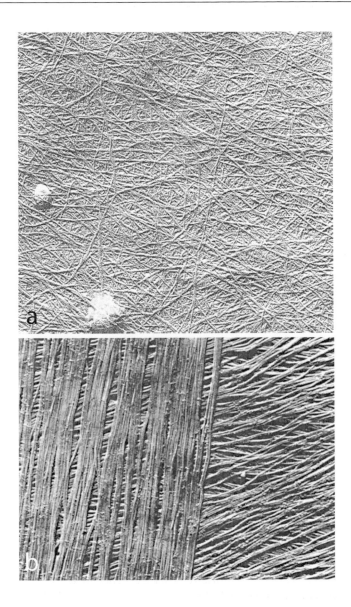

Figure 4.3 (a) Primary wall of a fiber from the secondary xylem of an angiosperm showing the arrangement of cellulose microfibrils. Magnification × 21 900. (b) Cell wall from the alga *Cladophora prolifera* showing different orientations of cellulose microfibrils in adjacent wall lamellae. Magnification × 15 230. (b) From Preston (1974). Used by permission of Springer-Verlag New York, Inc.

molecules; and according to Preston (1974) the molecular chains of the matrix are probably oriented parallel to the length of the microfibrils. Several models of the relationship between the cellulose microfibrils and the enclosing hemicelluloses and other compounds have been proposed. It was early suggested that the matrix polymers and the cellulose were bound together by covalent or hydrogen bonding (Keegstra *et al.*, 1973). In a more recent model (Fig. 4.5) the microfibrils are tethered by xyloglucan chains with the pectins and structural proteins filling the space between. The strength of a wall of this type would result from noncovalent bonding of the xyloglucan chains to the microfibril surfaces and the fact that some xyloglucan chains are embedded within the microfibrils (Fry, 1989; Hayashi, 1989; Cosgrove, 2001). A model of the developing primary wall proposed by

Figure 4.4 Diagrammatic representation of cell wall structure. (a) Part of a fiber. (b) A small fragment of the S2 layer of the wall of one cell consists of macrofibrils of cellulose between which are interfibrillar spaces filled with components of the wall matrix, shown in black. (c) Each macrofibril is composed of cellulose microfibrils embedded in non-cellulosic wall matrix. (d) Each microfribil consists of cellulose molecules. (e) Groups of microfibrils called micelles are arranged in a crystalline lattice. (f) A fragment of a cellulose molecule consisting of two glucose residues connected by an oxygen atom. From Esau (1977), slightly modified. Used by permission of John Wiley and Sons, Inc.

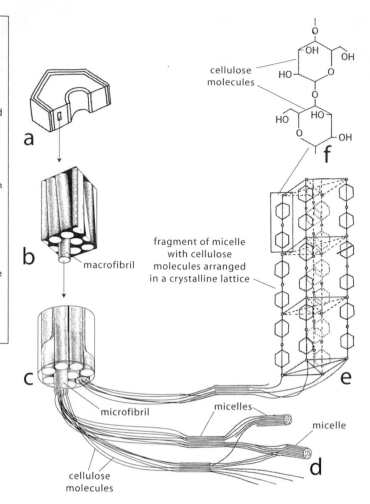

cellulose molecules

f

fragment of micelle with cellulose molecules arranged in a crystalline lattice

a

b

macrofibril

c

microfibril

e

micelles

micelle

d

cellulose molecules

Figure 4.5 A model of cell wall structure in which the cellulose microfibrils are tethered to each other by chains of xyloglucans (hemicelluloses). Pectins and structural proteins fill the space in between. A is a view in the plane of the wall, i.e., parallel to the plasmalemma. B is a sectional view, at right angles to that in A in which one is looking into the ends of the microfibrils of cellulose. ml, middle lamella; pl, plasmalemma. From Cosgrove (1999). Reprinted with permission from the *Annual Review of Plant Physiology and Plant Molecular Biology*, Volume 50, © 1999 Annual Reviews, http://www.annualreviews.org.

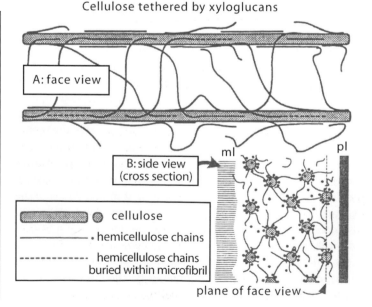

Cellulose tethered by xyloglucans

A: face view

B: side view (cross section)

ml

pl

cellulose
• hemicellulose chains
hemicellulose chains buried within microfibril

plane of face view

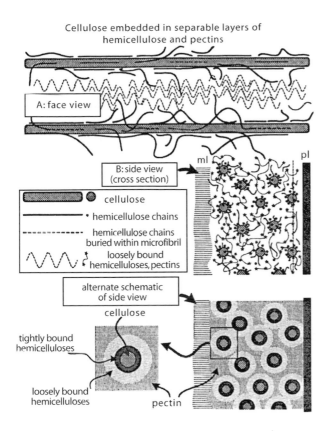

Cellulose embedded in separable layers of hemicellulose and pectins

A: face view

B: side view (cross section)

cellulose
• hemicellulose chains
hemicellulose chains buried within microfibril
loosely bound hemicelluloses, pectins

ml

pl

alternate schematic of side view

cellulose

tightly bound hemicelluloses

loosely bound hemicelluloses

pectin

Figure 4.6 A model of cell wall structure, adapted from Talbot and Ray (1992), in which cellulose microfibrils are enclosed by a layer of tightly bound hemicelluloses. The microfibrils and associated hemicelluloses are embedded in a matrix of pectin. ml, middle lamella; pl, plasmalemma. From Cosgrove (1999). Reprinted with permission from the *Annual Review of Plant Physiology and Plant Molecular Biology*, Volume 50, Ⓒ 1999 Annual Reviews, http://www.annualreviews.org.

Talbot and Ray (1992) envisions the microfibrils enclosed by a layer of tightly bound hemicelluloses such as xyloglucan ensheathed by a layer of loosely bound hemicelluloses (Fig. 4.6). The microfibrils and associated hemicelluloses are embedded in a matrix of pectin. Whereas the xyloglucan chains may be embedded in the microfibrils, they are not linked with other microfibrils. The strength of the walls is thought to result from non-covalent bonding between matrix polymers (Talbot and Ray, 1992; Cosgrove, 1999, 2001).

The relative proportions of the compounds in the matrix vary in accordance with the nature of the cell and its function. For example, the wall matrix in fibers that function in mechanical support contain large amounts of lignin which imparts both strength and flexibility to the cells. On the other hand, the wall matrix of epidermal parenchyma cells that prevent water loss from the plant contains large quantities of cutin, a compound impermeable to the passage of water. Except in walls containing suberin or cutin, walls are relatively porous. The interfibrillar capillaries in such walls are composed primarily of hydrated pectins that comprise the hydrophilic domain of the cell wall through which pass solutes of low moleclar weight and which allow the functioning of enzymes that facilitate the synthesis of wall components (Canny, 1995). The hydrophobic domain is formed by the crystalline cellulose/hemicellulose network where bonding between molecules of cellulose and hemicellulose lead to exclusion of water molecules from between the microfibrils (Wojtaszek, 2000).

One of the most important compounds in the cell wall matrix is **lignin** (a complex polymer of phenylpropane with *p*-coumarylic and synapylic acids) which may be added to the wall in various quantities, and at varying stages during development. It is often especially strongly concentrated in the primary wall and the S1 layer of the secondary wall. Its presence imparts strength to the wall, especially important in cells such as tracheids and vessel members through which water is transported. Were the walls of these cells unlignified they would probably collapse upon death of the protoplasts and the consequent loss of turgor pressure, thus impeding the transport of water through them. Another matrix compound of potentially great significance is the glycoprotein **expansin**, a component of the primary wall, which is thought to control wall extension during growth (Romberger *et al.*, 1993; Cosgrove, 2000; Cosgrove *et al.*, 2002).

Growth of the cell wall

Historically, two types of wall growth have been postulated, growth by intussusception and growth by apposition. Growth by **intussusception** as first proposed in 1858 by von Nägeli requires the synthesis of new wall substance within the primary wall during cell and wall expansion. This concept would require the transmission of precursor compounds of wall synthesis into the wall, and would seem to necessitate a loose, reticulate meshwork of microfibrils with room for the intermeshing of newly formed microfibrils. In the past, some workers have cited the fact that microfibrils in the primary wall *appear* to be intertwined as evidence for growth by intussusception. Although others have found it difficult to apply the concept of intussusception to the synthesis of new microfibrils within the wall, they have noted that it could easily be applied to the synthesis of new matrix compounds.

The concept of wall growth by **apposition** was suggested by Strasburger in 1882. Apposition of wall material requires the production of new layers of wall material, one upon another, like the application of plaster to the wall of a house. Strasburger as well as more recent workers considered this to be the pattern of secondary wall synthesis after the cell had stopped expanding. However, the Dutch botanists Roelofson and Houwink (1953) applied the concept of apposition to primary cell wall growth as well as expounded in their multinet hypothesis of wall growth. It has been observed that the outermost (oldest, and first-formed) primary wall layers contain longitudinally oriented microfibrils and that those in the most recently synthesized part of the wall (the youngest part) are more nearly transversely oriented. On this basis Roelofson and Houwink

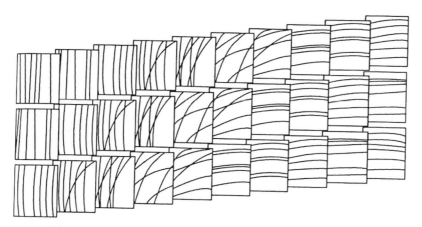

Figure 4.7 A diagram showing change of about 90 degrees in the orientation of cellulose microfibrils in the primary walls of meristematic and isodiametric parenchyma cells as wall synthesis occurs over time. The squares showing orientation of microfibrils (lines) represent a succession of inner wall sites. Note the gradual change in orientation from horizontal (transverse) on the right to vertical on the left. The changes in orientation of microfibrils do not occur simultaneously at all sites, but follow the same pattern. For more detail, please see the text. From Wolters-Arts *et al.* (1993). Used by permission of Springer-Verlag Wien.

concluded that, in cells that increase in length during growth, as layers of cellulose microfibrils (and associated matrix) are deposited one upon the other there is a passive change in orientation of microfibrils from nearly transverse (or sometimes random) to longitudinal resulting from elongation of the wall (see also Preston, 1982). Although this hypothesis has gained widespread acceptance, evidence of a helicoidal pattern of microfibrils in primary walls (Neville and Levy, 1984; Vian *et al.*, 1993; Wolters-Arts *et al.*, 1993) has suggested a different interpretation of the observations upon which it was based. For example, Wolters-Arts *et al.* (1993) observed a helicoidal change, over time, in orientation of newly deposited microfibrils of about 90 degrees in primary walls of cells of several types including cylindrical cells obtained from shoot apices, isodiametric cells obtained from mature bulbs and cells in suspension obtained from tobacco. They conclude that the change in orientation occurred gradually as successive layers of microfibrils were deposited at different angles (Fig. 4.7). They state that the change in direction of the microfibrils "could not have been due to passive reorientation" as proposed by the multinet hypothesis since the microfibrils of changed orientation at each site were newly synthesized and deposited. Romberger *et al.* (1993) believe that a helicoidal pattern of wall growth is "not necessarily irreconcilable with the multi-net growth hypothesis." They suggest further that "[this type of growth] allows the primary wall to be an active rather than a merely passive, mechanical participant in determining the direction and extent of growth."

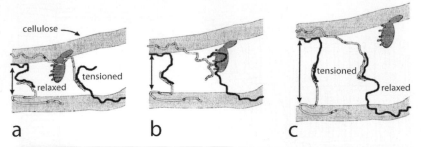

cellulose

tensioned

relaxed

tensioned

tensioned

relaxed

a   b   c

Figure 4.8 A diagrammatic illustration of wall loosening by expansin. It is hypothesized that the cellulose microfibrils become more widely separated when the expansin protein disrupts the bonding of the glycans (hemicellulose chains) to either the microfibril (a) or to each other (b) resulting in wall loosening (c). From Cosgrove (2000). Used with permission of Macmillan Magazines Limited.

Whereas growth in thickness of the secondary wall occurs largely after cell elongation has ceased, the primary wall which grows during cell elongation must expand in area predominantly in the direction of growth, and at the same time maintain its continuity. Several decades ago, resulting from the research of many workers, it was concluded that plant cells increase in size more rapidly in conditions of low pH, especially when it was below 5.5. It was hypothesized that in the early stages of wall growth auxin influences its acidification and stimulates the synthesis of one or more wall loosening enzymes (see Rayle and Cleland, 1992; Cosgrove, 2001). Subsequent efforts to identify wall-loosening enzymes resulted in the discovery of two proteins called **expansins** that mediate acid-induced wall extension (see Cosgrove, 2000, 2001). Consequently, it is widely accepted today that expansion of the primary wall involves wall loosening mediated by hormones such as auxin and the protein expansin, and the constant addition of new cell wall constituents required to maintain the continuity of the wall (see Lyndon, 1994; Fleming *et al.*, 1999; Cosgrove 2000). Although the mechanism of wall loosening is unknown, expansin is thought to "unlock" the system of wall polysaccharides, possibly through weakening glucan–glucan binding, thus allowing a separation of microfibrils (Fig. 4.8). With the continuing synthesis of new cellulose and under the influence of turgor pressure, cell walls grow (Cosgrove, 2000; Cosgrove *et al.*, 2002). As noted above, cellulose microfibrils (and associated microtubules lying just beneath the plasmalemma) are oriented more or less transversely to the long axis of the elongating cells. This orientation of microfibrils is believed to facilitate longitudinal expansion of the primary wall following wall loosening whereas a longitudinal orientation apparently restricts longitudinal wall growth (Seagull, 1986; Sauter *et al.*, 1993; Paolillo, 1995). It should be noted, however, that Paollilo (2000) has observed that in the outer epidermal walls of different parts of several taxa elongation continues after microfibrils become oriented longitudinally.

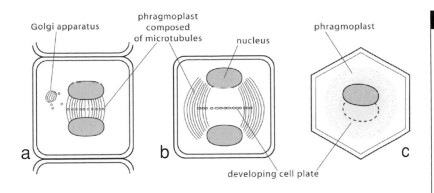

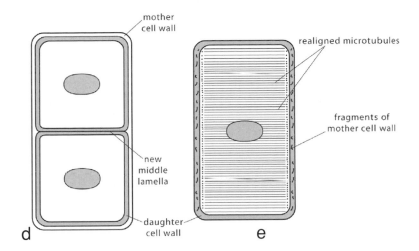

Figure 4.9 Formation of a new cell wall following mitosis. (a) Upon separation of the daughter nuclei, vesicles derived from Golgi bodies accumulate in a plane among the microtubules of the phragmoplast. (b) The phragmoplast begins to migrate toward the lateral walls of the cell. The Golgi vesicles begin to fuse, forming the cell plate. (c) The phragmoplast and the developing cell plate between the daughter nuclei, as seen from above. (d) The two new cells are separated by a new middle lamella, and each cell has a newly formed cell wall. (e) As each new cell begins to elongate, the microtubules become realigned transversely just beneath the plasmalemma. The mother cell wall disintegrates.

Cell wall development

The precursor compounds of both cellulose and matrix materials are synthesized in the protoplast and transported to the developing wall where synthesis is completed. Let us consider how this transfer of materials is accomplished. During the latter stages of mitosis, microtubules become arranged between the daughter nuclei, forming the phragmoplast, and Golgi vesicles aggregate in the "equatorial" plane between incipient daughter cells (Figs. 4.9a–c, 4.10). This disk-like aggregation of vesicles, called the cell plate and the surrounding phragmoplast, extend toward the original wall of the mother cell (Fig. 4.9b, c). Upon fusion the vesicle membranes become the plasmalemmas of contiguous protoplasts of the new cells, and the contents of the vesicles the new middle lamella that separates them (Fig. 4.9d). Golgi vesicles containing precursor compounds of wall material fuse with the plasmalemma of each new cell (Figs. 4.11, 4.12), adding to its area and at the same time ejecting their contents into the **periplast** immediately outside the plasmalemma where wall synthesis, a process including both enzymatic and non-enzymatic mechanisms, takes place (see Cosgrove, 1997; 1999). A new primary

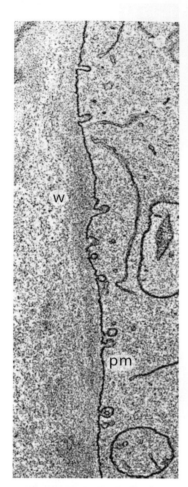

Figure 4.11 Fusion of Golgi vesicles with the plasmalemma during wall formation. Wall matrix compounds such as hemicelluloses and pectins contained within the vesicles are, thus, emptied into the region of wall formation. pm, plasma membrane; w, wall. Magnification × 23 520. From Whaley et al. (1960). Used by permission of the Botanical Society of America.

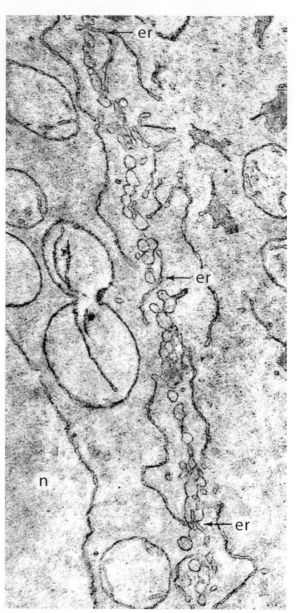

Figure 4.10 A developing cell plate consisting of numerous Golgi vesicles, some of which have fused with others. The vesicle membranes become the plasmalemmas of the new cells, the vesicle contents the new middle lamella. er, endoplasmic reticulum; n, nucleus. Magnification × 31 930. From Whaley et al. (1960). Used by permission of the Botanical Society of America.

wall is synthesized in both daughter cells. The original primary wall of the mother cell disintegrates (Fig. 4.9e), and the middle lamella between the daughter cells becomes continuous with that between the mother cell and contiguous cells. In preparation for elongation of a new cell, microtubules become transversely arranged just under the plasmalemma (Fig. 4.9e).

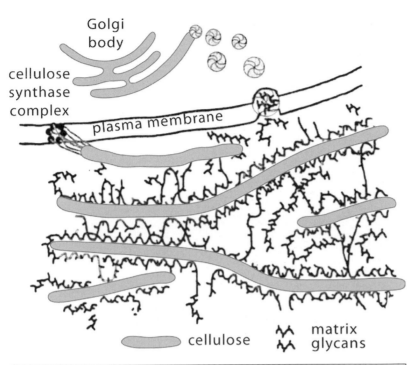

Figure 4.12 Diagram illustrating the biosynthesis of the cell wall. Cellulose microfibrils are generated by cellulose synthase complexes (large complexes of enzymes) in the plasmalemma (plasma membrane), and hemicelluloses and pectins (glycans) which comprise the wall matrix are synthesized in Golgi bodies and delivered to the wall by secretory vesicles. Within the developing wall, pectins form ionic gels and the hemicelluloses bind to the cellulose microfibrils. From Cosgrove (2000). Used by permission of Macmillan Magazines Limited.

The processes of cellulose microfibril formation and orientation during wall development have been the subjects of extensive research by many scientists during the past 30 years or so. Roelofson (1958) suggested that an enzyme located at the end of a microfibril might be responsible for the polymerization of glucose and its crystallization into a microfibril (see Brown, 1985). In 1976 Brown and Montezinos actually discovered cellulose microfibril synthesizing complexes in the alga *Oocystis*. With continuing research in many laboratories it is now well established that microfibrils are generated by terminal **cellulose synthase complexes** in the plasma membrane (Fig. 4.12) as earlier hypothesized by Heath (1974). These enzyme complexes are probably delivered to the plasmalemma in Golgi vesicles (Giddings and Staehelin, 1991). Two types of microfibril-generating complexes have been recognized, **linear complexes** of particles, commonly found in algae, and **rosette-shaped complexes** (Fig. 4.13) found in some algae and in higher plants. The rosette-shaped complexes consist of a cluster of six particles and sometimes contain a centrally located globular component (Fig. 4.14) (Brown, 1985; Delmer, 1987). Since these complexes occur at the ends of microfibrils, it has been hypothesized that the particles of which they are comprised are the enzymes

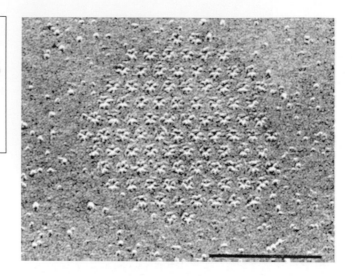

Figure 4.13 Cellulose synthase complexes, also called rosettes, in the plasmalemma of a cell of the alga *Microsterias denticulata*, as seen in a freeze-fracture electron micrograph. Bar = 0.1 μm. From Giddings *et al.* (1980). Used by copyright permission of the Rockefeller University Press.

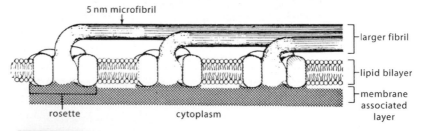

Figure 4.14 Diagram of a model of cellulose synthase complexes (rosettes) in the plasmalemma of *Microsterias*, and the microfibrils they have formed. From Giddings *et al.* (1980). Used by copyright permission of the Rockefeller University Press.

responsible for the synthesis of β-1,4-glucan chains of which cellulose is comprised (Giddings and Staehelin, 1988). These glucan chains "self-associate" through intra- and interchain hydrogen bonding forming insoluble, crystalline microfibrils (Delmer, 1987). Cellulose synthase complexes of similar form occur in algae, pteridophytes, and seed plants. The rosette type is especially common in pteridophytes and seed plants.

Extensive research during the past 20 years has demonstrated that there is frequently, but not always, a correlation between the orientation of microtubules and microfibrils. Consequently, it is believed that microtubules play an important role, perhaps assert some control, in aligning microfibrils (see, e.g., Lloyd, 1984; Seagull, 1990, 1991; Wymer and Lloyd, 1996; Baskin, 2001; Baskin *et al.*, 2004). Correlation between microtubules and microfibrils is especially apparent in regions of synthesis of the borders of circular bordered pits as well as the annular and helical thickenings in the primary walls of tracheary elements (Hogetsu, 1991). Immediately below the developing borders and wall thickenings lie microtubules, in contact with the plasmalemma, oriented in patterns identical to those of the developing pit borders and wall thickenings. Because there is not always a parallelism between microtubules and microfibrils, especially in walls with a helicoidal arrangement of microfibrils, and because in several cases depolymerization of microtubules did not affect the orientation of microfibrils, several authors have concluded that microtubules are not essential for the alignment of microfibrils (Roland and Vian, 1979; Boyd, 1985; Emons and Kieft, 1994; Emons and Mulder, 1998). Alternative explanations for microfibril orientation include self-assembly of cell wall components (Roland and Vian, 1979; Boyd, 1985), and the geometry of elongating cells (Emons and Kieft, 1994; Emons and Mulder, 1998). A recent model, proposed by Baskin (2001), postulates that self-assembly, cell geometry, and microtubule orientation all play a part in the development of cell wall structure. For more detailed discussions of the possible role of microtubules in orienting microfibrils, please see Chapters 7 and 11.

An interesting hypothesis for the mechanism of microfibril alignment, where strong evidence indicates a close relationship between microtubule and microfibril orientation, has been proposed by Giddings and Staehelin (1988). They hypothesize that the microtubules are connected to the plasmalemma by protein bridges. They suggest that the terminal cellulose synthase complexes from which microfibrils are synthesized are restricted to "channels" between the bridges of adjacent microtubules, and are pushed along these channels by the polymerization and crystallization of cellulose microfibrils. The microtubules are, themselves, possibly oriented by actin microfilaments (see Kobayashi et al., 1988). For more detail the interested student should consult the original papers cited above.

Plasmodesmata

Although plasmodesmata pass through cell walls they are not, strictly speaking, parts of the walls. Because, however, they play such an important role in development it is appropriate to provide here a detailed discussion of their structure and function. **Plasmodesmata** are highly specialized organelles that connect the protoplasts of adjacent protoplasts through the cell walls. They provide passageways for intercellular communication through the transport between cells of chemical substances ranging from small, soluble molecules to macromolecules such as proteins and nucleic acids (Ehlers and Kollmann, 2001). Although it is theoretically possible that all protoplasts may be connected by plasmodesmata, it is not certain that in large multicellular organisms there is ever any coordinated control throughout the entire organism. Recent research has shown, instead, that specific regions may be isolated from others by the absence of plasmodesmata, or even the modification of plasmodesmatal function (Ehlers et al., 1999), resulting in the formation of symplastic domains, localized sites of function, and/or development of certain tissue regions, tissues, or cell types.

Plasmodesmata originate following mitosis during cell plate formation. During cell division, Golgi and ER vesicles accumulate in the plane of the developing cell plate (Figs. 4.10, 4.15), and tubules of

Figure 4.15 Diagram illustrating the formation of primary and secondary plasmodesmata. Primary plasmodesmata develop from tubules of endoplasmic reticulum that are trapped between Golgi vesicles during cell plate formation just prior to the completion of cytokinesis. Secondary plasmodesmata develop in previously formed walls. Endoplasmic reticulum and Golgi vesicles aggregate on opposite sides of the wall which thins greatly, the endoplasmic reticulum tubules penetrate the wall, fuse, and differentiate, usually forming branched plasmodesmata. New wall is synthesized around the plasmodesmata and the wall regains its original thickness. From Kragler et al. (1998). Used by permission of Oxford University Press.

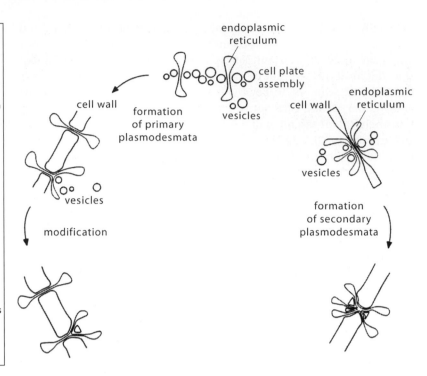

Figure 4.16 Primary plasmodesmata in the young walls of cells in a root tip of Zea mays, showing their relationship with endosplasmic reticulum (ER). Bar = 0.1 μm. From Lucas et al. (1993). Used by permission of Blackwell Publishing, Ltd.

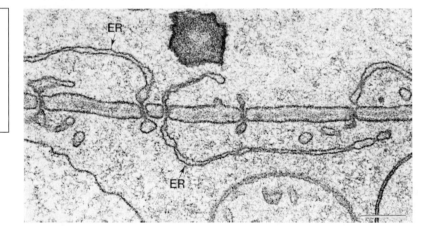

ER become trapped between them. As the vesicles fuse, forming the middle lamella, and primary wall is synthesized, these tubules of ER and associated cytoplasm (Fig. 4.16) become adpressed and differentiate into mature plasmodesmata (Figs. 4.17, 4.18). A plasmodesma consists of a **central cylinder** to which helically arranged proteinaceous particles are fused, comprising the **desmotubule** enclosed by plasmalemma (derived from vesicle membrane) which is continuous between adjacent cells (Figs. 4.17a, b, 4.18a, b). Centrally located within the desmotubule is a row of particles, embedded in the lipid of the ER, referred to as the **central column** (Fig. 4.18a, b). Embedded within its inner surface is a cylinder of protein particles (Fig. 4.18a, b). With expansion of the outer cylinder of plasmalemma and formation

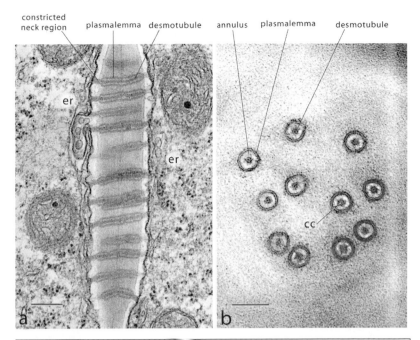

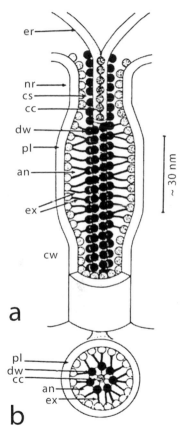

Figure 4.17 Plasmodesmata traversing the walls of adjacent phloem parenchyma cells in a leaf of *saccharum officinarum*. (a) Plasmodesmata as seen in longitudinal view. er, endoplasmic reticulum. Bar = 200 nm. (b) Plasmodesmata in sectional view. cc, central column. Bar = 100 nm. From Robinson-Beers and Evert (1991). Used by permission of Springer-Verlag GmbH and Co. KG. © Springer-Verlag Berlin Heidelberg.

of the **cytoplasmic annulus**, some of these particles become connected with those of the desmotubule by "spoke-like extensions," resulting in a system within the annulus of "microchannels" through which molecules are thought to pass (Fig. 4.18a, b). The narrow region at each end of the plasma membrane sheath is called the **orifice**, the region in which molecular transport is believed to be regulated. For more detail, see Ding *et al.* (1992) and Lucas *et al.* (1993) from which the foregoing description is taken.

Plasmodesmata that are formed during the process of cytokinesis are called **primary plasmodesmata** (Fig. 4.15). With increase in wall thickness, primary plasmodesmata may branch (Figs. 4.15, 4.19) (Kragler *et al.*, 1998; Ehlers and Kollmann, 2001). **Secondary plasmodesmata** form, *de novo*, in developing tissues across previously formed walls. At the sites of new secondary plasmodesmata the wall thins and ER cisternae and Golgi vesicles aggregate on either side of these regions (Fig. 4.15). The ER cisternae penetrate the thin wall, fuse, and differentiate, resulting in plasmodesmata, often branched, with a structure apparently identical to that of primary plasmodesmata. At these sites, following synthesis of new wall components (matrix from precursor compounds provided by the Golgi vesicles, and new cellulose microfibrils) the wall thickens around the secondary plasmodesmata (Monzer, 1991; see also Kragler *et al.*, 1998; Ehlers and Kollmann, 2001). Formation of secondary plasmodesmata is common

Figure 4.18 A model of a plasmodesma. (a) Longitudinal view. (b) Transverse (sectional) view. an, annulus; cc, central column; cs, cytoplasmic sleeve; cw, cell wall; dw, desmotubule wall; er, endoplasmic reticulum; ex, spoke-like extensions; nr, neck region; pl, plasmalemma. From Ding *et al.* (1992) (labeling slightly modified). Used by permission of Springer-Verlag Wien.

Figure 4.19 Modification of primary plasmodesmata during increase in cell wall thickness. Branches develop by the enclosure of branched tubules of endoplasmic reticulum. pl, plasmalemma; w, original cell wall; ml, middle lamella; nw, new cell wall; G, Golgi body; Gv, Golgi vesicle. From Ehlers and Kollman (2001). Used by permission of Springer-Verlag Wien.

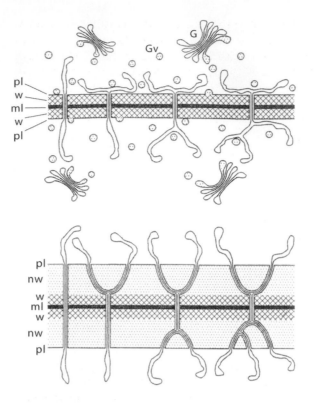

in the walls of cells that elongate greatly during development, thus compensating for the decrease in frequency of primary plasmodesmata. Secondary plasmodesmata are also thought commonly to occur in "non-division" walls as, e.g., the walls of cells that form the bounding layers between tissue regions of different ontogenetic origin such as the ground tissue (incipient mesophyll) and provascular tissue in developing leaves (Ehlers and Kollmann, 2001; see also Dengler *et al.*, 1985). Another good example is their presence in the periclinal walls between protodermal and underlying cells. Since, during development of the epidermis, protodermal cells divide only anticlinally, the walls between them and underlying cells are non-division walls. The non-division walls of cells in the carpel margins of some taxa which fuse only during gynoecium development also contain secondary plasmodesmata (van der Schoot *et al.*, 1995). The importance of secondary plasmodesmata is emphasized by the fact that they comprise a significant proportion of plasmodesmata that connect living cells in the plant (Ding *et al.*, 1992; see also Kragler *et al.*, 1998).

A major function of plasmodesmata is the interchange of information through the transmission of molecules from one protoplast into another, a process generally referred to as "trafficking." Molecular trafficking is essential to the regulation of physiological and developmental processes in the plant. Small molecules can diffuse freely through plasmodesmata, but large molecules may be restricted from passage by the molecular **size exclusion limit** of the plasmodesmata.

Some large molecules bind with certain proteins, called movement proteins, which facilitate the passage through the plasmodesma by increasing its size exclusion limit (see e.g. Kragler *et al.*, 2000; Oparka and Roberts, 2001; Aoki *et al.*, 2002). On the other hand, some plasmodesmata, such as those between companion cells and sieve tube members, have sufficiently large size exclusion limits which allow the direct passage of proteins that are essential to maintain the viability of the enucleate sieve tube members. Since the nutrients, hormones, and proteins essential to support the regulation of development and other physiological processes which characterize a particular developmental or functional domain vary, it is probable that specific mechanisms for the control of molecular trafficking in domains with different requirements must have evolved (Ehlers and Kollmann, 2001).

Recent studies (see Evert *et al.*, 1996) have demonstrated that the function of plasmodesmata may change during development. During early stages of development, leaves function as sinks since they utilize more photosynthate than they produce. Upon maturity, however, they become nutrient sources. During this transition from sink to source, plasmodesmata undergo a change in structure and function. Studies on tobacco by Oparka *et al.* (1999) indicate that during the stage in which developing leaves are nutrient sinks, the simple, unbranched plasmodesmata have a high size exclusion limit and photosynthate is transmitted from the phloem throughout the immature tissues of the leaf where it is utilized in the various processes of metabolism during development. As the leaf becomes a source and exporter of photosynthate, the plasmodesmata become much branched which increases their cross-sectional area, and which may enhance their capacity in phloem loading (Ehlers and Kollmann, 2001; see also Beebe and Russin, 1999; Oparka *et al.*, 1999). These changes in structure and function involve both primary and secondary plasmodesmata (Ehlers and Kollmann, 2001).

New plasmodesmata are not only formed during development, but also may become non-functional, or even eliminated, another way in which development can be controlled in the plant. For example, the protoplasts of developing tracheary cells are connected to adjacent parenchyma cells by primary pit fields that contain numerous plasmodesmata. With approaching maturity of the tracheary cells and just prior to autolysis of their protoplasts, both ends of the plasmodesmata are covered by newly synthesized layers of cell wall material, thus eliminating their functional ability (Lachaud and Maurousset, 1996; Kragler *et al.*, 1998).

REFERENCES

Aoki, K., Kragler, F., Xoconostle-Cazares, B., and W. J. Lucas. 2002. A subclass of plant heat shock cognate 70 chaperones carries a motif that facilitates trafficking through plasmodesmata. *Proc. Natl Acad. Sci. USA* **99**: 16342–16347.

Baskin, T. I. 2001. On the alignment of cellulose microfibrils by cortical microtubules: a review and a model. *Protoplasma* **215**: 150–171.

Baskin, T. I., Beemster, G. T. S., Judy-March, J. E., and F. Marga. 2004. Disorganization of cortical microtubules stimulates tangential expansion and reduces the uniformity of cellulose microfibril alignment among cells in the root of *Arabidopsis*. *Plant Physiol.* **135**: 2279–2290.

Beebe, D. U. and W. A. Russin. 1999. Plasmodesmata in the phloem-loading pathway. In A. J. E. van Bel and W. J. P. Kesteren, eds., *Plasmodesmata: Structure, Function, and Role in Cell Communication*. Berlin: Springer-Verlag, pp. 261–293.

Boyd, J. C. 1985. *Biophysical Control of Microfibril Orientation in Plant Cell Walls*. Dordrecht: Nijhoff.

Brown, R. M., Jr. 1985. Cellulose microfibril assembly and orientation: recent developments. *J. Cell Sci. (Suppl.)* **2**: 13–32.

Brown, R. M., Jr. and D. Montezinos. 1976. Cellulose microfibrils: visualization of biosynthetic and orienting complexes in association with the plasma membrane. *Proc. Natl Acad. Sci. USA* **73**: 143–147.

Canny, M. J. 1995. Apoplastic water and solute movement: new rules for an old space. *Annu. Rev. Plant Physiol. Plant Mol. Biol.* **46**: 215–236.

Chaffey, N., Barlow, P., and J. Barnett. 2000. A cytoskeletal basis for wood formation in angiosperm trees: the involvement of microfilaments. *Planta* **210**: 890–896.

Cosgrove, D. J. 1997. Assembly and enlargement of the primary cell wall in plants. *Annu. Rev. Cell Devel. Biol.* **13**: 171–201.

1999. Enzymes and other agents that enhance cell wall extensibility. *Annu. Rev. Plant Physiol. Plant Mol. Biol.* **50**: 391–417.

2000. Loosening of plant cell walls by expansins. *Nature* **407**: 321–326.

2001. Wall structure and wall loosening: a look backwards and forwards. *Plant Physiol.* **125**: 131–134.

Cosgrove, D. J., Li, L. C., Cho, H. T. *et al.* 2002. The growing world of expansins. *Plant Cell Physiol.* **43**: 1436–1444.

Côté, W. A. (ed.) 1967. *Wood Ultrastructrure: An Atlas of Electron Micrographs*. Seattle, WA: University of Washington Press.

Delmer, D. P. 1987. Cellulose biosynthesis. *Annu. Rev. Plant Physiol.* **38**: 259–290.

Dengler, N. G., Dengler, R. E., and P. W. Hattersley. 1985. Differing ontogenetic origins of PCR ("Kranz") sheaths in leaf blades of C4 grasses (Poaceae). *Am. J. Bot.* **72**: 284–302.

Ding, B., Turgeon, R., and M. V. Parthasarathy. 1992. Substructure of freeze-substituted plasmodesmata. *Protoplasma* **169**: 28–41.

Ehlers, K. and R. Kollmann. 2001. Primary and secondary plasmodesmata: structure, origin, and functioning. *Protoplasma* **216**: 1–30.

Ehlers, K., Binding, H., and R. Kollmann. 1999. The formation of symplasmic domains by plugging of plasmodesmata: a general event in plant morphogenesis? *Protoplasma* **209**: 181–192.

Emons, A. M. C. and H. Kieft. 1994. Winding threads around plant cells: applications of the geometrical model for microfibril deposition. *Protoplasma* **180**: 59–69.

Emons, A. M. C. and B. M. Mulder. 1998. The making of the architecture of the plant cell wall: how cells exploit geometry. *Proc. Natl Acad. Sci. USA* **95**: 7215–7219.

Esau, K. 1977. *Anatomy of Seed Plants*, 2nd edn. New York: John Wiley and Sons.

Evert, R. F., Russin, W. A., and A. M. Bosabalidis. 1996. Anatomical and ultrastructural changes associated with sink-to-source transition in developing maize leaves. *Int. J. Plant Sci.* **157**: 247–261.

Fleming, A. J., Caderas, D., Wehrli, E., McQueen-Mason, S., and C. Kuhlemeir. 1999. Analysis of expansin-induced morphogenesis on the apical meristem of tomato. *Planta* **208**: 166–174.

Frey-Wyssling, A. 1954. The fine structure of cellulose microfibrils. *Science* **119**: 80–82.

Fry, S. C. 1989. Cellulases, hemicelluloses and auxin-stimulated growth: a possible relationship. *Physiol. Plant.* **75**: 532–536.

Giddings, T. H., Jr. and L. A. Staehelin. 1988. Spatial relationships between microtubules and plasma-membrane rosettes during the deposition of primary wall microfibrils in *Closterium* sp. *Planta* **173**: 22–30.

1991. Microtubule-mediated control of microfibril deposition: a re-examintion of the hypothesis. In C. W. Lloyd, ed., *The Cytoskeletal Basis of Plant Growth and Form.* London: Academic Press, pp. 85–99.

Giddings, T. H., Brower, D. L., and L. A. Staehelin. 1980. Visualization of particle complexes in the plasma membrane of *Microsterias denticulata* associated with the formation of cellulose fibrils in primary and secondary walls. *J. Cell Biol.* **84**: 327–339.

Hayashi, T. 1989. Xyloglucans in the primary cell wall. *Annu. Rev. Plant Physiol. Plant Mol. Biol.* **40**: 139–168.

Heath, J. B. 1974. A unified hypothesis for the role of membrane bound enzyme complexes in plant cell wall synthesis. *J. Theor. Biol.* **48**: 445–449.

Hogetsu, T. 1991. Mechanism for formation of the secondary wall thickening in tracheary elements: microtubules and microfibrils of tracheary elements of *Pisum sativum* L. and *Commelina communis* L. and the effects of amiprophosmethyl. *Planta* **185**: 190–200.

Kaplan, D. R. 1992. The relationship of cells to organisms in plants: problems and implications of an organismal perspective. *Int. J. Plant Sci.* **153**: S28–S37.

Kaplan, D. R. and W. Hagemann. 1991. The relationship of cell and organism in vascular plants. *BioScience* **41**: 693–703.

Keegstra, K., Talmadge, K. W., Bauer, W. D., and P. Albersheim. 1973. The structure of cell walls. III. A model of the walls of suspension-cultured sycamore cells based on the interconnections of the macromolecular components. *Plant Physiol.* **51**: 188–196.

Kobayashi, H., Fukuda, H. and H. Shibaoka. 1988. Interrelation between the spatial disposition of actin filaments and microtubules during the differentiation of tracheary elements in cultured *Zinnia* cells. *Protoplasma* **143**: 29–37.

Kragler, F., Lucas, W. J., and J. Monzer. 1998. Plasmodesmata: dynamics, domains, and patterning. *Ann. Bot.* **81**: 1–10.

Kragler, F., Monzer, J., Xoconostle-Cazares, B., and W. J. Lucas. 2000. Peptide antagonists of the plasmodesmal macromolecular trafficking pathway. *EMBO J.* **19**: 2856–2868.

Lachaud, S. and L. Maurosset. 1996. Occurrence of plasmodesmata between differentiating vessels and other xylem cells in *Sorbus terminalis* L. Crantz and their fate during xylem maturation. *Protoplasma* **191**: 220–226.

Lloyd, C. W. 1984. Toward a dynamic helical model for the influence of microtubules on wall patterns in plants. *Int. Rev. Cytol.* **86**: 1–51.

Lucas, W. J., Ding. B., and C. van der Schoot. 1993. Plasmodesmata and the supracellular nature of plants. *New Phytol.* **125**: 435–476.

Lyndon, R. F. 1994. Control of organogenesis at the shoot apex. *New Phytol.* **128**: 1–18.

Monzer, J. 1991. Ultrastructure of secondary plasmodesmata formation in regenerating *Solanum nigrum* protoplast cultures. *Protoplasma* **165**: 86–95.

Neville, A. C. and S. Levy. 1984. Helicoidal orientation of cellulose microfibrils in *Nitella opaca* internode cells: ultrastructure and computed theoretical effects of strain reorientation during wall growth. *Planta* **162**: 370–384.

1985. The helicoidal concept in plant cell wall ultrastructure and morphogenesis. In C. T. Brett and J. R. Hillman, eds., *Biochemistry of Plant Cell Walls*. Cambridge, UK: Cambridge University Press, pp. 99–124.

Oparka, K. J. and A. G. Roberts. 2001. Plasmodesmata: a not so open-and-shut case. *Plant Physiol.* **125**: 123–126.

Oparka, K. J., Robards, A. G., Boewink, P. *et al.* 1999. Simple, but not branched, plasmodesmata allow the nonspecific trafficking of proteins in developing tobacco leaves. *Cell* **97**: 743–754.

Paolillo, D. J. 1995. The net orientation of wall microfibrils in the outer periclinal epidermal walls of seedling leaves of wheat. *Ann. Bot.* **76**: 589–596.

2000. Axis elongation can occur with net longitudinal orientation of wall microfibrils. *New Phytol.* **145**: 449–455.

Preston, R. D. 1974. *The Physical Biology of Plant Cell Walls*. London: Chapman and Hall.

1982. The case for multinet growth in growing walls of plant cells. *Planta* **155**: 356–363.

Rayle, D. L. and R. E. Cleland. 1992. The acid growth theory of auxin-induced cell elongation is alive and well. *Plant Physiol.*

Reuzeau, C. and R. F. Pont-Lezica. 1995. Comparing plant and animal extracellular matrix-cytoskeleton connections: are they alike? *Protoplasma* **186**: 113–121.

Robards, A. W. 1970. *Electron Microscopy and Plant Ultrastructure*. London: McGraw-Hill.

Robinson-Beers, K. and R. F. Evert. 1991. Fine structure of plasmodesmata in mature leaves of sugarcane. *Planta* **184**: 307–318.

Roelofson, P. A. 1958. Cell wall structure as related to surface growth. *Acta Bot. Neerl.* **7**: 77–89.

Roelofsen, P. A. and A. L. Houwink. 1953. Architecture and growth of the primary cell wall in some plant hairs and in the *Phycomyces* sporangiophores. *Acta Bot. Neerl.* **2**: 218–225.

Roland, J. C. 1981. Comparison of arced patterns in growing and non-growing polylamellate cell walls of higher plants. In D. B. Robinson and H. Quader, eds., *Cell Walls '81*. Stuttgart: Wissenschaftliche Verlagsgesellschaft, pp. 162–170.

Roland, J. C. and B. Vian. 1979. The wall of the growing plant cell: its three dimensional organization. *Int. Rev. Cytol.* **61**: 129–166.

Roland, J. C., Reis, D., Vian, B., Satiat-Jeunemaitre, B., and M. Mosiniak. 1987. Morphogenesis of plant cell walls at the supramolecular level: internal geometry and versatility of helicoidal expression. *Protoplasma* **140**: 75–91.

Romberger, J. A., Hejnowicz, Z., and Hill, J. F. 1993. *Plant Structure: Function and Development*. Berlin: Springer-Verlag.

Sauter, M., Seagull, R. W., and H. Kende. 1993. Internodal elongation and orientation of cellulose microfibrils and microtubules in deepwater rice. *Planta* **190**: 354–362.

Seagull, R. W. 1986. Changes in microtubule orientation and wall microfibril orientation during *in vitro* cotton fiber development: an immunofluorescent study. *Can. J. Bot.* **64**: 1373–1381.

——— 1990. The effects of microtubule and microfilament disrupting agents on cytoskeletal arrays and wall deposition in developing cotton fibers. *Protoplasma* **159**: 44–59.

——— 1991. Role of the cytoskeletal elements in organized wall microfibril deposition. In C. Haigler and P. J. Weimer, eds., *Biosynthesis and Biodegradation of Cellulose.* New York: Marcel Dekker, pp. 143–163.

van der Schoot, C., Dietrich, M. A., Storms, M., Verbeke, J. A., and J. A. Lucas. 1995. Establishment of cell-to-cell communication pathway between separate carpels during gynoecium development. *Planta* **195**: 450–455.

Vian, C., Roland, J. C., and D. Reis. 1993. Primary cell wall texture and its relation to surface expansion. *Int. J. Plant Sci.* **154**: 1–9.

Whaley, W. G., Mollenhauer, H. H., and J. H. Leech. 1960. The ultrastructure of the meristematic cell. *Am. J. Bot.* **47**: 319–399.

Wojtaszek, P. 2000. Genes and plant cell walls: a difficult relationship. *Biol. Rev.* **75**: 437–475.

Wolters-Arts, A. M. C., van Amstel, T., and J. Derksen. 1993. Tracing cellulose microfibril orientation in inner primary cell walls. *Protoplasma* **175**: 102–111.

Wyatt, S. E. and N. C. Carpita. 1993. The plant cytoskeleton–cell-wall continuum. *Trends in Cell Biology* **3**: 413–417.

Wymer, C. and C. Lloyd. 1996. Dynamic microtubules: implications for cell wall patterns. *Trends Plant Sci.* **1**: 222–228.

FURTHER READING

Bailey, I. W. 1957. Aggregations of microfibrils and their orientations in the secondary wall of coniferous tracheids. *Am. J. Bot.* **44**: 415–418.

Bailey, I. W. and M. R. Vestal. 1937. The orientation of cellulose in the secondary wall of tracheary cells. *J. Arnold Arbor.* **18**: 185–195.

Brown, R. M., Jr. 1999. Cellulose structure and biosynthesis. *Pure Appl. Chem.* **71**: 767–775.

Cook, M. E., Graham, L. E., Botha, C. E. J., and C. A. Lavin. 1997. Comparative ultrastructure of plasmodesmata of *Chara* and selected bryophytes: toward an elucidation of the evolutionary origin of plant plasmodesmata. *Am. J. Bot.* **84**: 1169–1178.

Cronshaw, J. 1965. Cytoplasmic fine structure and cell wall development in differentiating xylem elements. In W. A. Côté, Jr., ed., *Cellular Ultrastructure of Woody Plants.* Syracuse, NY: Syracuse University Press, pp. 99–124.

Ding, B. and W. J. Lucas. 1996. Secondary plasmodesmata: biogenesis, special functions and evolution. In M. Smallwood, J. P. Knox, and D. J. Bowles, eds., *Membranes: Specialized Functions in Plants.* Oxford, UK: BIOS Scientific Publishers, pp. 489–506.

Franceschi, V. R., Ding, B., and W. J. Lucas. 1994. Mechanism of plasmodesmata formation in characean algae in relation to evolution of intercellular communication in higher plants. *Planta* **192**: 347–358.

Frey-Wyssling, A. and K. Mühlethaler. 1965. *Ultrastructural Plant Cytology.* Amsterdam: Elsevier.

Harada, H. 1965. Ultrastructure and organization of gymnosperm cell walls. In W. A. Côté, Jr., ed., *Cellular Ultrastructure of Woody Plants*. Syracuse, NY: Syracuse University Press, pp. 215–233.

Houwink, A. L. and P. A. Roelofsen. 1954. Fibrillar architecture of growing plant cell walls. *Acta Bot. Neerl.* **3**: 385–395.

Itaya, A. and Y. M. Woo. 1999. Plasmodesmata and cell-to-cell communication in plants. *Int. Rev. Cytol.* **190**: 251–316.

Kerr, T. and I. W. Bailey. 1934. The cambium and its derivative tissues. X. Structure, optical properties, and chemical composition of the so-called middle lamella. *J. Arnold Arbor.* **15**: 327–349.

Lamport, D. T. A. 1965. The protein component of primary cell walls. *Adv. Bot. Res.* **2**: 151–218.

Liese, W. 1956. The fine structure of bordered pits in soft-woods. In W. A. Côté, Jr., ed., *Cellular Ultrastructure of Woody Plants*. Syracuse, NY: Syracuse University Press, pp. 271–290.

Linskens, H. F. and J. F. Jackson (eds.) 1996. *Plant Cell Wall Analysis*. Berlin: Springer-Verlag.

Majumdar, G. P. and R. D. Preston. 1941. The fine structure of collenchyma cells in *Heracleum spondylium* L. *Proc. Roy. Soc. London* B**130**: 201–217.

Manton, I. 1964. Morphology of microtubules in plant cells. *Science* **144**: 872–874.

Northcote, D. H. 1989. Control of plant-cell wall biogenesis: an overview. *Am. Chem. Soc. Symposium.* **399**: 1–15.

Northcote, D. H. and J. D. Pickett-Heaps. 1966. A function of the Golgi apparatus in polysaccharide synthesis and transport in root-cap cells of wheat. *Biochem. J.* **98**: 159–167.

Pickett-Heaps, J. D. 1968. Xylem wall deposition: radioautographic investigations using lignin precursors. *Protoplasma* **65**: 181–205.

Preston, R. D. 1952. *The Molecular Architecture of Plant Cell Walls*. New York: John Wiley and Sons.

Ray, P. M. 1967. Radioautographic study of cell wall deposition in growing plant cells. *J. Cell Biol.* **35**: 659–674.

Robards, A. W. 1968. A new interpretation of plasmodesmatal ultrastructure. *Planta* **82**: 200–210.

(ed.) 1974. *Dynamic Aspects of Plant Ultrastucture*. London: McGraw-Hill.

Roelofson, P. A. and Houwink, A. L. 1951. Cell wall structure of staminal hairs of *Tradescantia virginica* and its relation with growth. *Protoplasma* **40**: 1–22.

Srivastava, L. M. 1969. On the ultrastructure of cambium and its vascular derivatives. III. The secondary walls of the sieve elements of *Pinus strobus. Am. J. Bot.* **56**: 354–361.

Thimann, K. V. and R. Biradivolu. 1994. Actin and the elongation of plant cells. II. The role of divalent ions. *Protoplasma* **183**: 5–9.

Thimann, K. V., Reese, K., and V. T. Nachmias. 1992. Actin and the elongation of plant cells. *Protoplasma* **171**: 153–166.

Van Bel, A. J. E. and van Kesteren, W. J. P. (eds.) 1999. *Plasmodesmata: Structure, Function, and Role in Cell Communication*. Berlin: Springer-Verlag.

Wardrop, A. B. 1962. Cell wall organization in higher plants. I. The primary wall. *Bot. Rev.* **28**: 241–285.

1965. Cellular differentiation in xylem. In W. A. Côté, Jr., ed., *Cellular Ultrastructure of Woody Plants*. Syracuse, NY: Syracuse University Press, pp. 61–97.

Chapter 5

Meristems of the shoot and their role in plant growth and development

Perspective

Among the unusually interesting and unique aspects of plants is their indeterminate mode of growth. This results from the presence of apical meristems by which new cells and tissues are added to the plant body during every period of growth. As a consequence plants have the potential to increase in size at regular intervals throughout their lives. This accounts for the large size of some plants such as the redwoods of California as well as many hardwood tree species of temperate and tropical forests.

A **meristem** is a localized region of tissue which, by cell division, adds new cells to a plant or plant part. In the shoots of vascular plants the activity of meristems results in an increase in length and/or diameter, and following cell growth and differentiation, formation of the various mature tissue regions of the axes as well as the formation of organs such as leaves, cone scales, sporophylls, stipules, flower parts, etc. Some meristems are **self-perpetuating** and thus, can be considered to be "permanent" meristems. Most apical meristems and the vascular cambium are meristems of this type and, as a result of their activity, provide vascular plants with their mode of **indeterminate growth**. Others, such as the meristems that contribute to the formation of the petiole and blade of leaves, flower parts, and the various other lateral appendages of non-seed plants, cease functioning when these organs, characterized by **determinate growth**, reach their genetically predetermined size and form. For recent, in-depth discussions of the structure and function of apical and other meristems that cover material beyond the scope of this book, see Steeves and Sussex (1989) and Lyndon (1998).

Apical meristems

Apical meristems are borne at the tip of the **shoot apex**, the most distal region of a stem or lateral branch, and near the tip of roots (just behind the root cap). We shall defer consideration of apical

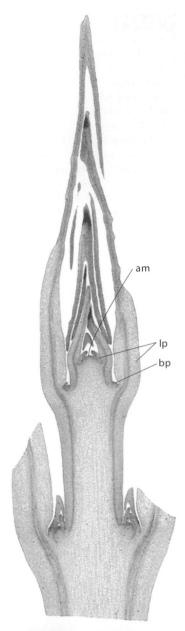

Figure 5.1 The shoot apex of a stem of *Syringa vulgaris* (lilac). am, apical meristem; lp, leaf primordium; bp, bud primordium

meristems of roots until the chapter on the root (Chapter 15). The shoot apex (Fig. 5.1) is comprised not only of the apical meristem but also of the subjacent transitional tissue regions in the axis, protoderm, ground meristem and provascular tissue (which we shall consider later in this chapter), and the leaf primordia near the stem tip. The term apical meristem, as used in this book, refers to that part of the shoot apex immediately distal to the first leaf primordium (Fig. 5.1), often referred to as the **apical dome**. The concept **promeristem** is essentially equivalent to apical meristem, as applied to the stems of seed plants. As generally defined it is considered to be the "self-perpetuating group of cells which does not undergo tissue differentiation but continually produces the cells which do differentiate" (Sussex and Steeves, 1967; see also Steeves and Sussex, 1989). Whereas apical meristem and promeristem are approximately equivalent terms when applied to seed plants, they differ strikingly when applied to some ferns. Ma and Steeves (1994, 1995a, b) provide evidence that in *Matteuccia struthiopteris* and *Osmunda cinnammomea* the first stages of differentiation of provascular tissue and parenchyma of the pith begin in tissues subjacent to the outer cell layer and distal to the first leaf primordia. Consequently, they conclude that in these taxa the apical cell and the single peripheral layer of cells produced by its cell divisional activity comprise the promeristem. Apical meristems vary in morphology and activity in the several major taxa of vascular plants, but all are characterized by some type of histological zonation. Whereas all cells in the apical meristem are meristematic, not all are **apical initials**, i.e., the cells that are the source, ultimately, of all other cells in the shoot system. Among extant vascular plants several of the more primitive taxa, Psilotaceae, *Equisetum*, some species of *Selaginella*, and most ferns are characterized by a single apical cell considered to be an **apical initial** (Fig. 5.2a, b) which resides at the tip of the apical meristem and from which all other cells therein are ultimately derived (see Bierhorst, 1971, 1977). In general, the apical cell of these plants is a four-sided tetrahedral cell with three triangular cutting faces (wall facets parallel to which new cells are produced by mitosis, cytokinesis, and cell wall deposition) and an outer, curved, triangular face (Fig. 5.3a). During some phases of development, however, the apical cell may be more irregular with more than three cutting faces. Typically, new cells are formed in a helical succession in three ranks (Fig. 5.3a). In a few ferns, the apical cell produces new cells from only two faces and, consequently, the histology of the stem has a bilateral symmetry as, e.g., in *Pteridium aquilinum* (Bierhorst, 1971). The apical meristem of extant lycophytes (except *Selaginella*) differs from that of ferns in the presence of a small cluster of apical initials at the tip of the apical dome.

In contrast to the concept of the promeristem proposed by Ma and Steeves (1994, 1995a, b), McAlpin and White (1974) suggest that the promeristem of many ferns consists of both the peripheral layer of the apical meristem between the most distal leaf primordia, and a subjacent group of cells. They also recognize a rib meristem from

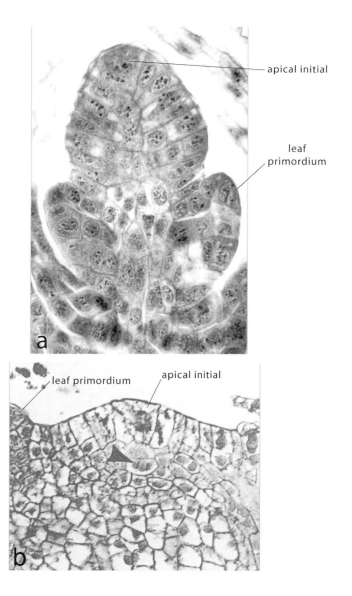

apical initial

leaf primordium

a

leaf primordium   apical initial

b

Figure 5.2 (a) Median longitudinal section of the shoot apex of *Equisetum*, showing the apical meristem containing a solitary apical initial. Magnification × 380. (b) Median longitudinal section of a shoot apex of *Quercifilix zeilanica*, a leptosporangiate fern. Note the broad apical meristem and the centrally located apical cell. Magnification × 158. (b) From McAlpin and White (1974). Used by permission of the Botanical Society of America.

which the pith is derived, and an internal peripheral zone (Fig. 5.3b). They note that this pattern of zonation is similar to that of seed plants. McAlpin and White (1974) precipitated a controversy by suggesting that ferns with apical meristems characterized by patterns of histological zonation do not contain a solitary apical initial in the classical sense, that is, one from which all other cells in the shoot are derived. Bierhorst (1977) strongly disagreed with this viewpoint and demonstrated the presence in 50 fern species of a solitary apical initial which he believed to be the source of all other cells in the shoot. For a detailed discussion of these two interpretations of the fern apical meristem, see Lyndon (1998: ch. 1).

The apical meristem of many gymnosperms, including cycads and *Ginkgo*, is characterized by distinctive cytohistological zonation (Fig. 5.4a, b). Typically, the apical meristem is bounded by a single

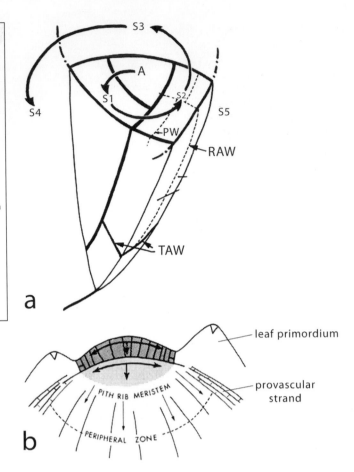

Figure 5.3 (a) Drawing of the apical cell (A) of a leptosporangiate fern and its derivatives. Note the helical succession of new cells (S1–S5). TAW, transverse anticlinal wall; RAW, radial anticlinal wall; PW, periclinal wall. (b) Diagram illustrating zonation in the apical meristem of a fern during active growth. Densely shaded surface cells and the less densely shaded subsurface zone comprise the promeristem according to McAlpin and White (1974). (a) Modified from Hébant-Mauri (1993). Used by permission of the National Research Council of Canada. (b) From McAlpin and White (1974). Used by permission of the Botanical Society of America.

surface layer below which is a small cluster of relatively large, vacuolate cells comprising the **central mother cell zone**. This zone is flanked by a **peripheral zone**, comprising small, conspicuously nucleate and densely staining cells characterized by active cell division which encloses a central region called the **transition zone** below which is a **rib meristem** consisting predominantly of longitudinal files of cells.

As we shall see below, many (possibly all) angiosperms are characterized by apical meristems with patterns of cytohistological zonation similar to those of gymnosperms. Unlike most gymnosperms, however, superimposed on this pattern of zonation is a pattern of tunica–corpus organization, based on the orientation of new cell walls as seen in median longitudinal sections (Fig. 5.5). The **tunica** is the outermost one to several (commonly two) layers of cells of the apical meristem distal to the first visible leaf primordium. During cell division in the tunica, new cell walls occur in **anticlinal planes**, i.e., in planes perpendicular to the surface of the apical dome. The **corpus**, the tissue internal to the tunica, comprises the major part of the apical meristem. During mitosis within the corpus new walls may be **periclinal**, i.e., in planes parallel to the surface, or in numerous other planes.

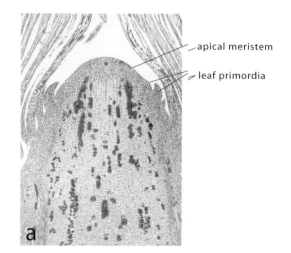

apical meristem

leaf primordia

Figure 5.4 (a) Median longitudinal section of the shoot apex of *Pinus strobus* (white pine). Magnification × 38. (b) The apical meristem of *Pinus strobus* showing histological zonation. Magnification × 107.

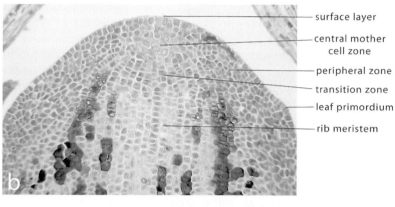

surface layer

central mother cell zone

peripheral zone

transition zone

leaf primordium

rib meristem

leaf primordium

L1 }
L2 } tunica layers

L3, corpus

Figure 5.5 Near median longitudinal section of the apical meristem of *Syringa vulgaris*. Magnification × 170.

In plants with two tunica layers, the designations, L1, L2, and L3, often used by researchers, correspond to the outer and inner tunica layers and the corpus (Lyndon, 1998). Cells of the tunica and corpus are connected by plasmodesmata which, as in other parts of the plant, are thought to control the formation of **symplastic domains** in which gene expression leads to the formation of protoderm, and ultimately to organs such as leaves and flower parts, peripherally, and in the more central part of the stem to production of cells that comprise the

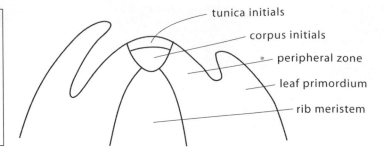

Figure 5.6 Diagram of a median longitudinal section of an apical meristem characteristic of many angiosperms, showing cytohistological zonation. Note the similarity of this apical meristem to that of *Pinus strobus* (Fig. 5.4b). Based on data and illustrations in Gifford (1950).

transitional tissue regions, ground meristem and provascular tissue (see, e.g., Clark, 2001; Ehlers and Kollmann, 2001). Such developmental domains are established by the differentiation of plasmodesmata with different size exclusion limits whereby molecules of certain sizes are allowed to be transported through them while passage of others is restricted. Domains may also be defined by an absence or very low frequency of plasmodesmata between certain tissue interfaces, or by the plugging of plasmodesmata by callose. Both the degree and mechanism of restriction of molecular transport through plasmodesmata can be modified by environmental conditions such as photoperiod (see Chapter 4 for more detail).

It is now well established that shoot apices of angiosperms exhibit, in addition to tunica–corpus zonation, patterns of **cytohistological zonation** similar to those of gymnosperms (see, e.g., Plantefol, 1947; Gifford, 1950; Popham and Chan, 1950; Buvat, 1952, 1955). Gifford (1950) recognized four zones in the shoot apices of a group of woody ranalean taxa (Fig. 5.6): an apical group of "tunica initials," a subtending zone of cells (which he called corpus initials) similar to the zone of central mother cells of gymnosperms, a peripheral zone, and a rib meristem. Similar patterns have been observed in many other angiosperms. In seed plants the apical initials and the central mother cells divide more slowly than those in the peripheral zone. Because of the conical nature of the apical meristem in most plants and the presence of these cells in its summit they can divide at a slow rate and still add cells to the more proximal zones in sufficient quantity to maintain the form and viability of the meristem. Even in plants with relatively flat or even concave apical meristems, the variation in frequency of cell division in the various zones is similar.

During periods of cytokinesis in angiosperms, the apical (tunica) initials contribute cells to the central mother cell zone and to the peripheral meristem. The central mother cell zone is the direct source of the cells of the rib meristem and the pith. The proximal region of the peripheral meristem is highly meristematic and is the site of formation of new leaf primordia and associated immature internodal tissue (Fig. 5.6). This region, often referred to as the organogenic region of the apical meristem, is approximately equivalent to that called the ring initial ("anneau initiale") in the French literature. (For detailed information on the anneau initiale and other

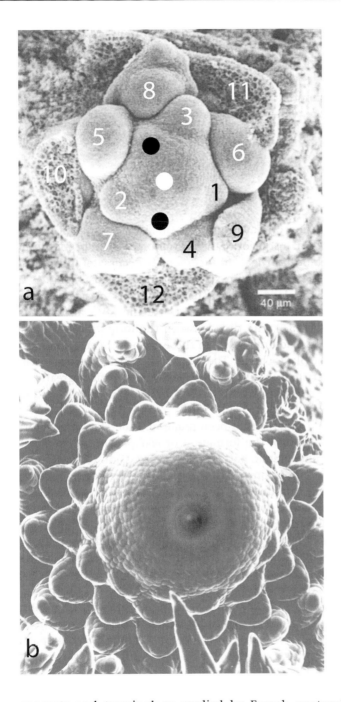

a

b

Figure 5.7 (a) Apical meristem and leaf primordia of *Arabidopsis thaliana* as seen from above. The primordia, arranged in a helix, are numbered from youngest to oldest. Positions of the next two primordia are indicated by dark dots, the apical meristem by a light dot. (b) The apical meristem of *Lycopodium* sp. and the numerous, helically arranged leaf primordia and young leaves it has produced. Magnification × 303. (a) From Clark (2001). Used by permission of Elsevier Science Ltd. (b) Photograph by P. Dayanandan.

concepts and terminology applied by French anatomists to apical meristems, see Plantefol (1947) and Buvat (1952, 1955).)

As viewed in a median longitudinal section, only a few leaf primordia are visible in the shoot apex (Figs. 5.1, 5.5a), thus imparting a misleading impression of its three-dimensional form and the actual number of leaf primordia. In fact, there are many leaf primordia, often tightly packed, in contact with one another (Fig. 5.7a, b), that vary in size in relation to their distance from the apical meristem, strand (see Larson, 1983; Nelson and Dengler, 1997) and it is possible, therefore, that the auxin signal which controls the initiation of leaf primordia is transported to the sites from a distant source through the provascular tissue (Dengler and Kang, 2001; see, also, Berleth and Sachs, 2001).

Research by Benková *et al.* (2003) indicates that auxin is probably supplied to the developing leaf primordia through the protoderm and accumulates at the primordium tip. From there it moves into the interior of the primordium where it controls the basipetal development of the leaf, and from which it is transported downward through provascular strands (Aloni *et al.*, 2003). Leaf traces differentiate along this pathway of auxin transport and ultimately connect to the acropetally differentiating vascular strands in the stem below supporting the viewpoint of earlier workers (see, e.g., Sachs 1969, 1984).

Transitional tissue regions

As new cells are being added to the more proximal regions of the apical meristem by the apical initials and central mother cells, the derivatives of these new cells are being added to the subjacent axial region of the shoot apex. The resulting transitional **tissue regions** (also called tissue domains), protoderm, ground meristem, and provascular tissue, are located between the apical meristem and mature tissues (Fig. 5.9). Cell division continues in upper parts of these regions and growth and differentiation take place more proximally. As a result of both cell division and growth, the shoot apex undergoes extensive elongation (Fig. 5.1). Steeves and Sussex (1989) characterize these activities as comprising the "expansion phase" of shoot development. Protoderm, ground meristem, and provascular tissue will ultimately develop into epidermis, cortex and pith, and the primary vascular system of the mature stem. Because, as development in these transitional tissue regions is proceeding, new cells are being added to them by the apical meristem, they are constant features of the shoot apex. Unlike the apical meristem upon which their presence is dependent, however, they are not self-perpetuating.

Cells closest to the apical meristem in these regions have been produced most recently and, are, of course, younger and thus, less mature, than those farther from it. As time passes, these cells and their derivatives elongate and differentiate (See a later section in this chapter on "Cell growth and development"). Since the cells more proximal to these have passed through the same stages of growth and differention even earlier, they are more mature. It is apparent, therefore, that the direction of differentiation in this region is acropetal, that is, in the direction of the shoot tip. Although this is generally true, as we shall see shortly, some tissues also differentiate basipetally, i.e., in the direction of the base of the plant.

In most vascular plants the **protoderm** is a single layer of cells which bounds the axis and its immature lateral appendages

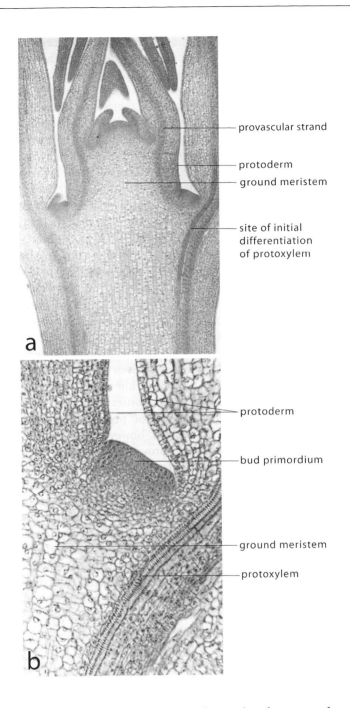

(primarily leaf primordia) in the region between the apical meristem and the proximal, mature part of the vegetative shoot (Fig. 5.9a, b). Over time, the various components of the epidermis (e.g., tabular cells, trichomes, guard cells, etc.) (see Chapter 8) develop concurrently with the addition to the shoot, distally, of new protoderm by the apical meristem. Differentiation within the protoderm is acropetal.

The protoderm encloses the **ground meristem** in which the provascular strands are embedded (Fig. 5.9). In general, cells of the fern *Adiantum capillis-veneris* seem to be formed in a similar manner. A network of wall thickenings consisting of parallel cellulose microfibrils develops external to a network of bundles of microtubules in the same position beneath the plasmalemma. The areas of thinner wall between the thicker regions of the reticulum expand laterally, presumably under the influence of turgor pressure in the cell, forming arm-like extensions and resulting in the multilobed form of these cells (Panteris *et al.*, 1993). There are also striking examples of the correlation between the orientation of microtubules and microfibrils in developing pit borders and annular and helical wall thickenings in tracheary elements. A discussion of the development of cell wall ornamentation will be deferred until Chapter 11 on the secondary xylem.

It is now clear that the orientation of microtubules beneath the plasmalemma is correlated with the synthesis of cellulose in certain patterns during an early stage in cell expansion. Following the formation of thickened wall regions, the alignment of the microtubules changes and they seem to spread out in a more uniform pattern. The mechanism of this realignment is unknown although it has been suggested that it might result from the activity of actin microfilaments (see Seagull, 1989). Finally, it is not clear just what role microtubules play in the synthesis of cellulose or by what mechanism they may influence or control the alignment of cellulose microfibrils. Recent research has resulted in several hypotheses regarding these significant aspects of development which we discuss in Chapter 4 (see, in particular, Baskin, 2001).

REFERENCES

Abbe, E. C., Phinney, B. O., and D. F. Baer. 1951. The growth of shoot apex in maize: internal features. *Am. J. Bot.* **38**: 744–751.

Aloni, R., Schwalm, K., Langhans, M., and C. I. Ullrich. 2003. Gradual shifts in sites of free-auxin production during leaf-primordium development and their role in vascular differentiation and leaf morphogenesis in *Arabidopsis*. *Planta* **216**: 841–853.

Anderhag, P., Hepler, P. K., and M. D. Lazzaro. 2000. Microtubules and microfilaments are both responsible for pollen tube elongation in the conifer *Picea abies* (Norway spruce). *Protoplasma* **214**: 141–157.

Baluska, F., Wojtaszek, P., Volkmann, D. and P. Barlow. 2003. The architecture of polarized cell growth: the unique status of elongating plant cells. *BioEssays* **25**: 569–576.

Baskin, T. I. 2001. On the alignment of cellulose microfibrils by cortical microtubules: a review and a model. *Protoplasma* **215**: 150–171.

Baskin, T. I., Beemster, G. T. S., Judy-March, J. E., and F. Marga. 2004. Disorganization of cortical microtubules stimulates tangential expansion and reduces the uniformity of cellulose microfibril alignment among cells in the root of *Arabidopsis*. *Plant Physiol.* **135**: 2279–2290.

Berleth, T. and T. Sachs. 2001. Plant morphogenesis: long-distance coordination and local patterning. *Curr. Opin. Plant Biol.* **4**: 57–62.

Benková, E., Michniewicz, M., Sauer, M. *et al.* 2003. Local, efflux-dependent auxin gradients as a common module for plant organ formation. *Cell* **115**: 591–602.

Bierhorst, D. W. 1971. *Morphology of Vascular Plants.* New York: Macmillan.

1977. On the stem apex, leaf initiation and early leaf ontogeny in filicalean ferns. *Am. J. Bot.* **64**: 125–152.

Buvat, R. 1952. Structure, évolution et fonctionnement du méristème apical de quelques dicotylédones. *Ann. Sci. Nat., Bot. Sér. ii* **13**: 199–300.

1955. Le méristème apical de la tige. *Ann. Biol.* **31**: 595–656.

Cho, H. T. and D. J. Cosgrove. 2000. Altered expression of expansin modulates leaf growth and pedicel abscission in *Arabidopsis thaliana. Proc. Natl. Acad. Sci. USA* **97**: 9783–9788.

Clark, S. E. 2001. Meristems: start your signaling. *Curr. Opin. Plant Biol.* **4**: 28–32.

Cosgrove, D. J. 1993. Wall extensibility: its nature, measurement and relationship to plant cell growth. *New Phytol.* **124**: 1–23.

1999. Enzymes and other agents that enhance cell wall extensibility. *Annu. Rev. Plant Physiol. Plant Mol. Biol.* **50**: 391–417.

2000. Loosening of plant cell walls by expansins. *Nature* **407**: 321–326.

Dengler, N. G. and J. Kang. 2001. Vascular patterning and leaf shape. *Curr. Opin. Plant Biol.* **4**: 50–56.

Devadas, C. and C. B. Beck. 1971. Development and morphology of stelar components in the stems of some members of the Leguminosae and Rosaceae. *Am. J. Bot.* **58**: 432–446.

Eckardt, T. 1941. Kritische Untersuchungen über das primäre Dickenwachstum bei Monokotylen, mit Ausblick auf dessen Verhältnis zur sekundären Verdickung. *Bot. Archiv* **42**: 289–334.

Edelmann, H. G. and U. Kutschera. 1993. Tissue pressure and cell-wall metabolism in auxin-mediated growth of sunflower hypocotyls. *J. Plant Physiol.* **142**: 467–473.

Ehlers, K. and R. Kollmann. 2001. Primary and secondary plasmodesmata: structure, origin, and functioning. *Protoplasma* **216**: 1–30.

Emons, A. M. C. and H. Kieft. 1994. Winding threads around plant cells. *Protoplasma* **180**: 59–69.

Evans, P. S. 1965. Intercalary growth in the aerial shoot of *Eleocharis acuta* R. Br. Prodr. I. Structure of the growing zone. *Ann. Bot.* **29**: 205–217.

Fleming, A. J., Caderas, D., Wehrli, E., McQuenn-Mason, S., and C. Kuhlemeir. 1999. Analysis of expansin-induced morphogenesis on the apical meristem of tomato. *Planta* **208**: 166–174.

Geldner, N., Friml, J., Stierhof, Y. D., Jürgens, G., and K. Palme. 2001. Auxin transport inhibitors block PIN1 cycling and vesicle trafficking. *Nature* **413**: 425–428.

Gifford, E. M., Jr. 1950. The structure and development of the shoot apex in certain woody Ranales. *Am. J. Bot.* **37**: 595–611.

Green, P. B. 1984. Shifts in plant cell axiality: histogenic influences on cellulose orientation in the succulent *Graptopetalum. Devel. Biol.* **103**: 18–27.

Hébant-Mauri, R. 1993. Cauline meristems in leptosporangiate ferns: structure, lateral appendages, and branching. *Can. J. Bot.* **71**: 1612–1624.

Hohl, M. and P. Schopfer. 1992. Growth at reduced turgor: irreversible and reversible cell-wall extension of maize coleoptiles and its implications for the theory of cell growth. *Planta* **187**: 209–217.

Jesuthasan, S. and P. B. Green. 1989. On the mechanism of decussate phyllotaxis: biophysical studies on the tunica layer of *Vinca major*. *Am. J. Bot.* **76**: 1152–1166.

Jones, N., Ougham, H., and H. Thomas. 1997. Markers and mapping: we are all geneticists now. *New Phytol.* **137**: 165–177.

Jung, G. and W. Wernicke. 1990. Cell shaping and microtubules in developing mesophyll of wheat (*Triticum aestivum* L.). *Protoplasma* **153**: 141–148.

Kaplan, D. R. 2001. Fundamental concepts of leaf morphology and morphogenesis: a contribution to the interpretation of molecular genetic mutants. *Int. J. Plant Sci.* **162**: 465–474.

Kutschera, U. 1992. The role of the epidermis in the control of elongation growth in stems and coleoptiles. *Bot. Acta* **105**: 227–242.

Larson, P. R. 1983. Primary vascularization and the siting of primordia. In D. J. E. Milthorpe, ed., *The Growth and Functioning of Leaves*. Cambridge, UK: Cambridge University Press, pp. 25–51.

Lyndon, R. F. 1994. Control of organogenesis at the shoot apex. *New Phytol.* **128**: 1–18.

1998. *The Shoot Apical Meristem, Its Growth and Development*. Cambridge, UK: Cambridge University Press.

Ma, Y. and T. A. Steeves. 1994. Vascular differentiation in the shoot apex of *Matteuccia struthiopteris*. *Ann. Bot.* **74**: 573–585.

1995a. Characterization of stelar initiation in shoot apices of ferns. *Ann. Bot.* **75**: 105–117.

1995b. Effects of developing leaves on stelar pattern development in the shoot apex of *Matteuccia struthiopteris*. *Ann. Bot.* **75**: 593–603.

Masuda, Y. 1990. Auxin-induced cell elongation and cell wall changes. *Bot. Mag. Tokyo* **103**: 345–370.

McAlpin, B. W. and R. A. White. 1974. Shoot organization in the Filicales: the promeristem. *Am. J. Bot.* **61**: 562–579.

Miller, D. D., Lancelle, S. A., and P. K. Hepler. 1996. Actin microfilaments do not form a dense meshwork in *Lilium longiflorum* pollen tube tips. *Protoplasma* **195**: 123–132.

Morrison, J. C., Greve, L. C., and P. A. Richmond. 1993. Cell wall synthesis during growth and maturation of *Nitella* internodal cells. *Planta* **189**: 321–328.

Muday, G. K. and A. DeLong. 2001. Polar auxin transport: controlling where and how much. *Trends Plant Sci.* **6**: 535–542.

Nelson, T. and N. G. Dengler. 1997. Leaf vascular pattern formation. *Plant Cell* **9**: 1121–1135.

Panteris, E., Apostolakos, P., and B. Galatis. 1993. Microtubule organization, mesophyll cell morphogenesis, and intercellular space formation in *Adiantum capillus veneris* leaflets. *Protoplasma* **172**: 97–110.

Paolillo, D. J., Jr. 1995. The net orientation of wall microfibrils in the outer periclinal epidermal walls of seedling leaves of wheat. *Ann. Bot.* **76**: 589–596.

Plantefol, L. 1947. Hélices foliaires, point végétatif et stèle chez les dicotylédonées: la notion d'anneau initial. *Rev. Gén. Bot.* **54**: 49–80.

Popham, R. A. and A. P. Chan. 1950. Zonation in the vegetative stem tip of *Chrysanthemum morifolium* Bailey. *Am. J. Bot.* **37**: 476–484.

Pyke, K. 1994. *Arabidopsis*: its use in genetic and molecular analysis of plant morphogenesis. *New Phytol.* **128**: 19–37.

Ray, P. M., Green, P. B., and R. E. Cleland. 1972. Role of turgor in plant cell growth. *Nature* **239**: 163–164.

Sachs, T. 1969. Polarity and induction of organized vascular tissues. *Ann. Bot.* **33**: 263–275.

1984. Axiality and polarity in vascular plants. In P. B. Barlow and D. J. Carr, eds., *Positional Controls in Plant Development*. Cambridge, UK: Cambridge University Press, pp. 193–224.

Sauter, M., Seagull, R. W., and H. Kende. 1993. Internodal elongation and orientation of cellulose microfibrils and microtubules in deepwater rice. *Planta* **190**: 354–362.

Schiefelbein, J. W. 2000. Constructing a plant cell: the genetic control of root hair development. *Plant Physiol.* **124**: 1525–1531.

Seagull, R. W. 1986. Changes in microtubule orientation and wall microfibril orientation during *in vitro* cotton fiber development: an immunofluorescent study. *Can. J. Bot.* **64**: 1373–1381.

1989. The plant cytoskeleton. *CRC Crit. Rev. Plant. Sci.* **8**: 131–167.

Snow, M. and R. Snow. 1947. On the determination of leaves. *New Phytol.* **46**: 5–19.

Steeves, T. A. and I. M. Sussex. 1989. *Patterns in Plant Development*, 2nd edn. Cambridge, UK: Cambridge University Press.

Steinmann, T., Geldner, N., Grebe, M. *et al.* 1999. Coordinated polar localization of auxin efflux crier PIN1 by GNOM ARF GEF. *Science* **286**: 316–318.

Sussex, I. M. and T. A. Steeves. 1967. Apical initials and the concept of promeristem. *Phytomorphology* **17**: 387–391.

Thimann, K. V. and R. Biradivolu. 1994. Actin and the elongation of plant cells. II. The role of divalent ions. *Protoplasma* **183**: 5–9.

Thimann, K. V., Resse, K., and V. T. Nachmias. 1992. Actin and the elongation of plant cells. *Protoplasma* **171**: 153–166.

Verbeke, J. A. 1992. Developmental principles of cell and tissue differentiation: cell–cell communication and induction. *Int. J. Plant Sci.* **153**: S86–S89.

Vidali, L., McKenna, S. T., and Hepler, P. K. 2001. Actin polymerization is essential for pollen tube growth. *Mol. Biol. Cell* **12**: 2534–2545.

Wang, H., Lockwood, S. K., Hoeltzel, M. F., and J. W. Schiefelbein. 1997. The *ROOT HAIR DEFECTIVE3* gene encodes an evolutionarily conserved protein with GTP-binding motifs and is required for regulated cell enlargement in *Arabidopsis*. *Genes Devel.* **11**: 799–811.

Wernicke, W., Gunther, P., and G. Jung. 1993. Microtubules and cell shaping in the mesophyll of *Nigella damascena* L. *Protoplasma* **173**: 8–12.

Wojtaszek, P. 2000. Genes and plant cell walls: a difficult relationship. *Biol. Rev.* **75**: 437–475.

Wolters-Arts, A. M. C., van Amstel, T., and J. Derksen. 1993. Tracing cellulose microfibril orientation in inner primary cell walls. *Protoplasma* **175**: 102–111.

FURTHER READING

Camefort, H. 1956. Étude de la structure du point végétatif et des variations phyllotaxiques chez quelques gymnospermes. *Ann. Sci. Nat., Bot. Sér. II.* **17**: 1–185.

Esau, K. 1977. *Anatomy of Seed Plants*, 2nd edn. New York: John Wiley and Sons.

Foster, A. S. 1938. Structure and growth of the shoot apex of *Ginkgo biloba*. *Bull. Torrey Bot. Club* **65**: 531–556.

1939. Problems of structure, growth and evolution in the shoot apex of seed plants. *Bot. Rev.* **5**: 454–470.

Gifford, E. M., Jr. and G. E. Corson, Jr. 1971. The shoot apex in seed plants. *Bot. Rev.* **37**: 143–229.

Kaufman, P. B., Cassel, S. J., and P. A. Adams. 1965. On nature of intercalary growth and cellular differentiation in internodes of *Avena sativa*. *Am. J. Bot.* **126**: 1–13.

Muday, G. K. and A. S. Murphy. 2002. An emerging model of auxin transport regulation. *Plant Cell* **14**: 293–299.

Popham, R. A. 1951. Principal types of vegetative shoot apex organization in vascular plants. *Ohio J. Sci.* **51**: 241–270.

Rinne, P. L. H. and C. van der Schoot. 1998. Symplasmic fields in the tunica of the shoot apical meristem coordinate morphogenetic events. *Development* **125**: 1477–1485.

Romberger, J. A., Hejnowicz, Z., and J. F. Hill. 1993. *Plant Structure: Function and Development*. Berlin: Springer-Verlag.

Schmidt, A. 1924. Histologische Studien an phanerogamen Vegetationspunkten. *Bot. Arch.* **8**: 345–404.

Sussex, I. M. 1955. Morphogenesis in *Solanum tuberosum* L: apical structure and developmental pattern of the juvenile shoot. *Phytomorphology* **5**: 253–273.

Wardlaw, C. W. 1957. The reactivity of the apical meristem as ascertained by cytological and other techniques. *New Phytol.* **56**: 221–229.

Williams, R. F. 1975. *The Shoot Apex and Leaf Growth: A Study in Quantitative Biology*. London: Cambridge University Press.

Chapter 6

Morphology and development of the primary vascular system of the stem

Perspective

The primary vascular system extends throughout the root system, the stem and its lateral branches, and appendages of the stem such as leaves, flowers, and fruits. The basic pattern of the primary vascular system is established initially by the arrangement of **provascular tissue** in the embryo. As development of the young plant proceeds, the provascular tissue becomes restricted to the shoot apex and to the root tip proximal to the root cap. Differentiation in the provascular tissue leads to the development of mature, functional primary xylem and primary phloem (Fig. 6.1). In primitive plants with central columns of primary vascular tissue (**protosteles**) (many pteridophytes as well as the roots of most plants), phloem surrounds the xylem (Fig. 6.1a). In those with tubular vascular systems (**siphonosteles**) this is usually also true, but in some taxa phloem may bound the xylem on the interior as well as on the exterior (Fig. 6.1b). In seed plants in which the primary vascular systems consist of discrete, or relatively discrete, vascular bundles (**eusteles**) (Fig. 6.1c, d), the spatial relationship of primary xylem and primary phloem varies according to the bundle type, i.e., whether collateral, bicollateral, amphicribral, or amphivasal. In **collateral bundles**, the primary xylem comprises the part of the bundle toward the inside of the stem and the primary phloem comprises the outer part (Figs. 6.1, 6.2, 6.4) whereas in **bicollateral bundles** phloem occurs both to the inside and to the outside of the primary xylem. In **amphicribral bundles** primary xylem is enclosed by primary phloem, and in **amphivasal bundles** primary phloem is enclosed by primary xylem (for illustrations, see Chapter 2). In some dicotyledons, in which vascular bundles are very close together, primary phloem may appear to form a nearly continuous cylinder (Fig. 6.2). In many monocotyledons, the primary phloem occurs in compact, well-defined strands enclosed by primary xylem to the inside and sclerenchyma fibers to the outside (Fig. 6.4).

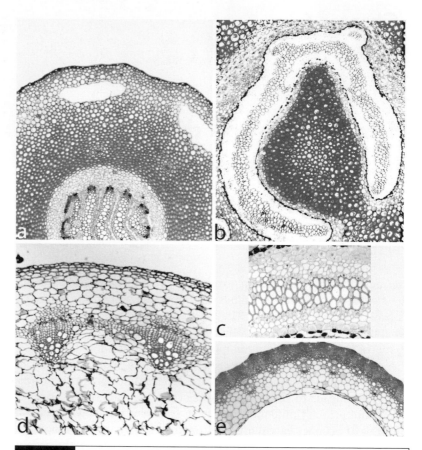

Figure 6.1 Some patterns of primary vascular tissues in pteridophytes (a, b) and seed plants (c, d) as seen in transverse sections. (a) A stem of *Lycopodium flabelliforme* with primary xylem and primary phloem arranged in a central column (a protostele). A protostele of this type in which primary phloem is interspersed in a system of interconnected plates of primary xylem is called a plectostele. Magnification × 44. (b) Primary xylem in the form of a cylinder is enclosed by primary phloem in the rhizome of the fern *Adiantum pedatum*. Magnification × 44. (c) Enlargement of a segment of (b) showing detail of the primary xylem and primary phloem (on both sides of the primary xylem). Magnification × 125. (d) Part of a cylinder of collateral vascular bundles composed of primary xylem and primary phloem in a stem of *Helianthus* (sunflower), a dicotyledon. Magnification × 48. (e) Peripheral collateral vascular bundles in the stem of *Triticum*, a monocotyledon. In many monocotyledons, vascular bundles are distributed throughout a parenchymatous ground tissue. Magnification × 43.

Cellular composition and patterns of development of primary xylem

Primary xylem is composed of both protoxylem and metaxylem; and primary phloem, likewise, is composed of protophloem and metaphloem (Fig. 6.2). In pteridophytes and gymnosperms, primary xylem consists of tracheids and parenchyma. In angiosperms it consists of vessel members, fibers, and parenchyma. The quantity and

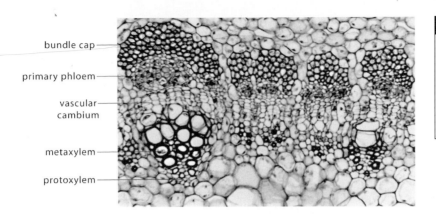

bundle cap

primary phloem

vascular cambium

metaxylem

protoxylem

Figure 6.2 Collateral vascular bundles of *Quercus* (oak), a dicotyledon. The vascular cambium had begun actively dividing prior to the preparation of this section. Note also the large bundle caps of sclerenchyma fibers. Magnification × 244.

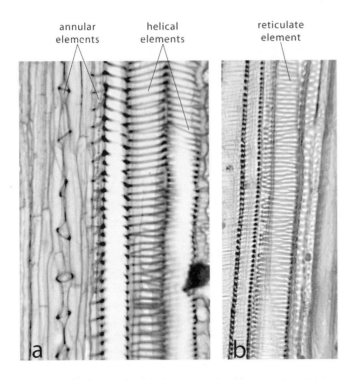

annular elements

helical elements

reticulate element

a

b

Figure 6.3 Tracheary elements in the primary xylem. Magnification × 275. (a) Annular and helical elements. (b) A reticulate element.

distribution of these cell types in the primary xylem varies greatly among different species. Conducting cells of the protoxylem and metaxylem are categorized on the basis of the morphology of their secondary wall deposits. In the **protoxylem**, conducting cells called **annular elements** are characterized by secondary wall in the form of rings, whereas the **helical elements** have a secondary wall deposited in the form of helices (Fig. 6.3a). The first tracheary elements of protoxylem to develop, the annular elements, usually have very small diameters, as viewed in transverse section, whereas those that develop subsequently are often progressively larger in diameter. In some monocots, especially grasses (Fig. 6.4) both annular and helical elements may have very large diameters.

Figure 6.4 A collateral vascular bundle of *Triticum*. Note the protoxylem lacuna which resulted from the break-up of one or more protoxylem elements during extension growth of the region. Magnification × 515.

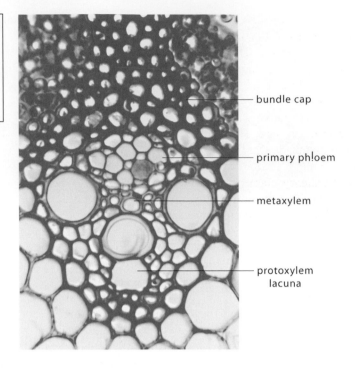

bundle cap

primary phloem

metaxylem

protoxylem lacuna

Scalariform, reticulate, and pitted tracheary elements, which comprise the metaxylem, vary widely in distribution among different taxa. **Scalariform elements** are characteristic of many pteridophytes, and are relatively less common in seed plants. Their secondary walls are deposited in the form of closely spaced, slightly inclined to nearly transverse bars which, in some taxa, seem to alternate in position on adjacent wall facets. It is often difficult to distinguish between some helical elements with secondary walls consisting of tight helices of low pitch, and scalariform elements. The secondary wall of **reticulate elements** is deposited in the form of a reticulum (Fig. 6.3b), and that of **pitted elements** is continuous except for the presence of bordered pits.

During development, annular elements differentiate first followed sequentially by helical, scalariform, reticulate, and pitted elements. Intergradation of wall characteristics in tracheary elements of primary xylem is common in some taxa resulting in cells that are appropriately labelled annular–helical, helical–scalariform, or scalariform–reticulate.

Although protoxylem and metaxylem elements are usually distinguished by their secondary wall characteristics, a more accurate basis for distinction of protoxylem and metaxylem is their time of development relative to that of other tissues, and their positions in the shoot or root. **Protoxylem** develops and becomes functional in shoot and root systems in apical regions undergoing elongation whereas **metaxylem** develops and becomes functional in more proximal regions that have ceased, or nearly ceased, to elongate.

Consequently, the morphology of cells of the protoxylem and metaxylem reflect the milieu of their development. Since vessel members in protoxylem mature (i.e., die and become functional) prior to cessation of elongation of surrounding tissues, they have evolved with a structure that allows them to remain functional while undergoing longitudinal (extension) growth. Their cell walls are largely thin, unlignified primary walls which are prevented from collapsing during rapid growth in early stages of development by the presence in annular elements of the closely spaced rings of secondary wall material or, in helical elements, the helical secondary wall thickenings. As the tracheary elements continue to elongate, the rings of secondary wall become more widely separated, and the pitch of secondary wall helices steeper, but, at the same time, they continue to provide support for the cells which thus maintain their function of transport. Ultimately, however, in many species, after death of the protoplast, continued elongation of the region and/or compression from the growth of surrounding cells results in the destruction of the protoxylem (Figs. 6.4, 6.10c) and, in some species, the formation of **protoxylem lacunae**. Lacunae formed in this way are especially prominent in many taxa of monocotyledons (Fig. 6.4). The tracheary elements of the metaxylem that differentiate in regions in which elongation has ceased are, on average, shorter than those of the protoxylem and have more continuous secondary walls, as we have seen in scalariform, reticulate, or pitted patterns.

Cellular composition and patterns of development of primary phloem

In pteridophytes and gymnosperms, the primary phloem consists of sieve cells and parenchyma. In angiosperms it consists of sieve tube members, companion cells, parenchyma, and fibers (Figs. 6.2, 6.4, 6.5). As in the primary xylem, different cell types in the phloem vary greatly in quantity and distribution in different species. Whereas recognition of protophloem and metaphloem on the basis of structural features is difficult if not impossible, they can be distinguished on the basis of time of development. Sieve elements in the protophloem, which differentiate early in regions of elongation, are subjected to stretching and ultimately are obliterated whereas the differentiation of sieve elements in the metaphloem occurs later in regions in which elongation has ceased.

In many angiosperms, the sieve tube members in protophloem are typically associated with phloem parenchyma cells (fiber primordia) that differentiate into fibers which comprise bundle caps (Fig. 6.2). Companion cells typically accompany sieve elements in the metaphloem (Figs. 6.4, 6.5), but may be absent in protophloem. As in secondary phloem, the sieve elements are characterized by protoplasts that lack nuclei, but which retain functional plasmalemmas; and as we shall see in Chapter 12, they are similar in many other ways.

Figure 6.5 Longitudinal section of primary phloem of *Cucurbita*. Note the companion cells associated with the sieve tube members, and the oblique and sectional views of simple sieve plates. Magnification × 389.

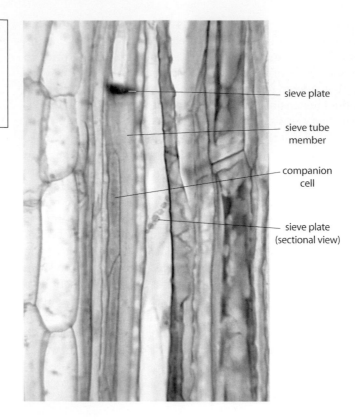

sieve plate

sieve tube member

companion cell

sieve plate (sectional view)

Sieve tube members have end walls that usually consist of simple sieve plates (Fig. 6.5) containing sieve pores which, during development, become enclosed by callose cylinders. The plasmalemma lines the pores and extends from cell to cell in the sieve tubes.

Differentiation of primary vascular tissues

In general, differentiation within provascular strands proceeds acropetally (i.e., toward the shoot apex) from sites of mature vascular tissues. Primary phloem which is continuous, and which differentiates acropetally, begins its development earlier than primary xylem and, thus, protophloem, the first primary phloem to differentiate (Fig. 6.6a), occurs closer to the apical meristem than protoxylem. The presence of functional protophloem in close proximity to the apical meristem is highly adaptive, providing to a region of very active metabolism and growth photosynthate, an essential energy source as well as the raw material from which protoplast and cell wall components are synthesized. Primary xylem (Fig. 6.6b, c) differentiates later than primary phloem and at more proximal levels in the same provascular strand. Unlike primary phloem the direction of its differentiation is acropetal in some regions and basipetal in others. Furthermore, it usually differentiates in several isolated regions. Initially, therefore, the most distal primary xylem to differentiate is commonly

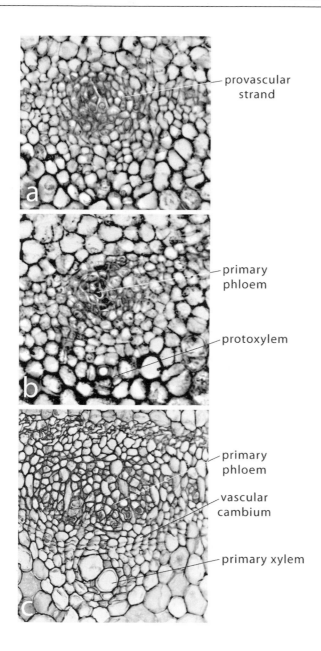

provascular
strand

primary
phloem

protoxylem

primary
phloem

vascular
cambium

primary xylem

Figure 6.6 Transverse sections of *Cassia didymobotrya* illustrating relative levels of differentiation of primary xylem and phloem in a developing vascular bundle. (a) A provascular strand in the stem with some mature primary phloem. Magnification × 439. (b) A developing vascular bundle with mature primary phloem and a single, recognizable protoxylem element. Magnification × 439. (c) A nearly mature vascular bundle. Note the protoxylem and an immature metaxylem element, lacking evidence of a secondary wall. Magnification × 244. From Devadas and Beck (1971).

discontinuous with that in more proximal regions (see Jacobs and Morrow, 1957; Larson, 1975).

Differentiation of primary vascular tissues in *Coleus blumei*, described by Jacobs and Morrow (1957), provides an excellent example which is generally applicable to other taxa of dicotyledons with collateral vascular bundles (Fig. 6.7). In *Coleus*, characterized by opposite and alternate pairs of leaves (decussate phyllotaxy), the initial site of differentiation of the first mature protoxylem in the shoot apex is usually in the base of a third visible leaf primordium below the apical meristem. From this site protoxylem differentiates acropetally into the leaf primordium and basipetally into the shoot axis, eventually

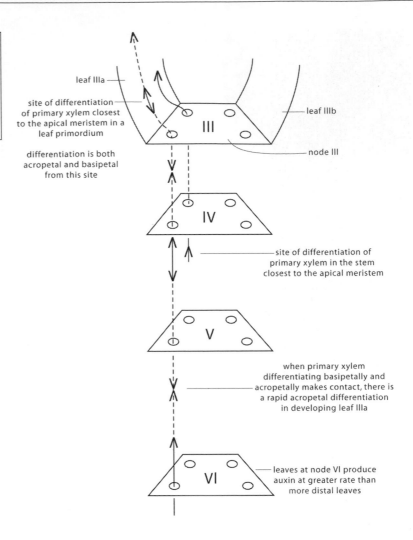

Figure 6.7 Diagrammatic representation of the sites of differentiation of primary xylem in *Coleus blumei*. Modified from Jacobs and Morrow (1957). Used by permission of the Botanical Society of America.

leaf IIIa

site of differentiation of primary xylem closest to the apical meristem in a leaf primordium

differentiation is both acropetal and basipetal from this site

leaf IIIb

node III

site of differentiation of primary xylem in the stem closest to the apical meristem

when primary xylem differentiating basipetally and acropetally makes contact, there is a rapid acropetal differentiation in developing leaf IIIa

leaves at node VI produce auxin at greater rate than more distal leaves

fusing with the acropetally differentiating protoxylem in the axis (see also Fig. 6.8 which shows a similar site of first initiation of protoxylem in a different taxon). In the shoot axis, the site of initial differentiation of protoxylem is just below node four. From this site, differentiation of primary xylem is also both acropetal and basipetal. The basipetally differentiating protoxylem will eventually contact that differentiating acropetally from the region of mature primary xylem, and that differentiating acropetally will contact that differentiating basipetally between nodes three and four. If the strands of basipetally and acropetally differentiating protoxylem are offset (i.e., fail to make direct contact), a bridge of protoxylem cells will differentiate between them (Jacobs and Morrow, 1957). Considering the fact that leaf arrangement in *Coleus* is decussate, one might expect the pattern of protoxylem differentiation in pairs of leaves to be very similar but, in fact, it is often quite different. It differs even in the two vascular bundles (leaf traces) that enter the leaf primordia (Jacobs and Morrow, 1957) (Fig. 6.7). For a description of the pattern of differentiation of

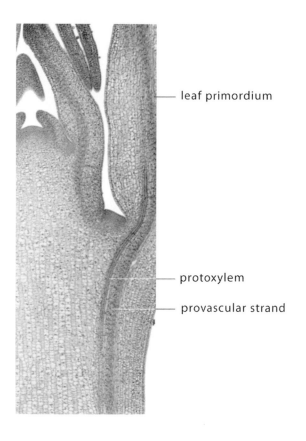

leaf primordium

protoxylem

provascular strand

Figure 6.8 Longitudinal section of part of the shoot apex of *Syringa vulgaris* showing a site of protoxylem differentiation in the base of a leaf primordium. From this site differentiation progresses both acropetally and basipetally. Magnification × 62.

primary vascular tissues in *Arabidopsis* which differs in some ways from that of *Coleus*, see Busse and Evert (1999).

During the development of the primary vascular system, tissues not only differentiate longitudinally, but also latitudinally (Fig. 6.9). This diagram represents a region in a developing vascular bundle in which primary xylem and phloem are differentiating acropetally in a collateral bundle of a seed plant (Fig. 6.9a–c). The primary phloem which develops closer to the apical meristem than the primary xylem (Fig. 6.9c, d) differentiates within the outer part of the provascular strand (i.e., the part adjacent to the ground meristem, or immature cortex) in regions of the shoot apex that are still elongating. Because of this elongation as well as compression resulting from the growth of surrounding cells, the sieve cells of the protophloem are ultimately obliterated, but not before sieve elements in the metaphloem have differentiated at levels adjacent to newly differentiated protophloem. Consequently, there is no discontinuity in the column of primary phloem. In the region of the protophloem, during, and to some degree, following obliteration of sieve elements, primary phloem fibers develop from parenchyma cells (fiber primordia) that comprise part of the protophloem, forming bundle caps. The fiber primordia often increase in number by division of parenchyma cells, resulting in large, conspicuous bundle caps (Fig. 6.2). Bundle cap fibers increase greatly in length during development and consequently

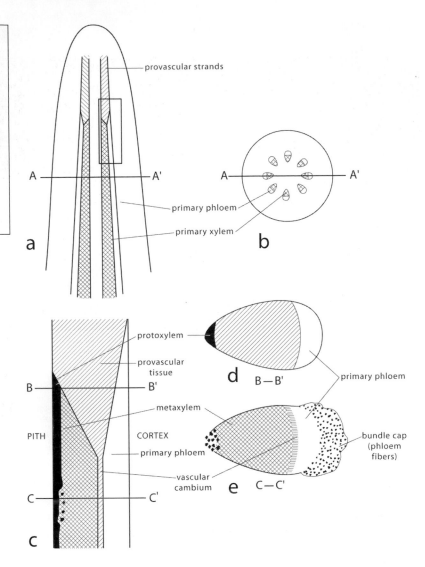

Figure 6.9 Diagrammatic representation of patterns of longitudinal and latitudinal (transverse) differentiation of primary xylem and primary phloem in a developing vascular bundle of a seed plant. Note that the bundle cap illustrated in (e) is not shown in (c). The distal, mature region of the vascular bundle in (a) and (c) are greatly foreshortened in relation to their width. See the text for descriptions.

intrude between surrounding cells. Since metaphloem develops to the inside of the protophloem after elongation has ceased in the region, its unity is maintained and it may remain functional for long periods of time, in herbaceous plants for as long as the plants live.

In most vascular plants primary xylem begins its differentiation later than primary phloem, that is, in more proximal regions of the provascular strands (Fig. 6.9c). Protoxylem differentiates within the inner part of the strands next to the immature pith and, as development proceeds, protoxylem and metaxylem gradually extend toward the exterior of the axis (Figs. 6.6b, c, 6.9d, e). Because, as noted above, protoxylem differentiates in a region of the stem still elongating, it is stretched and often fractured (Figs. 6.9e, 6.10c). In contrast, metaxylem differentiates after most, or all, elongation in the region has ceased and thus may maintain its function for some time, in

protoxylem

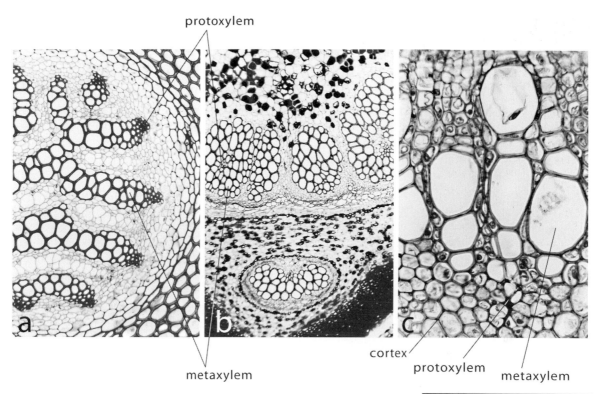

metaxylem

cortex
protoxylem
metaxylem

plants in which secondary xylem is not produced, for the life of the plant.

The patterns of latitudinal differentiation of primary vascular tissues are useful in plant taxonomy. The differentiation of primary phloem from the outer surface toward the inner surface of a provascular bundle (Fig. 6.9c) is referred to, especially in the older literature, as **centripetal development** whereas differentiation of the primary xylem from the inner toward the outer part of a bundle is referred to as **centrifugal development** (Figs. 6.6b, c, 6.9c, e, 6.10c). This pattern of development of protoxylem in relation to metaxylem is also called **endarch order of maturation** of primary xylem. If the sequence of development of primary xylem tracheary elements is reversed, that is, with annular elements to the exterior and helical, scalariform, and/or reticulate and pitted elements developing sequentially toward the center of the axis, the pattern of development, or order of maturation, is described as **exarch** (or centripetal) (Fig. 6.10a). This pattern is common in roots of vascular plants as well as in the roots and stems of pteridophytes which often, but do not always, contain central columns of primary vascular tissue (protosteles). If protoxylem elements differentiate more deeply within vascular bundles or central vascular columns and metaxylem elements develop around them, often in radiating files, the pattern of development or order of maturation is described as **mesarch** (Fig. 6.10b).

Figure 6.10 Transverse sections of stems showing order of maturation (latitudinal direction of development) of primary xylem. (a) Exarch (centripetal) development in the rhizome of *Lycopodium flabelliforme*. Magnification × 184. (b) Mesarch development in the fern *Osmunda*. Magnification × 45. (c) Endarch (centrifugal) development in the stem of *Quercus*. Magnification × 465.

The role of auxin in the development of the primary vascular system

It is widely accepted that differentiation of primary xylem and primary phloem from provascular tissue is directly related to the controlling influence of the hormone auxin (see, e.g., Sachs, 1981, 1991; Aloni, 1987; Roberts, 1988; Steeves and Sussex, 1989; Lyndon, 1990; Ma and Steeves, 1992; Stein, 1993). Auxin concentration and "flux" (= flow rate) (see Sachs, 1991; Stein, 1993) are thought to be primary factors influencing the type of cell or tissue that differentiates. For example, Wetmore and Rier (1963) showed that, in *Syringa* callus, primary xylem differented in high concentrations of auxin whereas primary phloem differentiatiated under lower concentrations. In the intact plant axis, auxin, synthesized in the shoot apex and/or leaf primordia, is transmitted through the symplast via plasmodesmata to the derivatives of meristems where it has a controlling influence on differentiation of protoderm, ground meristem, and provascular tissue, and subsequently, on differentiation of the mature tissue regions of the axis (i.e., epidermis, cortex, pith, primary vascular system, etc.).

Extensive evidence (see Sachs, 1981; Steeves and Sussex, 1989) indicates that developing leaf primordia are the primary source of auxin in angiosperms although the apical dome has been suggested as a source in some taxa (Sachs, 1981). Several workers have concluded that, in some ferns, auxin produced in the apical meristem controls the basic form of the primary vascular system (Wardlaw, 1946; Soe, 1959; Ma and Steeves, 1992). Auxin, apparently produced in leaf primordia, does, however, have a modifying influence on stelar form. For example, Ma and Steeves (1992, 1995) demonstrated that, in *Matteuccia struthiopteris*, leaf primordia control leaf gap formation as well as the differentiation of the pith. When, in their studies, leaf primordia were punctured (and, thereby, suppressed), the primary vascular system developed as a continuous cylinder without leaf gaps. If leaf primordia were prevented from forming over a period of several weeks, no pith was formed and the primary vascular system attained the form of a protostele. They observed, further, that when all leaf primordia were suppressed, no protoxylem or protophloem differentiated; but when a single primordium was left intact, protoxylem and protophloem differentiated in relation to that primordium.

Although extensive evidence indicates that auxin has a controlling influence on the development of leaf traces in seed plants, the mechanism of this control is unclear and obviously complex. We know that the primary phloem in leaf traces differentiates acropetally whereas the primary xylem differentiates both acropetally and basipetally in different sites in the shoot apex. These different patterns of development are thought to be related to differences in auxin concentration and flow rate (see Benková *et al.*, 2003; see also Chapter 5).

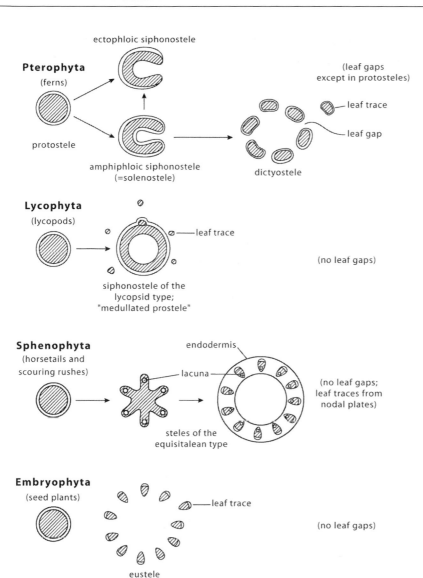

Figure 6.11 Diagrams illustrating some aspects of the structure and evolution of the steles in ferns, lycophytes, sphenophytes, and seed plants.

Early in the history of plant anatomy the primary vascular system of the plant axis (stem plus root) and certain associated tissues were recognized by the French botanists van Tieghem and Duliot as a unified system which they called the stele. Various types of steles were described by these and later workers (see Beck *et al.*, 1983). Many recent workers use the term "stele" as a synonym for the primary xylem and phloem in stems and roots. Since reference to stelar types is common in the anatomical and morphological literature, a summary of the most widely recognized types is provided in Fig. 6.11. The British botanist F. O. Bower (1930) proposed that the variation in stelar form was directly related to the size of the plant and its ability to provide, through the primary vascular system, the necessary nutrients and photosynthate to support the growth and development of the plant. A modern viewpoint, proposed by Stein (1993), suggests that these stelar types evolved not only in relation to the increasing size of the plant, but also, and perhaps more importantly, in relation to the size and complexity of its lateral branch systems and leaves, and the available auxin sources. Stein developed a computer model based on an initial hypothesis of Wight (1987) who proposed that the ribbed steles of some Devonian Aneurophytalean taxa developed under the influence of hormones produced by the lateral appendages. He suggested that hormones stimulated the differentiation of vascular tissue in the direction of the leaf traces which in these taxa diverged directly from the tips of the stelar ribs. Stein's model incorporates a detailed analysis of current knowledge of the induction of primary vascular tissues during development, and the application of this knowledge to stelar structure in both fossil and living plants, the size and structure of lateral branch systems and leaves, phyllotaxy, and the sources and sites of auxin production and transport as represented in mature axes by protoxylem and metaxylem strands. Predictions in the model of hormone availability and concentrations during development often conform to the actual structure of steles in members of the fossil Aneurophytales. Parameters of the model are also applied with equal success to other fossil taxa and to living seed plants. Stein's approach makes possible for the first time the interpretation of stelar evolution not only through a comparison of mature stelar patterns, but also through the developmental changes that accompanied, through time, the evolution of these patterns. His work also strongly supports the viewpoint that there is a causal relationship in several lines of evolution between the increasing size of plants during the Devonian and the increasing size and complexity of their lateral appendages (see also Niklas, 1984).

REFERENCES

Aloni, R. 1987. Differentiation of vascular tissues. *Annu. Rev. Plant Physiol.* **38**: 179–204.

Beck, C. B., Schmid, R., and G. W. Rothwell. 1983. Stelar morphology and the primary vascular system of seed plants. *Bot. Rev.* **48**: 691–815; 913–931.

Benková, E., Michniewicz, M., Sauer, M., *et al.* 2003. Local efflux-dependent auxin gradients as a common module for plant organ formation. *Cell* **115**: 591–602.

Bower, F. O. 1930. *Size and Form in Plants, with Special Reference to the Primary Conducting Tracts.* London: Macmillan.

Busse, J. S. and R. F. Evert. 1999. Pattern of differentiation of the first vascular elements in the embryo and seedling of *Arabidopsis thaliana. Int. J. Plant Sci.:* **160**: 1–13.

Devadas, C. and C. B. Beck. 1971. Development and morphology of stelar components in the stems of some members of the Leguminosae and Rosaceae. *Am. J. Bot.* **58**: 432–446.

Jacobs, W. P. and I. B. Morrow. 1957. A quantitative study of xylem development in the vegetative shoot apex of *Coleus. Am. J. Bot.* **44**: 823–842.

Larson, P. R. 1975. Development and organization of the primary vascular system of *Populus deltoides* according to phyllotaxy. *Am. J. Bot.* **62**: 1084–1099.

Lyndon, R. F. 1990. *Plant Development: The Cellular Basis*. London: Unwin Hyman.

Ma, Y. and T. A. Steeves. 1992. Auxin effects on vascular differentiation in ostrich fern. *Ann. Bot.* **70**: 277–282.

1995. Effects of developing leaves on stelar pattern development in the shoot apex of *Matteuccia struthiopteris*. *Ann. Bot.* **75**: 593–603.

Niklas, K. J. 1984. Size-related changes in the primary xylem anatomy of some early tracheophytes. *Paleobiology* **10**: 487–506.

Roberts, L. W. 1988. Hormonal aspects of vascular differentiation. In L. W. Roberts, P. B. Gahan, and R. Aloni, eds., *Vascular Differentiation and Plant Growth Regulators*. Berlin: Springer-Verlag, pp. 22–38.

Sachs, T. 1981. The control of the patterned differentiation of vascular tissue. *Adv. Bot. Res.* **9**: 151–262.

1991. *Pattern Formation in Plant Tissues*. Cambridge, UK: Cambridge University Press.

Soe, K. 1959. Morphogenetic Studies on *Onoclea sensibilis* L. Ph.D. thesis, Harvard University, Cambridge, MA.

Steeves, T. A. and I. M. Sussex. 1989. *Patterns in Plant Development*, 2nd edn. Cambridge, UK: Cambridge University Press.

Stein, W. 1993. Modeling the evolution of stelar architecture in vascular plants. *Int. J. Plant Sci.* **154**: 229–263.

Wardlaw, C. W. 1946. Experimental and analytical studies of pteridophytes. VII. Stelar morphology: the effect of defoliation on the stele of *Osmunda* and *Todea*. *Ann. Bot.* **9**: 97–107.

Wetmore, R. H. and J. P. Rier. 1963. Experimental induction of vascular tissues in callus of angiosperms. *Am. J. Bot.* **50**: 418–430.

Wight, D. C. 1987. Non-adaptive change in early land plant evolution. *Paleobiology* **13**: 208–214.

FURTHER READING

Esau, K. 1965. *Vascular Differentiation in Plants*. New York: Holt, Rinehart and Winston.

1977. *Anatomy of Seed Plants*, 2nd edn. New York: John Wiley and Sons.

Lyndon, R. F. 1998. *The Shoot Apical Meristem: Its Growth and Development*. Cambridge, UK: Cambridge University Press.

Northcote, D. H. 1995. Aspects of vascular tissue differentiation in plants: parameters that may be used to monitor the process. *Int. J. Plant Sci.* **156**: 245–256.

Shane, M. W., McCully, M. E., and M. J. Canny. 2000. The vascular system of maize stems revisited: implications for water transport and xylem safety. *Ann. Bot.* **86**: 245–258.

Chapter 7

Sympodial systems and patterns of nodal anatomy

Perspective: leaf traces

Vegetative shoots consist of stems bearing leaves. In order to develop, and to synthesize various necessary compounds required by the plant, leaves must have access to a source of water and essential minerals which are transported into them from the stem through the primary xylem. Photosynthate and other compounds synthesized in the leaves are, in turn, transported through the primary phloem into the stem and root system for storage and/or use. This transport of substances takes place in primary vascular connections between the stem vascular system and the base of leaves called **leaf traces**. Traces may diverge from the stem vascular system some distance below, or very near, the **nodes** (sites of attachment of leaves to stems) at which they enter the leaves. Leaf traces are composed of protoxylem, metaxylem, protophloem, and metaphloem, and typically contain **transfer cells** in both primary xylem and primary phloem. In seed plants, leaf traces are often larger and contain more tracheary cells than the vascular bundles from which they diverge, and they may increase in size distally. A leaf may be vascularized by only one or by several to many leaf traces.

In order to understand the morphology of nodal regions of shoots we must observe both transverse and median longitudinal sections through these regions. Remember that stems bear many leaves in various spatial distributions. Nevertheless in sections, depending on the distance between leaves and the size of the sections, only a single leaf and its vascular connection to the stem (leaf trace) may be visible.

Nodal structure of pteridophytes

The simplest nodal structure occurs in pteridophytes (Psilophyta, Lycophyta, Sphenophyta, and Pterophyta). In taxa with **protosteles** leaf traces simply diverge from the surface of the central vascular column near the level of the leaf and enter the leaf base. Commonly

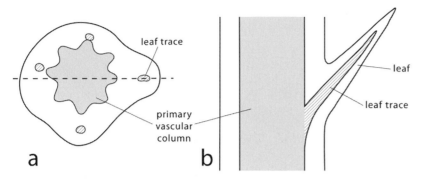

Figure 7.1 Nodal structure of a stem containing a protostele. (a) Transverse view showing a solid column of primary vascular tissue from which leaf traces diverge. (b) Longitudinal section along the plane indicated by the dashed line in (a).

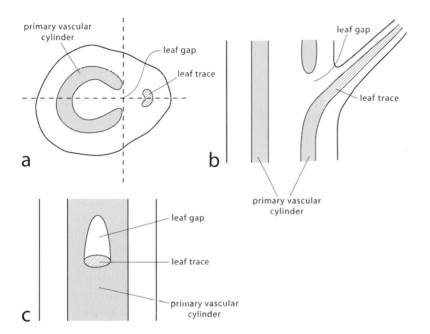

Figure 7.2 Nodal structure of a stem containing a siphonostele. (a) Transverse section. Primary xylem and primary phloem comprise a cylinder in which develop discontinuities called leaf gaps. (b) Longitudinal view of nodal structure as seen in a section along the plane indicated by the horizontal dashed line in (a). Note the leaf gap and its positional relationship to the leaf trace. (c) Longitudinal view of nodal structure along the plane indicated by the vertical dashed line in (a). Note the form of the leaf gap.

leaves of plants with protosteles are supplied by only one leaf trace (Fig. 7.1). In stems with **siphonosteles**, such as those of many ferns, the continuity of the cylinder of primary vascular tissue is broken immediately above the position of outward divergence of the leaf trace, leaving a discontinuity called a **leaf gap**, a region through which the parenchyma of the pith and cortex are continuous (Fig. 7.2). If the leaves of a particular species are closely spaced, the cylinder will be dissected by many such leaf gaps, and depending on their longitudinal extent (that is, their height as observed in face view), several to many of these gaps will be visible between regions of primary vascular tissue, giving, in transverse sections, the false impression of a system made up of separate vascular bundles. Such a system is termed a **dictyostele** (dissected stele) (Fig. 7.3). Whereas in many plants with dictyosteles, all discontinuites are, in fact leaf gaps, there are some plants in which some discontinuities in the dictyostele are not associated with leaves and, thus, are not leaf gaps. Studies by

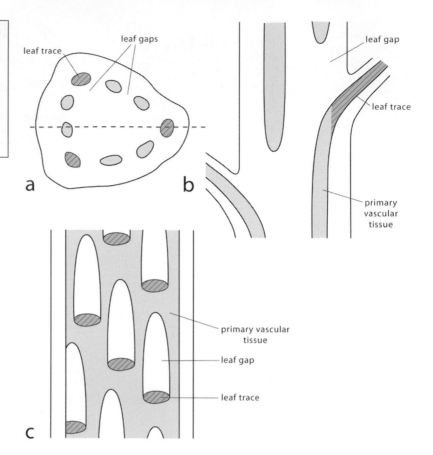

Figure 7.3 Nodal structure of a stem containing a dictyostele. (a) Transverse section. (b) Longitudinal section along the dashed line in (a). (c) Longitudinal view of the stelar surface showing dissection of the stele by numerous leaf gaps. Compare (b) and (c).

White and Weidlich (1995) demonstrate that the dictyosteles in the filicalean ferns *Diplazium* and *Blechnum* show a striking resemblence to the eusteles of some gymnosperms and dicotyledons with helically arranged leaves. Considering the wealth of information which suggests that ferns and seed plants are unrelated or, at best, only very distantly related (see Stewart and Rothwell, 1993), it seems likely that this similarity is the result of homoplasy (parallel evolution).

Sympodial systems of seed plants

The **eustele**, the primary vascular system in the stem of seed plants, consists of a system of vascular bundles plus the leaf traces that diverge from them. The stem vascular bundles, or **axial bundles**, and associated leaf traces comprise **sympodia**. The eustele, therefore, is a cylindrical system of sympodia. There are no leaf gaps in a eustele. The sympodia are considered to be discrete entities although there may be vascular connections between them. This is especially true among angiosperms, but in the more primitive angiosperms and in many gymnosperms the sympodia are not interconnected (see Beck *et al.*, 1983). As viewed in transverse section, a eustele appears as a cylinder

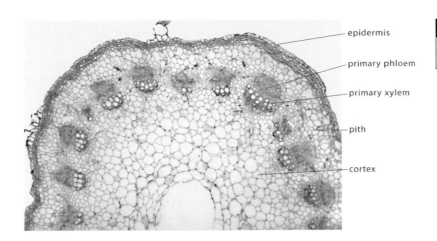

epidermis

primary phloem

primary xylem

pith

cortex

Figure 7.4 Transverse section of a eustele of a young stem of *Helianthus*. Magnification × 90.

of separate vascular bundles, some of which are axial bundles, others of which are leaf traces (Fig. 7.4).

The primary vascular systems of seed plants are often illustrated spread out in one plane. In order to thoroughly understand the eustele and some of its many manifestations, we shall illustrate it with a very simple system similar to that of a primitive gymnosperm (Fig. 7.5a). The system consists of five sympodia, labeled 1 through 5. Leaf traces diverge from every other axial bundle in a helical pattern. This helix, which represents the sequential development of leaf primordia by the apical meristem is called the **generative** (or **ontogenetic**) **spiral**. The levels at which traces enter the bases of leaves are indicated by triangles. Note that the oldest leaf trace visible in the diagram, that is, the lowest one (the one farthest from the apical meristem) diverged from axial bundle 4. During development, the next trace in the chronological sequence diverged from axial bundle 1, followed by a trace from axial bundle 3. The next youngest trace diverged from axial bundle 5, followed by traces from axial bundles 2 and 4. The last that we shall consider, and the most apical trace, diverged from axial bundle 1. You will note that in this region of the eustele, axial bundles 1 and 4 have each contributed a leaf trace to two different leaves. A helix connecting the two traces that diverged from one axial bundle would encircle the stem twice. A stem with a leaf arrangement in this pattern is referred to as having a **phyllotaxy** of 2/5, the numerator referring to the number of turns around the stem between two traces in the same sympodium (or, as seen on the stem surface, leaves in the same orthostichy) and the denominator indicating the number of sympodia in the system.

Figures 7.5b–e represent transverse sections at levels A-A through C-C in the primary vascular system. Axial bundles 1–5 are illustrated in all sections. Note that traces diverge laterally from axial bundles and then follow radial courses through the cortex. In section A the leaf trace that diverged from axial bundle 4 is shown within the cortex because the section was taken from a level below its entrance into the leaf. The next trace to diverge during development is shown

Figure 7.5 Diagrams illustrating the architecture of a simple eustele consisting of five sympodia. (a) A longitudinal view shown as if the cylinder of primary vascular bundles (1–5) had been spread out in one plane. (b–e) Transverse sections taken at the levels indicated by horizontal lines in (a). Please see the text for a detailed explanation.

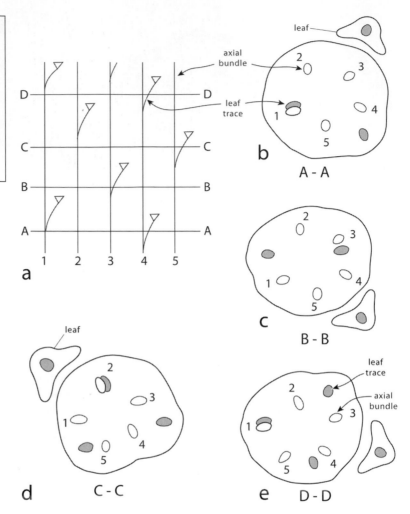

in contact with the axial bundle 1 because section A represents a level at which the separation of the trace from the axial bundle is not complete. In section B the trace that diverged from axial bundle 4 is in a leaf that has separated from the stem whereas the trace that diverged from axial bundle 1 is still in the stem cortex at this level. Note also the trace that diverged from axial bundle 3 just below the level of section B. It appears in section C in the outer cortex, and in section D in a leaf. Trace divergence from other axial bundles as shown at levels C and D follow similar patterns.

In summary, it is clear from these diagrams that, in the hypothetical plant represented, there was a eustele consisting of five sympodia, that during development leaf traces diverged in sequence along the generative spiral from every other axial bundle, that leaf traces (and, thus, leaves) were arranged in a helical pattern comprising a 2/5 phyllotaxy, and that each leaf was supplied by a single leaf trace that originated close to the node at which it entered the leaf. All eusteles are similar to the one just described, but may differ in being

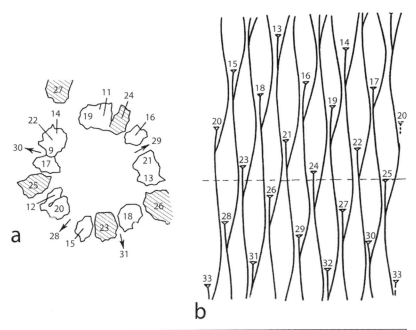

Figure 7.6 (a) The eustele of *Abies concolor* (fir) as seen in transverse section. An axial bundle is given the number of the leaf trace that will next diverge from it. Leaf traces are shaded. The approximate level of this section in the stele is indicated by the horizontal dashed line in (b). (b) Longitudinal diagrammatic representation of the pattern of the primary vascular system illustrated in (a). The triangles represent leaf traces near the levels of their entry into leaves. The traces are numbered in the order of their development, with the older traces indicated by higher numbers, the younger by lower numbers. This eustele consists of 13 sympodia (axial bundles bearing leaf traces). Note that the leaf traces follow a shallow helix called the ontogenetic spiral. From Namboodiri and Beck (1968a).

composed of fewer or more sympodia, and some, as we shall see, in being more complex in several other ways.

The eustele of fir (*Abies concolor*) is basically similar to the system we have described above, but differs in consisting of 13 sympodia and having a phyllotaxy of 5/13 (Fig. 7.6). The eustele of the ancient gymnosperm, *Ginkgo biloba* (Fig. 7.7) also consists of 13 sympodia and has a phyllotaxy of 5/13. It is distinctive, however, in that two leaf traces, each derived from a different, adjacent axial bundle, enter each leaf. Note also that the traces are quite long and vary greatly in length. Because some of the traces traverse, longitudinally, four or five nodes, a transverse section from any level of the mature region of a young stem will contain not just 13 vascular bundles (the number of axial bundles), but commonly 26, 13 axial bundles and 13 leaf traces.

Angiosperms have more complex eusteles, related primarily to the large size of their leaves. The leaves of many dicotyledons are supplied by three or five (occasionally seven) leaf traces. In such cases there will be a central trace, often larger than others, called the **median trace** flanked on either side by one or more smaller **lateral traces**, all of

Figure 7.7 Diagram of the eustele of *Ginkgo biloba*. Axial bundles are indicated by bold lines. Pairs of leaf traces are indicated by numbers in descending order along the ontogenetic spiral. See the text for more detail. Redrawn from Gunckel and Wetmore (1946). Used by permission of the Botanical Society of America.

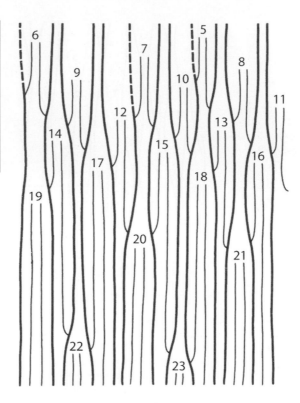

Figure 7.8 Transverse section of a seven-trace, multilacunate node of *Quercus* (oak). mltr, median leaf trace; ltr, lateral leaf traces. Magnification × 86.

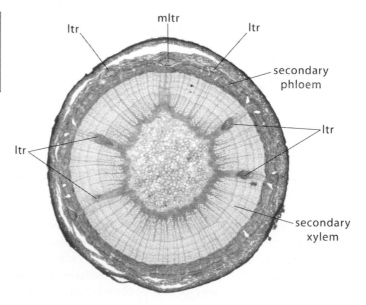

which enter the same leaf (Fig. 7.8). It is customary in diagrams of primary vascular systems, illustrated in one plane, to portray at the same level all traces that supply one leaf. These traces will usually be assigned the same number (indicating their position in the developmental sequence). This is clearly illustrated in the primary vascular systems of *Drimys* and *Potentilla* (Fig. 7.9a, b). Each leaf of these plants

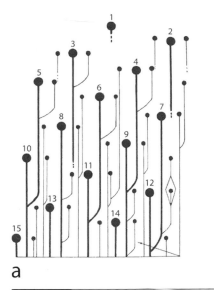
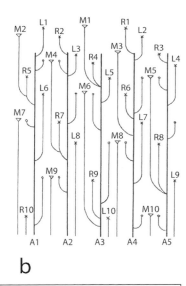

a   b

Figure 7.9 (a) Diagram of the eustele of *Drimys winteri* consisting of five sympodia. Note that each leaf receives three leaf traces, a median trace (large black dot) and right and left lateral traces (small black dots). The median and lateral traces branch from axial bundles of different sympodia. (b) Diagram of the eustele of *Potentilla fruticosa*. Median traces are indicated by triangles and laterals by Xs. This primary vascular system resembles that of *Drimys*, but the diagram differs in showing pairs of branch traces (small circles) flanking each median trace. Each pair of branch traces enters a lateral bud in the axil of the associated leaf. (a) From Benzing (1967b). Used by permission of the Botanical Society of America. (b) From Devadas and Beck (1972).

is provided with a median, and a right and a left lateral, trace. In *Potentilla* (Fig. 7.9b), the lateral and median traces have been assigned the same number as, for example M6 (median), L6 (left lateral), and R6 (right lateral). Note also that each of these three traces diverges from the axial bundle of a different sympodium, and that the median trace is much longer than the laterals. A final feature of importance in this diagram are the two traces, of variable length, on either side of the median that terminate (in the diagram) in two small circles. These are branch traces that enter the axillary bud just above the level of entry of the median trace into the leaf base. Note that in both *Drimys* (Fig. 7.9a) and *Potentilla* (Fig. 7.9b) there are five sympodia, that the phyllotaxy is 2/5, and that the large number of vascular bundles in the systems is related not only to the number of axial bundles, but also to the number of traces per leaf and to the fact that the traces extend longitudinally through several internodes and, thus, overlap those to other leaves.

The primary vascular systems described above are **open systems** in that the sympodia are not interconnected, a feature of many plants with helical phyllotaxy. **Closed systems**, characterized usually by an even number of sympodia, are common in plants with **verticillate** (leaves whorled), **distichous** (leaves two-ranked), or **decussate** (leaves opposite with alternating pairs at right angles) phyllotaxy.

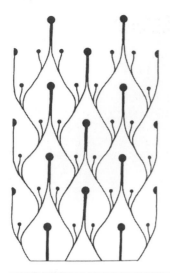

Figure 7.10 Diagram of the primary vascular system (eustele) of *Cercidiphyllum japonicum*. The large dots indicate median leaf traces, the small, laterals. Compare this closed system, consisting of four sympodia, with the open systems of *Drimys* and *Potentilla* (Fig. 7.9). For details please see the text. From Benzing (1967b). Used by permission of the Botanical Society of America.

Cercidiphyllum provides a good example of a closed primary vascular system in a plant with decussate phyllotaxy (Fig. 7.10). The four axial bundles are interconnected by pairs of leaf traces derived from adjacent axial bundles that fuse to form each median trace prior to its entrance into a leaf base. It is particularly interesting to note in this system, in which a median and two lateral traces enter each leaf, that the laterals branch from the traces that fuse to form the median traces of the next more distal pair of leaves. The eusteles of plants with distichous and decussate phyllotaxy are commonly composed of four sympodia; those with verticillate phyllotaxy, of six. For more information on the many architectural variations of eusteles in dicotyledons, please refer to Beck *et al.* (1983).

The primary vascular systems in the stems of monocotyledons, termed atactosteles, are considered by some to be fundamentally different from those of dicotyledons (Zimmermann and Tomlinson, 1972). The monocotyledon primary vascular system, however, may be simply a highly modified eustele (Beck *et al.*, 1983). These authors believe that the distinctive structural features of the primary vascular systems of monocotyledons are related primarily to the nature of the large leaves with broad leaf bases that, in some groups (e.g., grasses), overlap and encircle the stem, and which in many taxa are supplied by a very large number of leaf traces.

As viewed in transverse section the primary vascular system of many monocotyledons appears to consist of numerous, randomly scattered bundles (Fig. 7.11a, b). Those in a more peripheral position are often smaller and comprise a more compact zone than the larger, more central ones. Because some monocotyledons contain so many vascular bundles in their stems (hundreds or even thousands), development of an accurate understanding of the architecture of the system has been difficult. Major progress in understanding the system has resulted from the work of Zimmermann and Tomlinson (1972) and many other papers (see Beck *et al.* (1983) for more literature and a review). The description of the system below is based primarily on the work of Zimmermann and Tomlinson.

In monocotyledon seedlings the vascular system begins as a cylinder of relatively few bundles similar to that of dicotyledons. Furthermore, some mature monocotyledons (some members of the Liliaceae that do not have sheathing leaf bases) and several grasses such as *Avena* (oats) and *Triticum* (wheat) have a central pith surrounded by a cylinder of vascular bundles (Fig. 7.11c). In many monocotyledons, especially large ones, as development proceeds and the stem increases in diameter, the original axial bundles gradually increase in number by branching tangentially. Two types of axial bundles are recognized: **major bundles** which traverse, longitudinally, relatively great distances between the levels of divergence of consecutive leaf traces, and **minor bundles** which traverse much shorter distances between the levels of divergence of consecutive leaf traces (Fig. 7.12a). A unique feature of some monocotyledons is the presence of discontinuous cortical bundles which may accompany leaf traces as they enter leaf bases.

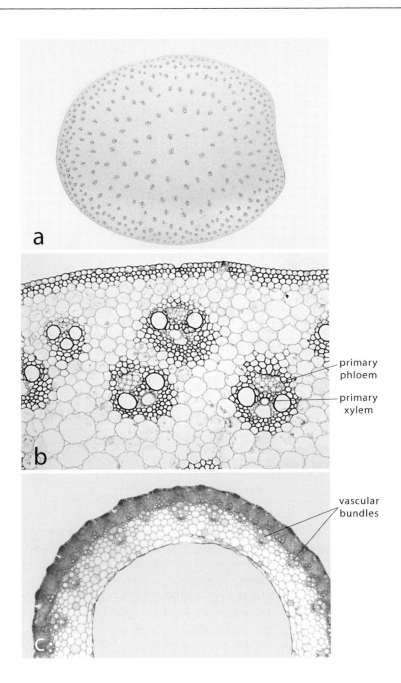

a

b

primary
phloem

primary
xylem

vascular
bundles

c

Figure 7.11 (a, b) Transverse sections of a stem of *Zea mays* illustrating the numerous vascular bundles of which the primary vascular system is comprised. Magnification (a) × 4.4, (b) × 182. (c) Transverse section of a stem of *Triticum* sp. Magnification × 49. characterized by an irregular cylinder of peripheral vascular bundles.

Following the divergence of a leaf trace an axial bundle follows an oblique (and often helical) course from a peripheral position toward the center of the stem, eventually turning sharply outward toward the periphery. After the divergence of another leaf trace, the course toward the center of the stem, and then outward, is repeated. Each axial bundle, consequently, follows an undulating, helical course through the stem with successive leaf traces diverging at regular intervals (Fig. 7.12a, b). Those axial bundles characterized by the greatest distances between divergence of successive leaf traces (major bundles) approach most closely the center of the stem

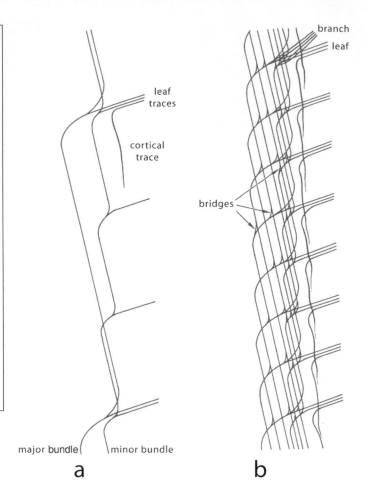

Figure 7.12 Diagrams illustrating the longitudinal course of vascular bundles in the primary vascular system of monocotyledons. (a) The basic components of the system are major bundles, minor bundles, and cortical bundles. Major and minor bundles are axial bundles from which leaf traces diverge. A leaf trace may also contain an outward extension of a cortical bundle. Following divergence of a leaf trace, both major and minor axial bundles follow oblique courses toward the center of the stem and then turn outward until another leaf traces diverges. (b) A part of the primary vascular system illustrating its complexity. For more detail, please refer to the text. (a, b) From Zimmerman and Tomlinson (1972). Used by permission of the University of Chicago Press. © 1972 University of Chicago. All rights reserved.

during their longitudinal, undulating course and may give rise to the median and near-median traces. Those characterized by shorter distances between levels of divergence of successive leaf traces (minor bundles) follow a similar course, but do not approach as close to the stem center (Fig. 7.12b). These may give rise to lateral traces. Adjacent axial bundles are often connected by bridge bundles which facilitate lateral transport in the vascular system. Branch traces that supply lateral appendages may branch from axial bundles and leaf traces (Fig. 7.12b).

With the information at hand, we can now interpret the pattern of zonation in a transverse section of a monocotyledon stem (Fig. 7.11a). The larger, more central bundles are major and minor axial bundles from which leaf and branch traces diverge. The peripheral zone of smaller bundles is a mixture of leaf and branch traces and in some taxa, cortical traces and bundles of fibers. Crowding in the peripheral zone is increased by the presence in it of segments of axial bundles distal to the levels of divergence of leaf traces (reduced in size in consequence of their branching in the formation of traces). The presence of these segments of axial bundles is related to the fact

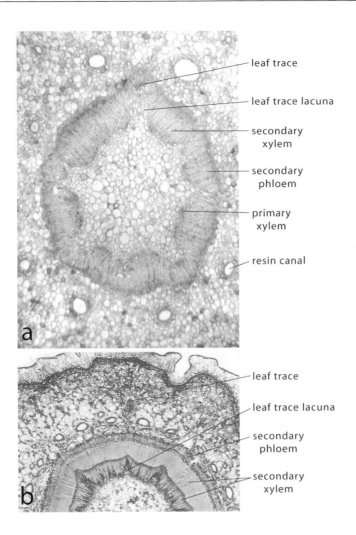

leaf trace

leaf trace lacuna

secondary
xylem

secondary
phloem

primary
xylem

resin canal

leaf trace

leaf trace lacuna

secondary
phloem

secondary
xylem

Figure 7.13 (a) Transverse section of a young stem of *Pinus* sp. (pine) showing a leaf trace lacuna across which the vascular cambium has not differentiated. Magnification × 47. (b) Transverse section of a 2-year-old stem of *Abies concolor*. The vascular cambium became continuous across the leaf trace lacuna toward the end of the first growing season. Magnification × 12.

that some parts of all axial bundles are peripheral by virtue of their undulating, longitudinal course through the stem.

Leaf trace lacunae

In contrast to most monocotyledons which produce only primary vascular tissues, most dicotyledons and all gymnosperms produce secondary vascular tissues as well which, as development proceeds, enclose the primary xylem in the stem and often make its analysis difficult. Typically, in these plants, however, the regions through which leaf traces pass are devoid of secondary vascular tissues (Fig. 7.13a) because, at this stage of development, the cambium has not differentiated across this region. These parenchymatous regions are called **leaf trace lacunae**. Unfortunately, in much of the older, and some of the recent botanical literature, these lacunae are confused with, and incorrectly labeled, leaf gaps. As we know, leaf gaps occur only in primary vascular cylinders, and among living plants are characteristic only of many ferns. Upon differentiation of vascular cambium across the region of the lacunae, often toward the end of the first season of growth, secondary xylem becomes continuous (Fig. 7.13b).

A widely used terminology applied to the nodal structure of woody dicotyledons and gymnosperms describes the relationship of leaf traces that supply a leaf and the lacunae in the secondary vascular cylinder through which they pass. If each leaf is supplied by only one leaf trace, the nodal structure is described as **one trace, unilacunar** or, simply, **unilacunar** (Fig. 7.13a). This relationship can also usually be accurately determined by examining spread out diagrams of the primary vascular system as, for example in Fig. 7.6b which illustrates the nodal structure of *Abies concolor* which is also unilacunar. If each leaf is supplied by two traces, as in *Ginkgo biloba*, the nodal structure is described as **two trace, unilacunar** (Fig. 7.7). In situations in which a single leaf is supplied by three traces, each traversing a separate lacuna, as in *Drimys winteri* (Fig. 7.9a) and *Potentilla fruticosa* (Fig. 7.9b), the nodal structure is described as **trilacunar**. In some taxa the median trace of a trilacunar node is comprised of several closely placed small bundles that may fuse in the leaf base. If a single leaf is supplied by five or more leaf traces, the nodal structure is described as **multilacunar** (Fig. 7.8).

The cauline vs. foliar nature of vascular bundles in the eustele

As emphasized by Stein (1993) the evolution of the various stelar architectures in plants is considered to be related to, and correlated with, the evolution of leaf and branch size and complexity, and to hormonal control of differentiation of primary xylem and phloem. Stein proposed a model of stelar evolution based on control of differentiation as it is presumed to be related to auxin sources (lateral branches and leaves and, in the most primitive, leafless vascular plants, the apical meristems of branches), the size of the apical meristem, longitudinal spacing of primordia, phyllotaxis, the rates of auxin synthesis in the sources, the concentration and rates of flow of the hormone, and the reaction to the hormone by the target tissue.

Stein's analysis throws interesting light on the controversy regarding the nature and evolution of the eustele. As he notes, "the currently popular evolutionary perspective" interprets the eustele as a system of discrete sympodia, each consisting of an axial bundle from which leaf traces diverge (see Namboodiri and Beck, 1968a, b, c; Beck *et al.*, 1983). The sympodia are considered to be cauline, that is, derived, evolutionarily, from stem vascular tissue. This viewpoint is based on comparisons of stelar morphology and leaf evolution of plants through time as demonstrated by the fossil record. The competing hypothesis considers the vascular bundles of the eustele to be composed solely of leaf traces and thus to be of foliar origin (see Esau,

1965). It is based on studies that show leaf primordia and young leaves to be primary sources of auxin that influence the differentiation of the primary vascular system. These two viewpoints may differ only in perspective, however. If one accepts that the most primitive vascular plants were leafless, and that leaves evolved from lateral branch systems (see Chapter 1), it follows that the hormonal source in leafless plants as well as those exhibiting intermediate stages in the evolution of the leaf was the apical meristems of individual lateral branches. One may speculate that upon evolution of leaf blades (according to the telome hypothesis, by a process of "webbing"of branch systems; see Stewart and Rothwell (1993)) the source of auxin became established in the leaf primordia, which accordingly would be, from an evolutionary standpoint, immature, reduced, and highly modified lateral branch systems. At various stages in the evolution of the stele, therefore, the source of the hormone controlling differentiation of the vascular tissues of the stem and its branches, whether apical meristems of lateral branch systems or leaf primordia and young leaves, probably varied with the level of evolutionary specialization of the plant.

Phyllotaxy

Let us now consider in more detail the nature of vegetative **phyllotaxy**, or leaf arrangement. Phyllotaxy in the mature plant is directly related to the position and size of leaf primordia produced on apical meristems of varying size and shape. During growth of the shoot apex, the leaf primordia produced nearest the tip of the apical meristem are gradually shifted laterally toward the periphery, and new primordia develop in sites above the older primordia. Auxin or some other morphogen is thought to influence the extensibility of cell walls in these sites on the surface of apical meristems destined to become leaf primordia, but the source of the morphogen is unclear (see Lyndon, 1994). Masuda (1990) and Kutschera (1992) have shown that auxin increases the extensibility of epidermal cells, and Lyndon (1994) believes that auxin is the hormone most likely to cause epidermal wall extensibility in the surface layers of the apical meristem. Green (1985) has shown that in sites of incipient leaf primordia, cellulose microfibrils are oriented more or less tangential to the surface which, with wall loosening, apparently facilitates the outward development of the primordium as a bulge from the surface of the meristem (see also Lyndon, 1994). However, the mechanism by which the microfibrils attain their tangential orientation is not clear at present. Jesuthasan and Green (1989) proposed that the microfibrils in the surface layer of the apical meristem of *Vinca* (periwinkle) are reoriented by being stretched by the growth of older, adjacent leaf bases. This, however, is at odds with the hypothesis that microfibril orientation is directly related to the orientation of microtubules in the outer protoplast (see Baskin, 2001). Marc and Hackett (1991, 1992) provide evidence that in *Hedera helix* (ivy) gibberellin influences rearrangement of cortical microtubules in cells of the surface layer in regions destined to become leaf primordia with the result that cellulose microfibrils become oriented tangentially in the cell walls. It has been suggested, also, that recently formed primordia may produce inhibitors that prevent new primordia from developing until the field of inhibition has diminished by virtue of the continued growth of the shoot apex and the lateral displacement of older primordia (Lyndon, 1994).

Factors such as hormonal influence on the extensibility of cell walls, orientation of microtubules and microfibrils that are directly related to the plane of cell division, and inhibition of new by older primordia all have a controlling influence on the position of leaf primordia in the shoot apex. Therefore, as suggested by Lyndon (1994), control of the sites of initiation of leaf primordia (and thus, control of phyllotaxy, the pattern of leaf arrangement) rests with "two cooperating systems, a chemical mechanism determining whether a primordium would form, and a bio-physical system determining precisely where it would form." The ultimate genetic control and details of the mechanism whereby genes influence the various factors involved in the establishment of phyllotaxy must await much additional research. (For more detail on the initiation and development of leaf primordia, please see Chapter 5).

It is clear that auxin (and/or other plant hormones, e.g., gibberellin, cytokinin) produced by leaf primordia and young leaves has a controlling influence on the differentiation of the primary vascular system, and that there is a direct relationship between phyllotaxy and the arrangement of vascular sympodia. We know, also, that phyllotaxy may change within a plant as, for example, the apical meristem enlarges during growth of a seedling. It may also differ in the main stem and lateral branches.

In plants with helical phyllotaxy, leaf primordia are produced in sequence along a helical pathway called the generative spiral. It is useful to number the primordia indicating their sequence of development. Many workers assign a high number to the oldest primordium on the apical dome and progressively lower numbers to younger primordia along the generative spiral. Others prefer to number the primordia in reverse order. As we have seen in the spread-out diagrams of primary vascular systems, leaf traces can be similarly numbered. It is also apparent that the leaf trace pattern in the diagrams of primary vascular systems we have studied earlier in this chapter directly reflect the pattern of leaf traces and leaves on the stem surface.

The angle between radii drawn to the center of successively produced primordia is called the **angle of divergence**. On average, in plants with only one generative spiral (**monojugate plants**), the angle of divergence is approximately 137.5 degrees. In **bijugate plants**, with two generative spirals, one clockwise, the other counter-clockwise, the angle of divergence is 68.7 degrees.

A line connecting the primordia in the generative spiral is called a **parastichy**. In fact this term is applied to any helical line connecting

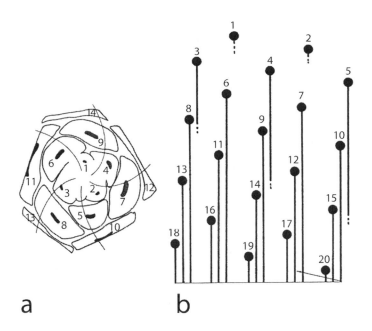

a   **b**

Figure 7.14 Diagrams of *Illicium parviflorum*. (a) A transverse view of a shoot apex showing leaf primordia arranged in two sets of contact parastichies, one set containing five parastichies spiraling to the left and the other containing three parastichies, spiraling to the right. (b) Diagram of the primary vascular system in longitudinal view. Please refer to the text for a detailed description. (a, b) From Benzing (1967a). Used by permission of the Botanical Society of America.

leaf primordia. If the primordia in a helical series are in physical contact in a bud the parastichy is referred to as a **contact parastichy**. The generative spiral is a non-contact parastichy, and the parastichy characterized by the shallowest helix on an apical dome.

The leaf primordia in a shoot apex, as viewed from above, comprise two conspicuous sets of contact parastichies, one set spiraling to the right, the other to the left, and one steeper than the other (Fig. 7.14a). The numerical difference between the primordia in these two sets is used to characterize the phyllotaxy of particular plants. The steeper parastichies (Fig. 7.14b), which include leaf primordia 1, 6, 11; 4, 9, 14; 2, 7, 12, etc., (indicated by curved lines in Fig. 7.14a), are characterized by primordia with numbers that differ by 5. The other set of parastichies which include leaf primordia 1, 4, 7, 10; 2, 5, 8, 11, etc., are characterized by numbers that differ by 3 (Fig. 7.14a, b). The phyllotaxy of this plant can, therefore, be indicated as 3:5 which tells us that the leaf primordia in each of the steep parastichies were initiated by the apical meristem along the generative spiral five plastochrons apart (a plastochron is the time between the initiation of successive primordia in a parastichy) and that leaf primordia in the other, shallower, parastichy were initiated three plastochrons apart. We can also conclude that every leaf primordium produced along the generative spiral occurs in one of the steep parastichies. The same logic can be applied to the shallower parastichies.

One can often recognize more than two sets of parastichies in the shoot apices of most plants with helical phyllotaxy. The steeper the parastichy, the higher the characterizing number. The steepest parastichies, in which mature leaves would be nearly directly above each other, are called **orthostichies**. If in the example above, the steep parastichies were, in fact, orthostichies, we would observe that, in each, leaves would be separated by five internodes (Fig. 7.14b).

Most commonly, leaf primordia occur in 1, 2, 3, 5, 8, or 13 parastichies. These numbers comprise the mathematical series known as the **primary Fibonacci series**, named after the Italian mathematician Leonardo Fibonacci. Addition in sequence of any two numbers of this series will give the number following the preceding pair. The denominator of common phyllotactic fractions, 5, 8, or 13, indicates the number of orthostichies and, as we know, also the number of vascular sympodia in the primary vascular system. The numerator represents the number of turns around the stem and along the generative spiral between successive leaves in an orthostichy, usually 2 as in 2/5, 3 as in 3/8 and 5 as in 5/13.

In plants with decussate phyllotaxy, the leaf primordia develop in pairs 180 degrees apart, and successive pairs occur perpendicular to each other. In plants with distichous phyllotaxy, primordia are also 180 degrees apart, but occur in only one plane. The concept "angle of divergence" is not usually applied to primordia of plants with decussate or distichous phyllotaxy since during development the leaf primordia are not clearly arranged in a helical pattern.

The concepts of the generative spiral, parastichies, and orthostichies that have been applied to phyllotactic systems are important because they not only have a mathematical, but also a biological basis. For more detail, and in-depth discussions of the mathematical and biological bases for phyllotaxy, see Richards (1951) and Romberger *et al.* (1993).

REFERENCES

Baskin, T. I. 2001. On the alignment of cellulose microfibrils by cortical microtubules: a review and a model. *Protoplasma* **215**: 150–171.

Beck, C. B., Schmid, R., and G. W. Rothwell. 1983. Stelar morphology and the primary vascular system of seed plants. *Bot. Rev.* **48**: 691–816; 913–931.

Benzing, D. H. 1967a. Developmental patterns in stem primary xylem of woody Ranales. I. Species with unilacunar nodes. *Am. J. Bot.* **54**: 805–813.
  1967b. Developmental patterns in stem primary xylem of woody Ranales. II. Species with trilacunar and multilacunar nodes. *Am. J. Bot.* **54**: 813–820.

Devadas, C. and C. B. Beck. 1972. Comparative morphology of the primary vascular systems in some species of Rosaceae and Leguminosae. *Am. J. Bot.* **59**: 557–567.

Esau, K. 1965. *Vascular Differentiation in Plants.* New York: Holt, Rinehart and Winston.

Green, P. B. 1985. Surface of the shoot apex: a reinforcement-field theory for phyllotaxis. *J. Cell Sci.* (Suppl.) **2**: 181–201.

Gunckel, J. E. and R. H. Wetmore. 1946. Studies of development in long shoots and short shoots of *Ginkgo biloba* L. II. Phyllotaxis and the organization of the primary vascular system: primary phloem and primary xylem. *Am. J. Bot.* **33**: 532–543.

Jesuthasan, S. and P. B. Green. 1989. On the mechanism of decussate phyllotaxis: biophysical studies on the tunica layer of *Vinca major*. *Am. J. Bot.* **76**: 1152–1166.

Kutschera, U. 1992. The role of the epidermis in the control of elongation growth in stems and coleoptiles. *Bot. Acta* **105**: 227–242.

Lyndon, R. F. 1994. Control of organogenesis at the shoot apex. *New Phytol.* **128**: 1–18.

Marc, J. and W. P. Hackett. 1991. Gibberellin-induced reorganization of spatial relationships of emerging leaf primordia at the shoot apical meristem in *Hedera helix* L. *Planta* **185**: 171–178.

1992. Changes in the pattern of cell arrangement at the surface of the shoot apical meristem in *Hedera helix* L. following gibberellin treatment. *Planta* **186**: 503–510.

Masuda, Y. 1990. Auxin-induced cell elongation and cell wall changes. *Bot. Mag. Tokyo* **103**: 345–370.

Namboodiri, K. K. and C. B. Beck. 1968a. A comparative study of the primary vascular system of conifers. I. Genera with helical phyllotaxis. *Am. J. Bot.* **55**: 447–457.

1968b. A comparative study of the primary vascular system of conifers. II. Genera with opposite and whorled phyllotaxis. *Am. J. Bot.* **55**: 458–463.

1968c. A comparative study of the primary vascular system of conifers. III. Stelar evolution in gymnosperms. *Am. J. Bot.* **55**: 464–472.

Richards, F. J. 1951. Phyllotaxis: its quantitative expression and relation to growth in the apex. *Phil. Trans. Roy. Soc.* London **B235**: 509–564.

Romberger, J. A., Hejnowicz, Z., and J. F. Hill. 1993. *Plant Structure: Function and Development*. Berlin: Springer-Verlag.

Stein, W. 1993. Modeling the evolution of stelar architecture in vascular plants. *Int. J. Plant Sci.* **154**: 229–263.

Stewart, W. N. and G. W. Rothwell. 1993. *Palaeobotany and the Evolution of Plants*, 2nd edn. Cambridge, UK: Cambridge University Press.

White, R. A. and W. H. Weidlich. 1995. Organization of the vascular system in the stems of *Diplazium* and *Blechnum* (Filicales). *Am. J. Bot.* **82**: 982–991.

Zimmermann, M. H. and P. B. Tomlinson. 1972. The vascular system of monocotyledonous stems. *Bot. Gaz.* **133**: 141–155.

FURTHER READING

Balfour, E. E. and W. R. Philipson. 1962. The development of the primary vascular system of certain dicotyledons. *Phytomorphology* **12**: 110–143.

Dormer, K. J. 1972. *Shoot Organization in Vascular Plants*. London: Chapman and Hall.

Esau, K. 1943. Vascular differentiation in the vegetative shoot of *Linum*. II. The first phloem and xylem. *Am. J. Bot.* **30**: 248–255.

1977. *Anatomy of Seed Plants*, 2nd edn. New York: John Wiley and Sons.

Jensen, L. C. W. 1968. Primary stem vascular patterns in three subfamilies of the Crassulaceae. *Am. J. Bot.* **55**: 553–563.

Larson, P. R. 1975. Development and organization of the primary vascular system in *Populus deltoides* according to phyllotaxy. *Am. J. Bot.* **62**: 1084–1099.

Philipson, W. R. and E. E. Balfour. 1963. Vascular patterns in dicotyledons. *Bot. Rev.* **29**: 382–404.

Zimmermann, M. H. and P. B. Tomlinson. 1967. Anatomy of the palm *Rhapis excelsa*. IV. Vascular development in apex of vegetative aerial axis and rhizome. *J. Arnold Arbor.* **46**: 122–142.

Chapter 8

The epidermis

Perspective

Most terrestrial plants live in a highly evaporative environment and one in which they are constantly exposed to toxic substances, to attack and invasion by various small insects and pathogens, to the potentially damaging effects of solar radiation, and to potential damage from high winds. Consequently, several protective tissues have evolved that reduce water loss from the plant, restrict the entry of organisms and toxic substances into the plant body, mitigate the effects of radiation, and strengthen and support the plant thereby reducing its susceptibility to damage from rapid air movement. These include the epidermis of shoot and root systems (sometimes called rhizodermis in the root), the periderm and the rhytidome. These tissues, while providing these functions, must also under certain conditions allow oxygen used in respiration to enter the plant and carbon dioxide utilized in photosynthesis to exit the plant. Consequently, the epidermis and other surficial, protective tissues represent both structural and functional compromises. As the bounding tissue of all young parts of a plant, and of the aerial parts of plants that are comprised solely or largely of primary tissues, the epidermis also provides an important supporting function. In the stem of *Tulipa* (tulip), for example, the epidermis plus a layer of subepidermal collenchyma can contribute as much as 50% to overall stem stiffness (Niklas and Paolillo, 1997). We shall consider the epidermis in some detail in this chapter, and periderm and rhytidome in Chapter 13.

Epidermis of the shoot

In the shoot, the **epidermis** develops directly from the protoderm (Fig. 8.1). In most plants it is composed of a single layer of specialized cells that comprises the outer layer of all parts of the shoot system in regions distal to the development of periderm. The epidermis consists largely of living parenchyma cells, usually tabular (i.e., with a

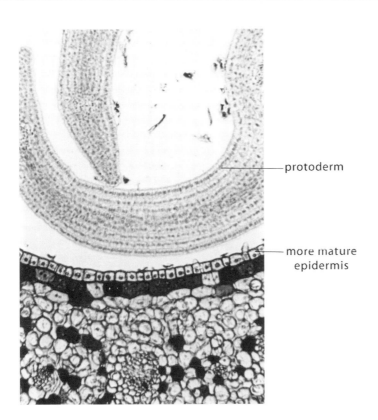

protoderm

more mature epidermis

narrow radial dimension and comparatively large inner and outer surface areas), of various shapes and frequently with wavy anticlinal walls (especially in leaves and petals). Interspersed in the parenchymatous groundmass of leaves are stomata (Fig. 8.2) (most commonly in the lower epidermis) and associated subsidiary cells. In grasses and many other monocotyledons, and on elongate parts (e.g., stems, petioles, midribs, etc.) of most vascular plants, epidermal cells are usually elongate (Fig. 8.3) with the long axis parallel to the long axis of the plant structures. The sinuous nature of the walls in some gymnosperms (Fig. 8.3a) and most grasses (Figs. 8.3b, 8.4) results from the synthesis during development of thickened bands of cellulose microfibrils that extend across the anticlinal walls between inner and outer periclinal walls. During growth, the epidermal cell walls expand between the bands (Panteris *et al.*, 1994). Epidermal cells of grasses may contain **idioblasts** (single or small groups of specialized cells) such as fibers, as well as cork and silica cells (Fig. 8.4) which often occur in pairs. The function of the non-living **cork cells**, characterized by suberized walls, is unknown. The SiO_2 in the **silica cells**, often deposited in laminae, is the basis of the abrasive nature of many grasses. Specialized epidermal cells may also contain pigments, oils, crystals, tannins, etc. Grasses are also characterized by the presence of **bulliform cells** (Fig. 8.5), large thin-walled cells which, during expansion growth, contribute to the unrolling of young leaves. In

Figure 8.5 Section of a leaf of *Stipa robusta*, a grass, containing bulliform cells in grooves in the adaxial surface.

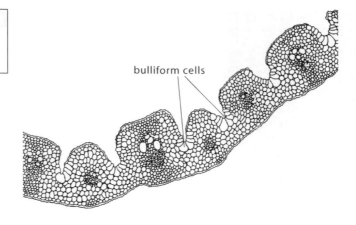

bulliform cells

Figure 8.6 (a) A multiple epidermis of the leaf of *Ficus*. Note the thin cuticle. The large cells below the outer layer, lacking chloroplasts, are considered to function in water storage. Magnification × 193. (b) The epidermis of a stem of *Smilax*. The outer periclinal walls of the epidermal cells are impregnated with lignin and cutin, and have become greatly expanded. Overlying the expanded cell walls is a thick cuticle. Magnification × 167.

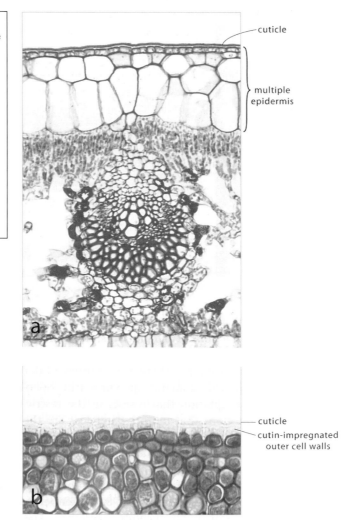

cuticle

multiple epidermis

cuticle
cutin-impregnated outer cell walls

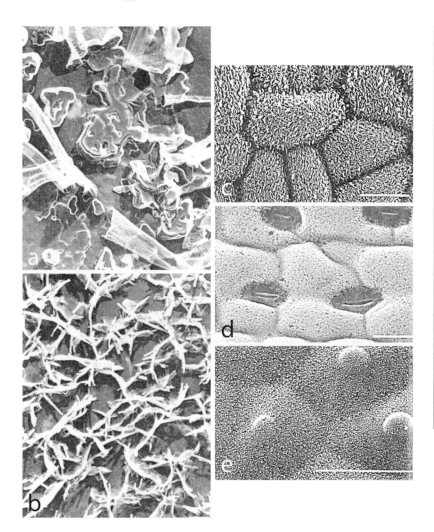

Figure 8.7 (a, b) Scale-like wax platelets on the adaxial surface of the leaf of *Brassica oleracea* (a), and *Lupinus albus* (b). Magnification × 9 206. (c–e) Characteristics of water-repellent leaf surfaces. (c, d) Epidermal cells with convex outer walls covered by a dense layer of epicuticular waxes in *Hypericum aegypticum* (c) and *Marsilea mutica* (d). Note the stomata in *Hypericum*. Bars = 10 μm. (e) Epidermal cells with convex outer walls with conspicuous papillae covered by epicuticular waxes in *Lupinus polyphyllos*. Bar = 50 μm. The covering of fine wax platelets makes the leaf surfaces hydrophobic. Consequently, water and contaminating particles flow off because of the reduction in adhesion to the leaf surfaces. (a, b) From Juniper (1959). Used by permission of Elsevier. (c–e) From Neinhuis and Barthlott (1997). Used by permission of Oxford University Press.

impermeable to the passage of water and other small molecules. It varies in thickness from less than 1 μm to 15 μm or more, and its thickness seems to be related to the environment in which the plant lives. For example, plants living in moist conditions usually have thin cuticles whereas those that live in arid conditions often have thick cuticles.

In some plants the innermost layer of the cuticle has a high content of **pectin** (easily detected by staining with ruthenium red) which often impregnates the outer laminae of the epidermal cell walls. Pectin decreases in concentration outwardly whereas the wax content of the cuticle increases outwardly, and the outer surface layer consists solely of **epicuticular wax**. Highly magnified, the wax can be seen to consist of scale-like platelets (Fig. 8.7a, b) of various forms, but without magnification it may appear smooth and glossy. How the precursor compounds of cutin and wax move from the plasmalemma through the cell wall and to the exterior is not clearly understood. It has been suggested that, during development, minute **cuticular pores**

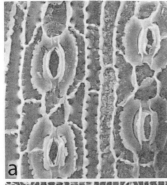

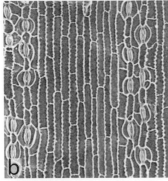

Figure 8.8 Variation in epidermal cell patterns on the inner surface of the cuticle of species of *Dacrydium*, a gymnosperm (Podocarpaceae), considered useful taxonomic characters in this family. (a) *D. guillauminii*. Magnification × 278. (b) *D. lycopodioides*. Magnification × 118. From Stockey and Ko (1990). Used by permission of the University of Chicago Press. © 1990 The University of Chicago. All rights reserved.

(often called microchannels) facilitate the transfer of wax to the surface of the cuticle, but Neinhuis *et al.* (2001) provide indirect evidence that molecules of wax and water simply move together through the cuticle under the influence of cuticular transpiration.

Whereas the major role of the cuticle and intracuticular waxes is restriction of water loss from the plant, the epicuticular waxes play a significant additional role in reflecting light, thus reducing the possibility of overheating and excessive water loss through transpiration (see Barnes and Cadoso-Vilhena, 1996; Neinhuis *et al.*, 2001). The epicuticular wax also "waterproofs" the plant, i.e., makes it water-repellent (Fig. 8.7c–e). Consequently, leaves are "self-cleaning." According to Neinhuis and Barthlott (1997), during rainfall, water and contaminating particles flow off the leaf because of the consequent reduction in adhesion to the hydrophobic surfaces. Variations in the structure of epicuticular wax are thought to correlate with specific epicuticular wax alleles in *Sorghum bicolor* (Jenks *et al.*, 1992) and therefore could be useful taxonomic characters. Epidermal cell patterns, imprinted in great detail on the inner surface of the cuticle, are also used as taxonomic characters in some gymnosperms as, for example, in *Dacrydium* (Podocarpaceae) (Fig. 8.8a, b) (Stockey *et al.*, 1998).

In some ferns (e.g., members of the Polypodiaceae which thrive in conditions of low light intensity) as well as in a few aquatic angiosperms epidermal cells contain chloroplasts and thus function in photosynthesis.

Since the epidermis comprises the outer covering of plants and plant parts in contact with the atmosphere, cell development is more complex than that of internal parenchymatous cells. During differentiation, because of their generally tabular form the anticlinal walls reach their greatest extent sooner than the periclinal walls which have a considerably greater surface area and grow over a longer period of time. Furthermore, the outer periclinal walls, containing cutin and waxes, differ in chemical composition, and are often thicker than inner walls. Consequently, control mechanisms must not only direct the differential synthesis of precursor compounds as well as the production of secretory vesicles but must also control the transfer of the correct compounds to the appropriate sites of wall synthesis (see Wojtaszek, 2000).

Trichomes in the shoot system are appendages of the epidermis. They may remain alive or may die following development. Unlike root hairs, which are extensions of single epidermal cells, trichomes of the shoot consist of one or more cells. They may be simple, consisting of only a single non-epidermal cell, or remarkably complex and multicellular. Some are glandular, others non-glandular. We shall defer consideration of glandular hairs to Chapter 15 on secretion in plants and emphasize non-glandular hairs in this chapter. Some non-glandular trichomes develop from cells of the protoderm as in *Gossypium* (cotton), others are initiated by cell division in a mature epidermal cell. They vary in both function and morphology, and may

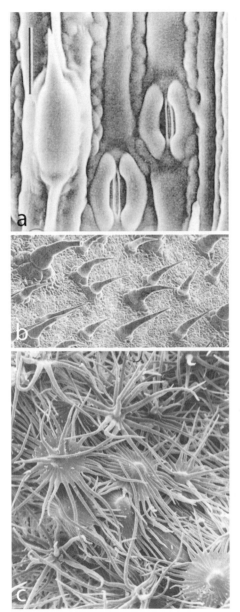

Figure 8.9 Non-glandular epidermal trichomes. (a) A unicellular trichome of *Sorghastrum contractum*. Bar = 20 μm. (b) Multicellular, unbranched trichomes on the abaxial surface of a leaf of *Cucumis sativus*. Magnification × 136. (c) Multicellular, branched trichomes on a leaf of *Verbascum* (mullein). Magnification × 50. (a) From Dávilla and Clark (1990). Used by permission of the Botanical Society of America. (b) From Troughton and Donaldson (1972). Used by permission of the New Zealand Ministry of Research, Science and Technology. (c) Photograph by P. Dayanandan.

be branched, unbranched, or peltate. **Unbranched hairs** are either unicellular (Fig. 8.9a) or multicellular (Fig. 8.9b) and sometimes spine-like. **Branched trichomes** are often digitate or stellate (Fig. 8.9c). **Peltate hairs** are scale-like consisting of a group of cells forming a plate oriented parallel to the epidermal surface and attached by a very short stalk, often consisting of a single cell. Hairs occur on all parts of the plant including floral organs. The function of non-glandular trichomes is poorly understood. Dense coverings of non-living, air-filled, branched hairs such as those of *Verbascum* (mullein) tend to hold a layer of vapor-filled air on the surface of the epidermis that is thought to reduce transpiration, and thus desiccation, in dry and/or windy conditions. The presence of such pubescent layers, especially on leaf surfaces may, under certain conditions, be a very important supplement to the cuticle in inhibiting water loss from the plant although there is little experimental evidence to support this viewpoint. It seems very likely, however, that a dense pubescence would reflect light and thus, prevent overheating of the plant. It has been suggested that trichomes may also prevent or restrict insect predation. The vesiculate hairs of saltmarsh plants such as *Atriplex* (saltbush) are repositories of salt which prevent the build-up of toxic levels of salt in the plant. Epidermal trichomes are also useful in taxonomy (see, e.g., Amarasinghe *et al.*, 1991).

Epidermis of the root

The **root epidermis**, often called the rhizodermis, differs from that of the shoot in development as well as in structure and function. In monocotyledons the epidermis originates from a tier of meristematic cells at the root tip which also gives rise to the cortex. In dicotyledons the epidermis originates from an apical tier of cells which also gives rise to the root cap. In each case, the outermost layer of cells derived from these meristematic regions comprises a protoderm from which the epidermis ultimately develops.

In the absence of a cuticle, the epidermis of the root in most plants differs significantly from that of the shoot. The presence of a very thin, cuticle-like layer, however, has been observed in a few taxa. Furthermore, there are usually no, or only a very few, stomata in the root epidermis. These structural differences are, of course, directly related to the absorptive function of the root, and to the fact that roots of most plants are subterranean, non-photosynthetic, and not subject to the desiccating effects of the aerial environment. Root hairs, unlike the trichomes of the shoot, are simple extensions from single epidermal cells. Root hairs develop from specialized cells in the epidermis called **trichoblasts**, and occupy a position between the root tip and the level of the root at which periderm develops. Absorption occurs through the root hairs which greatly expand the absorptive surface of the root, but absorption also occurs through all outer surfaces of the epidermal cells. The percentage of absorption that occurs directly through roots hairs and other root surface areas is unknown. In most plants, however, it is clear that a great part of all absorption occurs through the fungal hyphae of mycorrhizae that may extend out from the root in all directions over a large area. As the root elongates, the more proximal root hairs cease to function and new root hairs develop acropetally in the distal part of the root hair zone. We shall discuss in more detail the development of root hairs as well as the role of mycorrhizae in root absorption in the chapter on the root (Chapter 16).

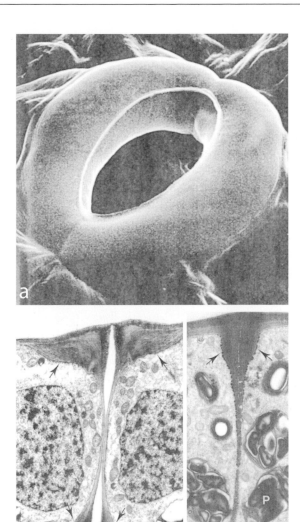

Figure 8.10 (a) Scanning electron micrograph of a stoma from the epidermis of *Cucumis sativus* showing two expanded, kidney-shaped guard cells enclosing an open stomatal aperture. Magnification × 224. (b) Transverse section of a stoma of the fern *Asplenium nidus*, passing through the closed stomatal aperture. Note the thickened cell wall regions (arrows), the large nuclei, and the numerous mitochondria. Magnification × 4260. (c) Section through one of the ends of a stoma of *Asplenium nidus* where the guard cells are connected to each other. Note the wall thickenings (arrows) and the large plastids (p) containing starch grains. Magnification × 4714. (a) From Troughton and Donaldson (1972). Used by permission of the New Zealand Ministry of Research, Science and Technology. (b, c) From Apostolakos and Galatis (1999). Used by permission of Blackwell Publishing, Ltd.

Stomata

Stomata (Figs. 8.2a, b, 8.8a, b, 8.10a, b) occur in the epidermis of all parts of the shoot system, even in flower parts such as stamens and pistils. They occur in great frequency in leaves where they are most abundant in the lower epidermis. A stoma consists of the **stomatal aperture** and two enclosing **guard cells**. Some workers prefer to use the term "stoma" for the aperture or pore only, and "stomatal apparatus" for the stoma plus the guard cells. Because stoma (singular) and stomata (plural) are used so extensively in the plant physiology and plant development literature to mean aperture plus guard cells, they will be used that way in this book.

With the exception of those in grasses and sedges, the stomata of all plants are very similar. The guard cells are kidney-shaped, and the cell walls in the polar regions (where the ends of the pair of cells

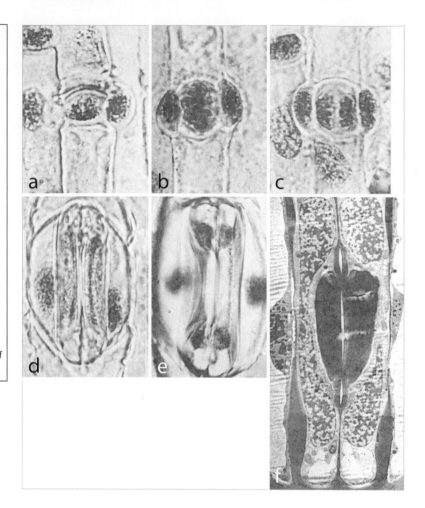

Figure 8.11 Perigenous development of a stoma in the epidermis of the internode of *Avena*, a grass. (a–c) Guard cells and subsidiary cells originate independently from different mother cells. (d) Nearly mature stoma and associated subsidiary cells. (e) Mature stoma. On the basis of the arrangement of guard cells and subsidiary cells in the mature state, and without knowledge of early developmental stages, one would probably conclude, incorrectly, that development of the grass stoma was mesogenous. (f) Enlargement of a mature stoma of *Avena*. Note the elongate nucleus and the thick walls opposite the stomatal aperture. Magnification (a–e) × 1333, (f) × 4425. From Kaufman *et al.* (1970). Used by permission of the Botanical Society of America.

are attached) are thicker than other parts of the cell wall (Fig. 8.10b). Unlike other cells in the epidermis of most plants, the guard cells contain chloroplasts in which starch grains may develop (Fig. 8.10c). At maturity the guard cells may be at the same level in the epidermis as the surrounding subsidiary cells, raised above them (Fig. 8.10a), or sunken below them (Fig. 8.2b). Commonly, immediately below the stoma there is a large air space in the leaf mesophyll called the **substomatal chamber** (Fig. 8.2b).

In contrast to the kidney-shaped guard cells of most plants, grasses and sedges have guard cells of very different morphology. The ends of the guard cells are thin-walled and bulbous whereas the central region is narrow and thicker-walled (Fig. 8.11e, f). Consequently, the cell resembles somewhat a dumb-bell. When the bulbous ends are turgid and expanded, the aperture is open; when not expanded, the aperture is closed. The shape of the nucleus conforms to that of the cell with the ends enlarged and connected by a slender strand (Fig. 8.11f). An especially interesting aspect of the grass stoma is the fact that the protoplasts of the guard cells are connected through a pore in the common walls of the bulbous ends (polar regions). Consequently changes in turgor pressure occur nearly simultaneously in the two guard cells.

The mechanism of movement in guard cells

Until relatively recently, the cause of increase in turgor pressure in guard cells had been thought to be the result of an increase in concentration of photosynthate in the protoplast. It is now thought, however, that photosynthate is utilized primarily as a source of energy for the guard cells. Excellent evidence indicates that the solute resulting in increase in turgor in the guard cells is potassium ions (K^+). As anions of malic acid and cloride accumulate in the guard cell vacuoles, potassium ions migrate into the vacuoles, neutralizing the anions, and with an increase in concentration of K^+ in the cells osmosis occurs, resulting in increase in turgor pressure. Consequently, the cells swell; but since the ends of the guard cells are attached they bulge away from each other resulting in opening of the aperture. It has also been proposed that, in some species, the cellulose microfibrils are radially arranged in the walls of the guard cells. Consequently with increasing turgor pressure, the cells are able to increase in length (though not in diameter) which results in expansion of the aperture.

The source of water and ions during the process of stomatal opening is the surrounding subsidiary cells. Interestingly, there are very few plasmodesmata, in some cases none at all, between the subsidiary cells and the guard cells. Consequently, the movement of water and ions into and out of the guard cells is through the cell wall apoplast. As in the case of transfer cells, the cell walls of the subsidiary cells contiguous with those of the guard cells often have extensive infoldings.

Whether the stoma is open or closed also seems to be correlated with CO_2 concentration in the substomatal chamber. During periods of photosynthesis when CO_2 is utilized and the concentration in the substomatal chamber is low, the stomata are open, but during periods of very low light intensity when the CO_2 concentration is high, the stomata close. They also close under conditions of water stress which usually occurs when the temperature is high and air movement is great, thus, reducing water loss through the stomatal apertures. During such conditions, although photosynthesis may be occurring and CO_2 is being utilized, the concentration in the substomatal chambers may also be high since the air movement increases the available supply of CO_2 thus steepening the gradient of CO_2 into the leaf.

Development of stomata

As a leaf primordium grows, stomata develop in one of three ways. Following an unequal cell division in the protoderm, the smaller cell may divide followed by another division in each daughter cell to form four cells. The two inner cells will become guard cells, the two outer, subsidiary cells. This type of development, in which cells of the stoma and subsidiary cells are derived from the same mother cell, is referred to as **mesogenous development** (for some variations, see Carr and Carr, 1991). It is characteristic of many gymnosperms, in which this pattern of development is also termed **syndetocheilic**. If the guard cells and the subsidiary cells are derived from different protodermal initials, development is referred to as **perigenous**. This pattern is common in many dicotyledons. In gymnosperms this pattern of development is also called **haplocheilic**. In some plants in which one of two subsidiary cells has the same origin as the guard cells, development is termed **mesoperigenous**.

One is not always able accurately to determine these patterns of development by observing mature stomata. In grasses, for example, the guard cells and the two adjacent subsidiary cells appear to have had a common origin and to have developed mesogenously. In fact, the guard cells and subsidiary cells originate from separate initials and, thus, the development of stomata is perigenous (Fig. 8.11a–e) (Kaufman *et al.*, 1970).

As in other cells within the plant, the form of guard cells is directly related to the presence, during development, of microtubules (often associated with actin microfilaments) immediately below the plasmalemma that mirror the synthesis of cellulose microfibrils in the cell walls. Both shape and function are related to variable thickness of the cell walls and the effect this has on the expansion of the cells during growth and during fluctuations in turgor pressure. This association of microtubules with cell wall synthesis in the contiguous walls adjacent to the developing pore in the stomata of the fern *Asplenium nidus* (Fig. 8.12a, b) has been beautifully demonstrated by Apostolakos and Galatis (1999). An association of microtubules with wall synthesis in developing stomata of *Avena sativa* (oats) was demonstrated as early as 1970 by Kaufman and co-workers.

Several different patterns of guard cells and subsidiary cells have been recognized, and used for taxonomic purposes (see Metcalfe and Chalk, 1950: vol. I, p. xiv). We shall consider the four most common types (Fig. 8.13) referred to as anomocytic (irregular), also called "ranunculaceous" (because this type is of frequent occurrence in the Ranunculaceae); paracytic (parallel), also "rubiaceous"; anisocytic (unequal), also "cruciferous"; and diacytic (cross-celled), also "caryophyllaceous". In the **anomocytic** type the guard cells and subsidiary cells have no well-defined pattern; or, from a developmental standpoint, there may be no subsidiary cells. The **paracytic** type is characterized by subsidiary cells and guard cells having parallel long axes. The **anisocytic** type is distinctive in having three subsidiary cells each of a different size. In the **diacytic** type, there are two subsidiary cells that have contiguous walls that are at right angles to the long axis of the guard cells, and which enclose the guard cells.

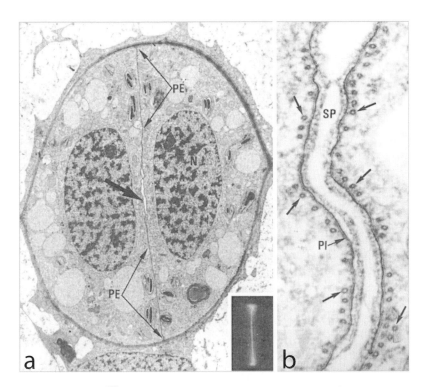

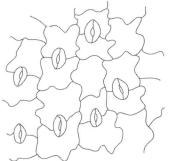

Figure 8.12 (a) Paradermal section of an immature stoma of the fern *Asplenium nidus*. Thin arrows delimit polar ends (PE) between the developing aperture (large, thick arrow). Magnification × 743. (b) Enlargement showing stomatal aperture at approximately the same developmental stage as that shown in (a) and the associated microtubules (arrows). PL, plasmalemma; Magnification × 66 340. SP, developing aperture. From Apostolakos and Galatis (1999). Used by permission of Blackwell Publishing, Ltd.

Figure 8.13 Diagrams illustrating the relationship between guard cells and associated subsidiary cells. Please see the text for a detailed explanation.

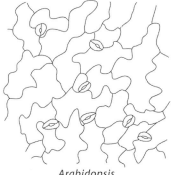

Arabidopsis
anomocytic
(ranunculaceous)

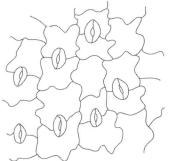

Viscaria
diacytic
(caryophyllaceous)

Vigna
paracytic
(rubiaceous)

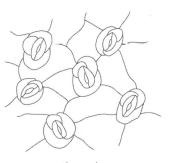

Crassula
anisocytic
(cruciferous)

REFERENCES

Amarsinghe, V. Graham, S. A., and A. Graham. 1991. Trichome morphology in the genus *Cuphea* (Lythraceae). *Bot. Gaz.* **152**: 77–90.

Apostolakos, P. and B. Galatis. 1999. Microtubule and actin filament organization during stomatal morphogenesis in the fern *Asplenium nidus*. II. *Guard cells. New Phytol.* **141**: 209–223.

Barnes, J. D. and J. Cadoso-Vilhena. 1996. Interactions between electromagnetic radiation and the plant cuticle. In G. Kerstiens, ed., *Plant Cuticles: An Integrated Functional Approach*. Oxford, UK: Bios Scientific Publishers, pp. 157–174.

Bowman, J. 1994. Arabidopsis: *An Atlas of Morphology and Development*. Heidelberg: Springer-Verlag.

Carr, D. J. and S. G. M. Carr. 1991. Development of the stomatal complexes during ontogeny in *Eucalyptus* and *Angophora* (Myrtaceae). *Austral. J. Bot.* **39**: 43–58.

Dávilla, P. and L. G. Clark. 1990. Scanning electron microscopy survey of leaf epidermis of *Sorghastrum* (Poaceae: Andropogoneae). *Am. J. Bot.* **77**: 499–511.

Jenks, M. A., Rich, P. J., Peters, P. J., Axtell, J. D., and E. N. Ashworth. 1992. Epicuticular wax morphology of *bloomless* (*bm*) mutants in *Sorghum bicolor*. *Int. J. Plant Sci.* **153**: 311–319.

Juniper, B. E. 1959. The surfaces of plants. *Endeavour* **18**: 20–25.

Kaufman, P. B., Petering, L. B., Yocum, C. S., and D. Baic. 1970. Ultrastructural studies on stomata development in internodes of *Avena sativa. Am. J. Bot.* **57**: 33–49.

Metcalfe, C. R. and L. Chalk. 1950. *Anatomy of the Dicotyledons*, 2 vols. Oxford, UK: Clarendon Press.

Neinhuis, C. and W. Barthlott. 1997. Characterization and distribution of water-repellent, self-cleaning plant surfaces. *Ann. Bot.* **79**: 667–677.

Neinhuis, C., Koch, K., and W. Barthlott. 2001. Movement and regeneration of epicuticular waxes through plant cuticles. *Planta* **213**: 427–434.

Niklas, K. J. and D. J. Paolillo, Jr. 1997. The role of the epidermis as a stiffening agent in *Tulipa* (Liliaceae) stems. *Am. J. Bot.* **84**: 735–744.

Panteris, E., Apostolakos, P., and B. Galatis. 1994. Sinuous ordinary epidermal cells: behind several patterns of waviness, a common morphogenetic mechanism. *New Phytol.* **127**: 771–780.

Stockey, R. and H. Ko. 1990. Cuticle micromorphology of *Dacrydium* (Podocarpaceae) from New Caledonia. *Bot. Gaz.* **151**: 138–149.

Stockey, R., Frevel, B. J., and P. Woltz. 1998. Cuticle micromorphology of *Podocarpus*, subgenus *Podocarpus*, section *Scytopodium* (Podocarpaceae) of Madagascar and South Africa. *Int. J. Plant Sci.* **159**: 923–940.

Tenberge, K. B. 1991. Ultrastructure and development of the outer epidermal wall of spruce (*Picea abies*) needles. *Can. J. Bot.* **70**: 1467–1487.

Troughton, D. and L. A. Donaldson. 1972. *Probing Plant Structure*. New York: McGraw-Hill.

Wojtaszek, P. 2000. Genes and plant cell walls: a difficult relationship. *Biol. Rev.* **75**: 437–475.

FURTHER READING

Behnke, H.-D. 1984. Plant trichomes – structure and ultrastructure: general terminology, taxonomic applications, and aspects of trichome-bacteria interaction in leaf tips of *Dioscorea*. In E. Rodriguez, P. L. Healey, and I.

Mehta, eds., *Biology and Chemistry of Plant Trichomes*. New York: Plenum Press, pp. 1–21.

Cutler, D. F., Alvin, K. L., and C. E. Price (eds.) 1982. *The Plant Cuticle*. London: Academic Press.

Davis, D. G. 1971. Scanning electron microscopic studies of wax formation on leaves of higher plants. *Can. J. Bot.* **49**: 643–546.

Eglington, G. and R. J. Hamilton. 1967. Leaf epicuticular waxes. *Science* **156**: 1322–1335.

Esau, K. 1965. *Plant Anatomy*, 2nd edn. New York: John Wiley and Sons.

Juniper, B. E. and C. E. Jeffree. 1983. *Plant Surfaces*. London: Edward Arnold.

Kerstiens, G. (ed.) 1996. *Plant Cuticles: An Integrated Functional Approach*. Oxford, UK: BIOS Scientific Publishers.

Koch, K., Neinhuis, C., Ensikat, H. J., and W. Barthlott. 2004. Self assembly of epicuticular waxes on living plant surfaces imaged by atomic force microscopy (AFM). *J. Exp. Bot.* **55**: 711–718.

Levitt, J. 1974. The mechanism of stomatal movement: once more. *Protoplasma* **82**: 1–17.

Mansfield, T. A., Hetherington, A. M., and C. J. Atkinson. 1990. Some current aspects of stomatal physiology. *Annu. Rev. Plant Physiol. Plant Mol. Biol.* **41**: 55–75.

Martin, J. T. and B. E. Juniper. 1970. *The Cuticles of Plants*. London: Edward Arnold.

Pant, D. D. 1965. On the ontogeny of stomata and other homologous structures. *Plant Sci. Ser. Allahabad* **1**: 1–24.

Raschke, K. 1975. Stomatal action. *Annu. Rev. Plant Physiol.* **26**: 309–340.

Romberger, J. A., Hejnowicz, Z., and J. F. Hill. 1993. *Plant Structure: Function and Development*. Berlin: Springer-Verlag.

Sachs, T. and N. Novoplansky. 1993. The development and patterning of stomata and glands in the epidermis of *Peperomia*. *New Phytol.* **123**: 567–574.

Srivastava, L. M. and A. P. Singh. 1972. Stomatal structure in corn leaves. *J. Ultrastruct. Res.* **39**: 345–363.

Uphof, J. C. T. 1962. *Handbuch der Pflanzenanatomie*, vol. 4, part 5, *Plant hairs*. Berlin: Gebrüder Borntraeger.

Chapter 9

The origin of secondary tissue systems and the effect of their formation on the primary body in seed plants

Perspective: role of the vascular cambium

As the vascular cambium becomes active and secondary tissues are formed, the consequent increase in diameter of the stem may have profound effects on the primary body. This is especially true in woody, arborescent taxa among conifers and dicotyledons. The vascular cambium is an extensive, permanent secondary meristem, one cell thick, conical in form, often described as cylindrical, that begins its development between primary xylem and primary phloem. In most gymnosperms and dicotyledons it is present in all main stems and roots and their branches, extending from near their tips to the bases of stems and roots. In some woody plants it even extends into leaf petioles. In most woody taxa it differentiates first in developing vascular bundles at about the same time as metaxylem begins its development (Figs. 9.1a, b, 9.2), that is, after elongation in the provascular strands has ceased. This **fascicular cambium** may become active, producing some secondary xylem and phloem before cambial differentiation occurs between the bundles (Fig. 9.1b), that is, in the **interfascicular regions**. In many woody, arborescent taxa, additional provascular strands differentiate between the initial vascular bundles, often so close together that they may contact each other laterally (Fig. 9.1b, c). The vascular cambium then becomes continuous across the vascular bundles (Fig. 9.1c, d).

If the **interfascicular cambium** differentiates relatively early, as in many woody herbaceous taxa, it may develop from cells of ground meristem (Fig. 9.3a); if later, its interfascicular development may actually result from dedifferentiation of mature interfascicular parenchyma. Cambial activity begins initially in the vascular bundles (Fig. 9.4a). Upon the establishment of continuity of the cambium, its cell divisional activity results in continuous increments of secondary xylem and secondary phloem (Figs. 9.3b, 9.4b). In many woody vines, only the fascicular cambium produces typical secondary xylem and phloem whereas the interfascicular cambium produces secondary parenchyma (Fig. 9.3c) (see Chapter 14 for more information on the structure of woody vines).

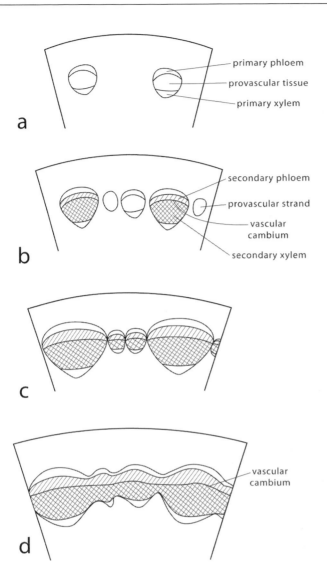

a

primary phloem
provascular tissue
primary xylem

b

secondary phloem
provascular strand
vascular cambium
secondary xylem

c

d

vascular cambium

Figure 9.1 Diagrams illustrating the sites of development of the vascular cambium and its formation as a continuous cylinder in woody, arborescent seed plants. Please see the text for a detailed explanation.

In herbaceous taxa that produce only meager amounts of secondary tissues, the fascicular cambium is the most, or in some plants, the only, active region (Fig. 9.3d, e). Thus secondary vascular tissues are restricted to the sites of the original primary vascular bundles. In non-woody herbs, the cambium, of course, does not develop, or if it develops will not become very active (Fig. 9.3e). As you might expect, the degree of development and activity of the vascular cambium intergrades between these several conditions. In plants characterized by a residual meristem, fascicular cambium develops from provascular tissue (which has developed from residual meristem) and interfascicular cambium develops from any intervening residual meristem; or if intervening residual meristem differentiates into interfascicular parenchyma, the interfascicular cambium will develop by dedifferentiation of cells in this tissue. (For more detail, see Chapter 5 on meristems of the shoot).

Figure 9.2 Transverse section of vascular bundles in *Cassia didymobotrya* (Leguminosae) showing the presence of fascicular cambium within individual bundles. Fascicular cambium becomes functional at the approximate level at which metaxylem elements are differentiating. Magnification × 275. From Devadas and Beck (1971).

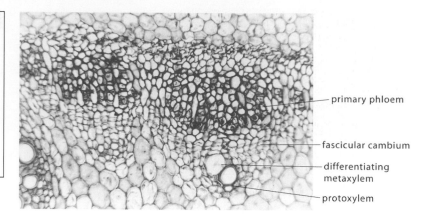

primary phloem

fascicular cambium

differentiating metaxylem

protoxylem

Figure 9.3 Diagrams showing origin of the vascular cambium in woody, herbaceous dicotyledons (a, b), vines (c), and herbaceous dicotyledons (d, e). Please see the text for descriptions.

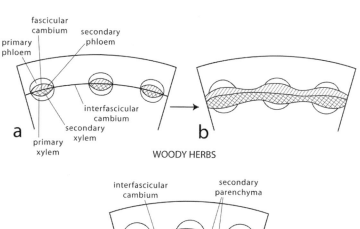

fascicular cambium
primary phloem
secondary phloem
interfascicular cambium
secondary xylem
primary xylem

a   b

WOODY HERBS

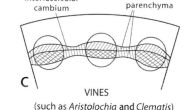

interfascicular cambium
secondary parenchyma

c

VINES
(such as *Aristolochia* and *Clematis*)

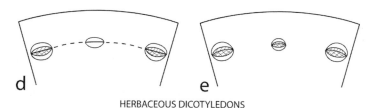

d   e

HERBACEOUS DICOTYLEDONS
(such as *Geum* and *Agrimonia*)   (such as *Ranunculus* and *Impatiens*)

In arborescent forms and woody perennials the vascular cambium functions throughout the life of the plants, in some for hundreds of years. In annuals, the cambium persists only during the growth of the plant. It may cease its activity and develop into xylem and/or phloem before the plant dies. There is no vascular cambium in some very herbaceous dicotyledons, most monocotyledons, and most pteridophytes.

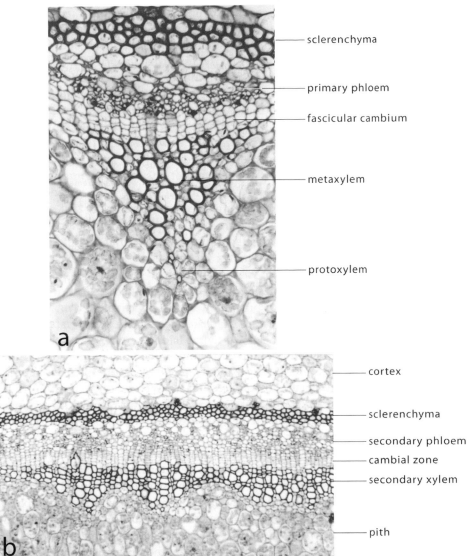

sclerenchyma

primary phloem

fascicular cambium

metaxylem

protoxylem

a

cortex

sclerenchyma

secondary phloem

cambial zone

secondary xylem

pith

b

The effect of secondary growth on the primary body

Once the cambium has differentiated as a continuous layer in the most distal regions of the stem and in lateral branches, its activity results, as we now know, in the production of a layer of secondary xylem to the inside and a layer of secondary phloem to the outside of itself. Since, of course, the newly formed cambium is simply the most distal part of the vascular cambium that extends to the base of the stem, these new layers of vascular tissue also extend to the base of the stem (Fig. 9.5). Each year, therefore, in any one region of the stem, an additional increment of secondary xylem and an additional increment of secondary phloem will be produced by activity of the vascular cambium.

Figure 9.4 Transverse sections of *Pelargonium*. (a) An active fascicular vascular cambium has produced some secondary xylem. Magnification × 263. (b) Fascicular and interfascicular cambia have formed a continuous cylinder, and have produced continuous increments of secondary xylem and secondary phloem Magnification × 100.

Figure 9.6 (a–c) Diagrams illustrating the divergence of a leaf trace from the cylinder of primary vascular bundles and its entrance into a leaf. (a, b) Transverse sections. (c) Radial section along the plane x–x′ in (a). (d, e) Transverse sections illustrating the divergence of a leaf trace at levels of one (a) and three (e) increments of secondary vascular tissues, and illustrating the presence of a lacuna (a discontinuity in the secondary xylem) just above the leaf trace. (f) Tangential section along the plane y–y′ in (e).

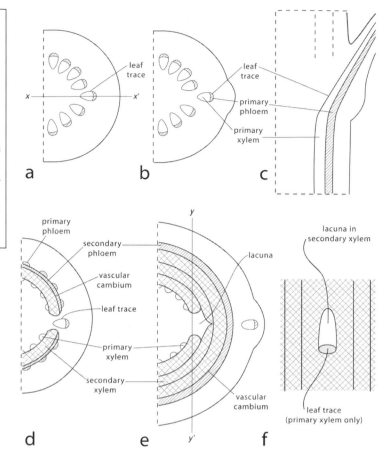

Figure 9.7 Diagrams of radial sections showing (a) a severing of the leaf trace following differentiation of the vascular cambium through the trace, a common occurrence in deciduous taxa, and (b) a persistent trace which increases in length over a period of several years by production of additional conducting tissue by the "armpit" cambium. Please see the text for more detail.

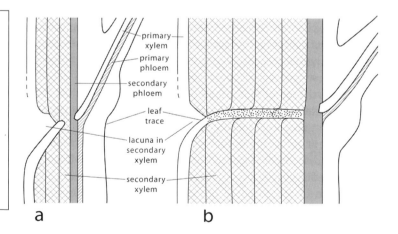

the activity of a small, specialized meristem within the vascular cambium in contact with the ends of the traces. These, small, generally circular regions are called **armpit cambia**. They produce new conducting tissues that are composed of cells that simulate those of the primary xylem, and that add to the length of the traces (Fig. 9.7b).

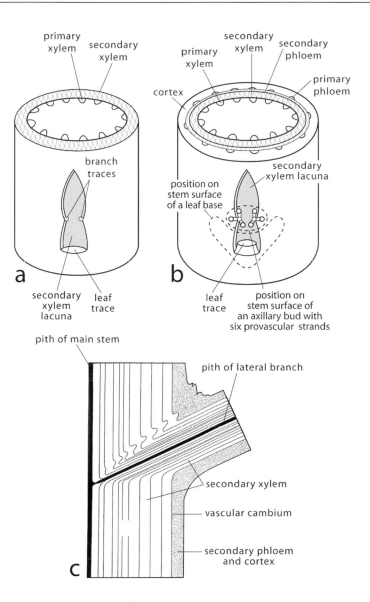

Figure 9.8 (a) Diagram showing the positional relationship of branch traces and a leaf trace in a woody dicotyledon stem with one increment of secondary xylem. (b) Diagram showing the positional relationship of a leaf base and an axial bud on the surface of the stem; also the origin of the cylinder of provascular strands in the bud. (c) Diagram of a radial section illustrating the enclosure by secondary xylem of the base of a lateral branch. Note the continuity of the vascular cambium in the main stem and lateral branch; also the continuation of the pith and primary xylem (black) from the stem into the lateral branch.

As a bud primordium develops, commonly in the axil of a leaf primordium, two incipient branch traces consisting of provascular tissue extend into the base of the young bud (Fig. 9.8a, b). Each branch trace divides several times forming a new, cylindrical, system of provascular strands (Fig. 9.8b). In a subsequent year, as the bud begins to grow, this will differentiate into the primary vascular system of a lateral branch. As development continues, transitional meristems, including provascular tissue, are produced by activity of the apical meristem in the bud; the vascular cambium differentiates within these provascular strands, as described above, which produces concentric increments of secondary xylem and phloem in the lateral branch (Fig. 9.8c). A lacuna, usually in the first increment of secondary xylem in the lateral branch, is associated with each pair of branch traces (Fig. 9.8a, b). Since the vascular cambium of the main axis becomes continuous with that of the lateral branch, the pith and primary xylem of the main axis are continuous through the lacuna, extending into the lateral (Fig. 9.8c). Thus, we see that the innermost primary tissue regions (pith and primary xylem) and secondary tissues are continuous in all branches of the plant body.

REFERENCE

Devadas, C. and C. B. Beck. 1971. Development and morphology of stelar components in the stems of some members of the Leguminosae and Rosaceae. *Am. J. Bot.* **58**: 432–446.

FURTHER READING

Beck, C. B., Schmid, R., and G. W. Rothwell. 1983. Stelar morphology and the primary vascular system of seed plants. *Bot. Rev.* **48**: 691–815.

Eames, A. J. and L. H. McDaniels. 1947. *An Introduction to Plant Anatomy*, 2nd edn. New York: McGraw-Hill.

Esau, K. 1965. *Plant Anatomy*, 2nd edn. New York: John Wiley and Sons.

  1977. *Anatomy of Seed Plants*, 2nd edn. New York: John Wiley and Sons.

Chapter 10

The vascular cambium: structure and function

Perspective

It is difficult to overemphasize the importance of the vascular cambium which produces secondary xylem and secondary phloem. In the following two chapters we shall discuss in detail the structure, functions, and the importance to the plant of these tissues which also have great significance for mankind. Wood (i.e., secondary xylem) is a material of which the buildings in which we live and work are constructed. It is the source of the paper on which we write, on which newspapers, magazines, and books are printed, and of many synthetic fabrics such as rayon and nylon of which our clothes are made, to name only a few of its many uses. The phloem is of the utmost importance as the tissue through which photosynthate is transported from the leaves to sites of utilization or storage in the plant. It is the availability of photosynthate which makes possible the development of nutritious, edible parts of plants, such as fruits, nuts and grains, bulbs, tubers, other edible roots, and leaves, etc., the source of so much of the food supply of humans and other organisms. It is important, therefore, that we know more about the detailed structure and activity of the vascular cambium, a lateral meristem of such great significance.

Structure of the vascular cambium

It is generally agreed that the vascular cambium is composed of a layer of cells only one cell thick, and that all of these cells are meristematic cambial initials from which cells of the secondary xylem and phloem are derived. Cambial initials may be displaced in relation to one another, however, because of the differential growth of immature cambial derivatives and the resultant forces generated. Furthermore, in most plants, the derivatives of cambial initials divide further, forming a zone of immature cells, the **cambial zone** (Fig. 10.1a). The cells of this zone to the inside of the cambium will ultimately differentiate into cells of the secondary xylem, and those to the outside into cells of the secondary phloem. The width of the cambial zone reflects

Figure 10.1 (a) Transverse section of part of a young stem of *Quercus* sp. (oak) showing an active cambial zone and its derivatives. Magnification × 238. (b) Transverse section showing an inactive vascular cambium of *Tilia americana* (basswood) bounded by secondary xylem and secondary phloem produced in the previous growing season. Magnification × 476.

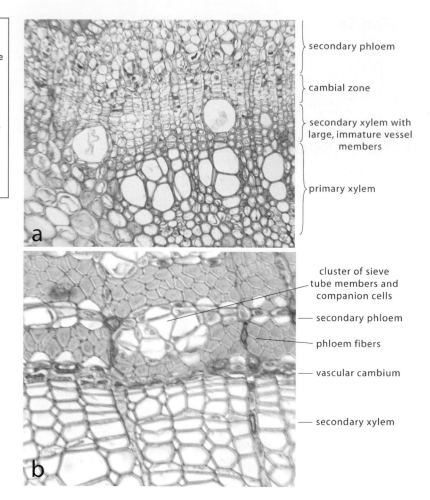

secondary phloem

cambial zone

secondary xylem with large, immature vessel members

primary xylem

a

cluster of sieve tube members and companion cells

secondary phloem

phloem fibers

vascular cambium

secondary xylem

b

the ratio of the rate of production of cambial derivatives and the rate of their differentiation into mature cells of the xylem and phloem. When the rate of cell division exceeds that of differentiation (at the beginning of growth) the cambial zone is broad (Fig. 10.1a), whereas when the rate of differentiation exceeds that of cell division, the zone becomes narrow (Fig. 10.1b). During periods of active growth the cambial initials are difficult, often impossible, to distinguish from their recently formed derivatives. During periods of dormancy, however, at least in trees of temperate zones, the cambium can often be recognized as the layer of cells immediately adjacent to the boundary layer of the secondary xylem (Fig. 10.1b), i.e., the last layer of cells to differentiate prior to cessation of cambial activity (see, e.g., Barnett, 1992). (See Larson (1994) for a detailed, historical review of the nature and activity of the vascular cambium, and Lachaud *et al.* (1999) for a comprehensive review of the structure and function of the cambium.)

Two classes of cambial initials are recognized: **fusiform initials** from which the longitudinally elongate conducting, and other associated cells of the secondary vascular tissues are derived, and **ray**

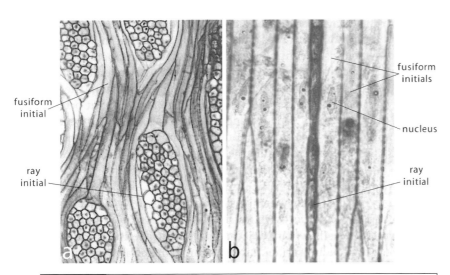

Figure 10.2 Tangential sections showing fusiform and ray initials in the fascicular cambium. (a) *Kalopanax pictus*. The broad ray initials, comprising numerous ray cell initials, will give rise to multiseriate rays. Magnification × 132. (b) *Tilia americana*. Note the uniseriate ray initial, the large nuclei, the tapered ends of the fusiform initials, and the beaded appearance of the walls of contiguous cells which results from the presence of primary pit fields. Magnification × 371. (a) From Kitin *et al.* (1999). Used by permission of Oxford University Press.

initials from which vascular rays are derived (Fig. 10.2). Cells derived from fusiform initials comprise the axial system of secondary vascular tissues and those derived from the ray initials, the radial system of vascular rays. Fusiform initials are longitudinally elongate and, typically, characterized by tapered ends (Fig. 10.2). They vary in length from less than 0.2 mm (about 170 µm in *Robinia pseudoacacia* (black locust)) to nearly 7 mm in the gymnosperm *Agathis robusta*. They are uninucleate and have unlignified primary walls, are quite narrow radially and much wider in tangential dimension (Fig. 10.3). Although basically tabular, these cells often have many wall facets. Within the

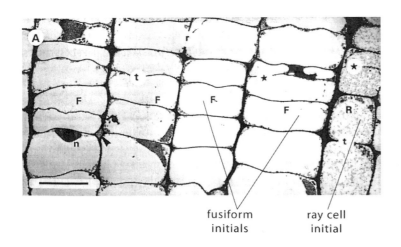

Figure 10.3 Transverse section of the cambial zone of the root of *Aesculus hippocastanum* (horse chestnut) showing the short radial and longer tangential dimensions of the fusiform initials (F). Note how the dimensions of the ray cell initial differ. Bar = 10 µm. From Chaffey *et al.* (1997). Used by permission of the University of Chicago Press. © 1997 The University of Chicago. All rights reserved.

Figure 10.5 Arrangement of cambial derivatives. (a) Transverse section of *Pinus strobus* (white pine) illustrating radial files of tracheids in the secondary xylem, and sieve cells in the functional secondary phloem. Magnification × 218. (b) Transverse section of secondary xylem of *Quercus rubra* (red oak) illustrating distortion of radial files of cambial derivatives which resulted from great increase in size of vessel members. Magnification × 160.

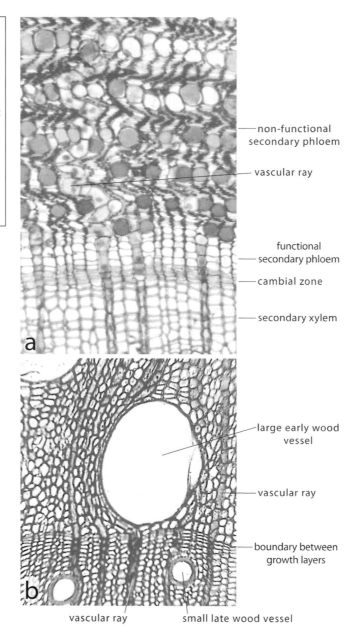

non-functional secondary phloem

vascular ray

functional secondary phloem

cambial zone

secondary xylem

large early wood vessel

vascular ray

boundary between growth layers

vascular ray   small late wood vessel

divisions in fusiform initials which result in the addition of new initials to the cambium. In relatively short fusiform initials these divisions are radial anticlinal divisions in which the resulting cell walls between daughter cells are parallel to their long axes (Fig. 10.7a). The production of new initials from long fusiform initials is by oblique anticlinal divisions, followed by intrusive growth of the overlapping ends of the daughter cells (Fig. 10.7b). New fusiform initials can also be formed by elongation (parallel to the long axis of the stem) and intrusive growth of ray cell initials (Lachaud *et al.*, 1999).

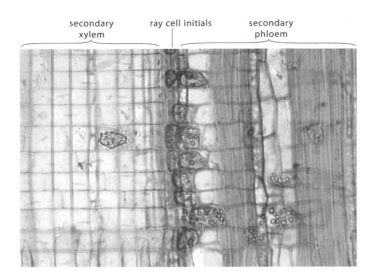

secondary xylem   ray cell initials   secondary phloem

Figure 10.6 Radial longitudinal section of *Tilia americana* showing ray cell initials. Note variation in the radial dimensions of ray cell initials and derivatives. Magnification × 253.

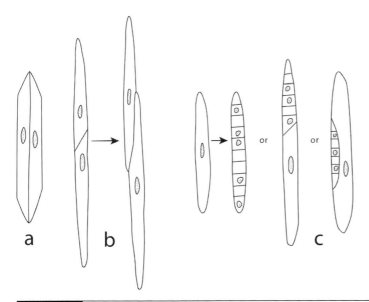

a   b   or   or   c

Figure 10.7 Diagrams illustrating increase in circumference of the cambium through production of new cambial initials by anticlinal divisions in fusiform initials. (a) A short fusiform initial may divide by a radial anticlinal division. (b) Long fusiform initials usually divide by oblique anticlinal divisions followed by intrusive tip growth of the daughter initials. (c) A new ray initial may be formed by subdivision of an entire short fusiform initial by a series of transverse anticlinal divisions, by segmentation of the end of a fusiform initial, or by subdivision of a short fusiform initial formed, following mitosis, by the development of an arcuate wall in a longer initial (right).

New ray initials are formed from fusiform initials in three ways (Fig. 10.7c). In plants with relatively short fusiform initials, the entire initial becomes subdivided into ray cell initials by a series of transverse anticlinal divisions. In longer fusiform initials, new ray initials may be formed by the development of an oblique anticlinal wall near one end followed by a series of transverse anticlinal divisions in the part of the initial thus cut off. The formation in a fusiform initial of a more centrally located arcuate wall followed by a series of transverse anticlinal divisions will also result in the formation of a new ray initial. The result in each case is the formation of a uniseriate ray initial. A multiseriate ray initial is formed by a series of longitudinal, anticlinal divisions in its constituent cells which increases its tangential width.

As the cambium increases in circumference during secondary growth, a balance must be maintained between the number and distribution of fusiform and ray cell initials. It is especially important to maintain the ratio of ray initials to fusiform initials because rays provide passageways for the transport of both nutrients and messenger molecules to the cambium and the actively differentiating cambial derivatives, incipient secondary xylem and phloem cells (Chaffey and Barlow, 2001; Van Bel, 1990; Van Bel and Ehlers, 2000).

Although divisions that result in the production of new fusiform and ray cell initials and that contribute toward an increase in the girth of the vascular cambium occur throughout a period of cambial activity, the greatest number occur toward the end of a period of growth. Thus, on average, fusiform cambial initials are shorter at the end of a growth period and gradually increase in length as growth begins in the next season. New ray initials are usually shorter (longitudinally) than the average height of ray initials, and this is reflected in the height of the rays that result from their cytokinetic activity. During development, however, the initials and the derivative rays will increase in height. Over the life of the plant there is also an increase in the length of fusiform initials and in the height of ray initials from the first year of cambial activity through successive years until stabilization occurs. Consequently, the longest tracheids, fibers, and vessel members as well as the tallest rays are usually found in the most recently formed secondary wood.

An interesting characteristic of cambial activity, especially common in trees of both gymnosperms and angiosperms, is the loss of initials from the cambium. This loss of both fusiform and ray cell initials occurs during the production of new initials which increase the girth of the cambium during secondary growth of stems and roots. If more new cambial initials are produced than required to maintain the continuity of the cambium, some are eliminated and mature into cells of the secondary xylem or phloem. (For more detail, see Bannan (1956, 1968), Evert (1961), and Cheadle and Esau (1964).)

Plant hormones and cambial activity

It is widely accepted that the cambium is the major pathway for polar auxin transport, with auxin moving primarily, but not exclusively, basipetally from sources of production in apical meristems and young leaves. The basipetal flux of auxin is correlated with the basipetal reactivation of the dormant cambium. Furthermore, much research indicates that the activity of the cambium is regulated by plant hormones, in particular, auxins and gibberellins, both of which are known to stimulate cell division in cambial initials and young cambial derivatives and may have a significant role in regulating the frequency and distribution of fusiform and ray cell initials (Little and Savidge, 1987; Lev-Yadun and Aloni, 1991). It has been shown that the fate of derivatives of the apical meristems of both root and shoot is determined more by position than by cell lineage (see Chapters 4 and 15). Although the production of radial files of cambial derivatives seems to indicate a significant role for cell lineage, Lachaud et al. (1999) conclude that the identity of cambial daughter cells is also regulated through positional information and intercellular communication. This conclusion is based in part on the studies of auxin concentrations in the cambium. In *Pinus sylvestris* (Scots pine) there is a radial transport of auxin (derived from the polar flow) across the cambium through the rays (Uggla et al., 1996, 1998). In the cambial zone, the highest concentration of auxin occurs in the region of greatest frequency of cell division, and decreases toward both the mature xylem and mature phloem. Uggla et al. (1998) conclude, therefore, that the concentration of auxin along the radial concentration gradient across the cambial zone might provide positional information that controls both rate of cell division and the identity of the differentiating cambial derivatives as either xylem or phloem cells. Other hormones also have significant roles in the regulation of cambial activity. For example, cytokinins increase the sensitivity of cambial cells to auxin, whereas abscisic acid tends to inhibit periclinal divisions (see Lachaud et al., 1999). Some workers believe that sucrose in cambial cells is important in the initiation of growth during reactivation of dormant cambium (Iqbal, 1995; see also Krabel et al., 1994). Physical factors such as short day length and water stress are also known to influence cambial activity. Both short days and water stress induce dormancy by stimulating the synthesis of inhibitors of cytokinesis. Whereas abscisic acid synthesis is known to result from water stress, some evidence indicates that there is no correlation between its synthesis and the advent of short days (see, e.g., Alvim et al., 1979).

Submicroscopic structure of cambial initials

In a general sense, the cells of the cambium resemble those of other meristematic cells in their submicroscopic structure (see, e.g., Rao, 1985; Catesson, 1994; Chaffey et al., 1997; Lachaud et al., 1999). Each contains a nucleus, mitochondria, endoplasmic reticulum, ribosomes, Golgi bodies, microtubules, microfilaments, storage products including starch grains, and spherosomes (membrane-bound oil globules).

Variations in cell organelles can be seasonal or taxonomic. For example, plastids containing starch grains are abundant in some species during the early phases of dormancy, and occur rarely or not at all in some other taxa. During periods of cytokinesis, cyclosis (streaming of cytoplasm) is conspicuous, especially in fusiform initials, somewhat less so in ray cell initials.

Cambial initials are characterized by thin, unlignified, primary walls, containing numerous primary pit fields traversed by plasmodesmata which connect the protoplasts of adjacent cambial cells or cambial derivatives. It is through these symplastic connections that hormone signals are transmitted to the sites of developing xylem and phloem cells. One of the most intriguing questions is how this essential symplastic continuity between cells is maintained during growth and elongation of fusiform initials and cambial derivatives such as incipient fibers or phloem parenchyma cells. We shall discuss this problem in a later section of this chapter.

The onset of dormancy and the reactivation of dormant cambium

During the period leading up to the cessation of cell division (in temperate zones, usually in mid-autumn), many changes occur in cambial initials and immature derivatives. Perhaps the most conspicuous is an increase in thickness of the cell walls, especially the radial walls (Fig. 10.8). Increase in wall thickening is correlated with the presence of a helical array of microtubules beneath the plasmalemma which are thought to influence the synthesis of additional cellulose microfibrils, thus increasing the wall thickness (Chaffey *et al.*, 1998). Other dramatic changes include the division of the large central vacuole into many small ones (Fig. 10.8), cessation of cyclosis, and the synthesis of starch grains and spherosomes which, according to Catesson (1994), serve as nutrient sources necessary during the resumption of cell divisional activity. Hydrolysis of starch during the cold-hardening phase of dormancy is of great importance since it increases the osmotic potential in the cell, thus increasing frost resistance (Krabel *et al.*, 1994). Additional changes include an increase in the number of mitochondria, reduction in the amount of rough ER, an increase in the number of free ribosomes, deposition in the vacuoles of sugars and amino acids as well as (in some taxa) tannins, and cessation of Golgi activity (see Rao and Dave, 1983; Rao, 1985; Catesson, 1994; Farrar and Evert, 1997).

The period of dormancy in *Abies balsamea* has been characterized as consisting of phases of "rest" and "quiescence" (Little and Bonga, 1974) followed by transition to meristematic activity. Some workers have applied this terminology generally (e.g., Catesson, 1994). It is clear, however, that although cytokinesis ceases during the early, "rest" phase of dormancy and does not resume until late winter or early spring, metabolic activity in fusiform initials may continue at

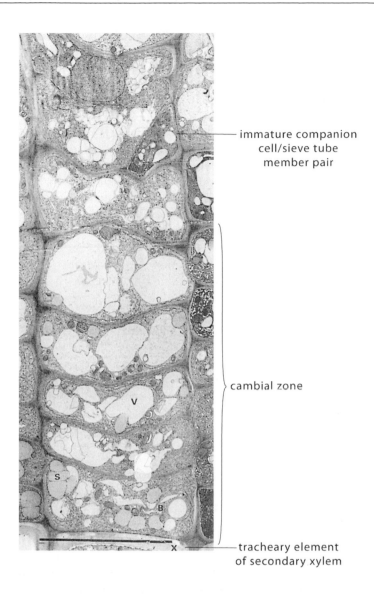

immature companion
cell/sieve tube
member pair

cambial zone

tracheary element
of secondary xylem

Figure 10.8 A transverse section through the dormant cambial zone of *Aesculus hippocastanum*. The fusiform initials are characterized by thickened radial walls and numerous vacuoles (V). Note also the spherosome (S) in the layer adjacent to the mature secondary xylem, and the immature companion cell–sieve tube member pair produced during the previous year's growth. Bar = 10 μm. From Barnett (1992). Used by permission of Oxford University Press.

a relatively high level. Catesson (1994) suggests that this metabolic activity might be related to the development of frost hardiness, but Lachaud *et al.* (1999) warn that distinguishing between changes related to cold hardening and those linked to cell division could be difficult.

In late winter or early spring, dramatic changes occur in cambial initials that lead to a resumption of cell division by the cambium. The cells swell and their walls undergo a process of thinning. Chaffey *et al.* (1998) suggest that this reduction in cell wall thickness results from a process of enzymatic dissolution of the inner cell wall produced prior to dormancy, but leaving intact the original, outer primary cell wall. They and others have observed that as cambial activity is initiated cortical (i.e., peripheral) microtubules are randomly arranged in the cambial initials and their immature derivatives. During differentiation of cambial derivatives, the orientation of microtubules typically changes from random to helical, as secondary wall material is synthesized. This change in orientation characterizes differentiating cells of both the secondary xylem and secondary phloem (see, e.g., Fukuda, 1997; Chaffey *et al.*, 2000, 2002). Other changes include the resumption of cyclosis, fusion of small vacuoles into a single, large central vacuole, hydrolysis of metabolic products such as starch and lipids, increase of Golgi bodies and resumption of their activity, and an increase of rough ER (Farrar and Evert, 1997).

As cell division begins in the cambial zone, several different patterns have been observed in different species. For example, Evert (1963) and Derr and Evert (1967) observed cell divisions throughout the cambial zone beginning at about the same time in *Pyrus* and *Robinia*. However, Grillos and Smith (1959) concluded that in *Pseudotsuga taxifolia* (Douglas fir) the first divisions occur adjacent to the xylem. In a study of reactivation of the cambium of roots of *Aesculus* (horse chestnut), Barnett (1992) demonstrated the first production of new cells in a zone of incipient phloem cells immediately outside the cambium, produced prior to dormancy, which complete their development upon resumption of growth in early spring. Similar observations were made by Catesson (1964) and Tucker and Evert (1969) in *Acer* (maple). In *Aesculus* the cambial initials did not begin to divide for two weeks after the begining of cytokinesis and cell differentiation in the phloem region, and not until four weeks later did division begin to produce cambial derivatives that developed into xylem cells. It is not clear whether or not this pattern is characteristic of other temperate, hardwood species.

Cytokinesis in fusiform initials

Cytokinesis in fusiform initials in which the new cell wall is parallel to the long axis of the cell is of considerable interest because of the great distance the phragmoplast must traverse following nuclear division (Fig. 10.9a, b). Prior to mitosis a pre-prophase band composed of cortical microtubules (see Lambert *et al.*, 1991) and associated actin microfilaments appears, which marks the site of the future cell plate (Lloyd, 1991; Wick, 1991). In early prophase the nucleus migrates to the center of the cell supported by an aggregation of microtubules and microfilaments which become arranged in the plane of division. This microskeletal sheet comprises the **phragmosome**. The phragmosomal microtubules then disappear through depolymerization leaving only microfilaments which, with others, comprise the mitotic spindle. Upon completion of mitosis the **phragmoplast** consisting of newly formed microtubules and microfilaments begins its expansion in the division plane toward the walls of the parent cell (Fig. 10.9b) (see Lloyd, 1989, 1991). In radial view (Fig. 10.9a), the cell plate, consisting of numerous Golgi and/or ER vesicles gradually extends along the division plane behind the phragmoplast as it migrates toward the cell

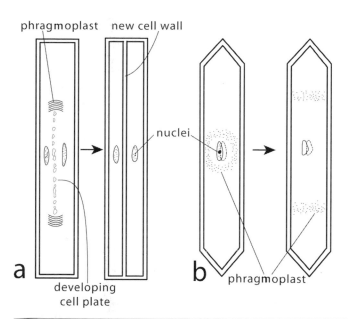

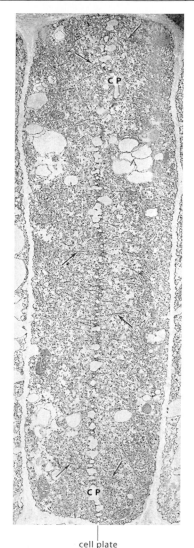

Figure 10.9 (a) Diagrams of radial views of fusiform initials illustrating the developing cell plate as the phragmoplast moves toward each end of the cell followed by formation of the middle lamella and cell walls. (b) Tangential views illustrating the early circular form of the phragmoplast and its subsequent bar-like appearance.

Figure 10.10 Transmission electron micrograph of a transverse section of the developing cell plate (CP) of a dividing fusiform initial of *Ulmus americana* (elm). The phragmoplast and cell plate have reached the radial walls prior to completion of vesicle fusion. Note the phragmoplast microtubules (arrows). Magnification × 4749. From Evert and Deshpande (1970). Used by permission of the Botanical Society of America.

wall. Initially the phragmoplast, as seen in tangential view, appears as a circle of microtubules and microfilaments (Fig. 10.9b). Subsequently, however, as it proceeds toward the ends of the cell it appears as two transverse bands. By this stage, the cell plate has expanded to such an extent that it is in contact with both radial walls, as can be seen in a transverse section (Fig. 10.10). It is interesting to note that as the results of some unknown control system, the phragmosome and, as a result, the cell plate do not become aligned with the walls in adjacent cells. Consequently, the daughter cells resulting from cytokinesis overlap the ends of other cells in the developing tissue like bricks in a wall (see Flanders *et al.*, 1990; Lloyd, 1991). For extensive bibliography and more detailed discussions of the role of the cytoskeleton in cell division including the origin and role of the pre-prophase band, the architecture of the phragmoplast, origin of the microtubules, their role and the role of actin microfilaments in cytokinesis, see Lloyd (1988, 1989), Seagull (1989), Baskin and Cande (1990), Flanders *et al.* (1990), Lambert *et al.* (1991) and Wick (1991).

The problem of differential growth of cambial cells and immature cambial derivatives

A long-standing problem of cell growth in the cambium and its immature derivatives involves the method whereby two adjacent cells, or parts of cells, can grow differentially, i.e., at different rates, and maintain their symplastic continuity. It is well established that when a new cambial initial is formed following an oblique, anticlinal division, the cell elongates by growth of its tips. A cambial derivative such as an incipient fiber may elongate greatly during its development adjacent to an incipient vessel member that may elongate only slightly, or not at all. It has been assumed that elongation of contiguous cells at different rates would, of necessity, disrupt both developing pit pairs and plasmodesmata. Several explanations have been proposed by which differential growth of contiguous cells could occur without such disruptions (see Larson (1994) for a detailed, historical account). Earlier workers strongly supported the view that cambial fusiform initials and certain cells in the cambial zone such as incipient fibers elongated primarily by symplastic growth. Although symplastic growth in the main body of immature cells may play a contributing role in their elongation, several studies (e.g., Evert, 1960; Barnett and Harris, 1975; Wenham and Cusick, 1975; Barnett, 1981; Robards and Lucas, 1990) have shown that new cambial initials and immature cambial derivatives increase in length primarily by intrusive growth of the cell tips which extend between other cells in the tissue. This process may be facilitated by digestion of the middle lamella by enzymes produced by the immature growing tip which allows the thin-walled tip more effectively to push its way between contiguous cells (Hejnowicz, 1980). Evidence of tip growth comes from the thin-walled nature of the tips, the presence in them of numerous Golgi bodies, mitochondria, and ribosomes, and the increase in thickness of the cell walls toward the cell tips during development. It has been shown that cambial fusiform initials in several conifers and angiosperms seem to have no plasmodesmata in their radial walls (Barnett and Harris, 1975; Rao and Dave, 1983). Others have observed plasmodesmata in both radial and tangential walls (e.g., Kidwai and Robards, 1969). It is widely accepted, however, that early derivatives of fusiform initials lack both pits and plasmodesmata in the walls of cell tips that are undergoing intrusive growth (Wardrop, 1954; Bannan, 1956; Barnett and Harris, 1975). If this were true also of fusiform cambial initials as suggested by Larson (1994), the potential problem of disruption of symplasmic connections would become moot. Prior to completion of the development of cambial initials and their derivatives, new pits pairs and plasmodesmatal connections (secondary plasmodesmata; see Chapter 4) form between contiguous cells.

REFERENCES

Alvim, R., Saunders, P. F., and R. S. Barros. 1979. Abscisic acid and the photoperiod induction of dormancy in *Salix viminalis* L. *Plant Physiol.* **63**: 774–777.

Bannan, M. W. 1956. Some aspects of the elongation of fusiform cambial cells in *Thuja occidentalis*. *Can. J. Bot.* **34**: 175–196.

1968. Anticlinal divisions and the organization of conifer cambium. *Bot. Gaz.* **129**: 107–113.

Barnett, J. R. 1981. Secondary xylem cell development. In J. R. Barnett, ed., *Xylem Cell Development*. Tunbridge Wells, UK: Castle House Publications, pp. 47–95.

1992. Reactivation of the cambium in *Aesculus hippocastanum* L.: a transmission electron microscope study. *Ann. Bot.* **70**: 169–177.

Barnett, J. R. and J. M. Harris. 1975. Early stages of bordered pit formation in radiata pine. *Wood Sci. Technol.* **9**: 233–241.

Baskin, T. I. and W. Z. Cande. 1990. The structure and function of the mitotic spindle in flowering plants. *Annu. Rev. Plant Physiol. Plant Mol. Biol.* **41**: 277–315.

Catesson, A. M. 1964. Origine, fonctionnement et variations cytologiques saisonnières du cambium de l'*Acer pseudoplatanus* L. (Acéracées). *Ann. Sci. Nat., Bot., Sér. xii* **5**: 229–498.

1994. Cambial ultrastructure and biochemistry: changes in relation to vascular tissue differentiation and the seasonal cycle. *Int. J. Plant Sci.* **155**: 251–261.

Chaffey, N. J. and P. W. Barlow. 2001. The cytoskeleton facilitates a three-dimensional symplasmic continuum in the long-lived ray and axial parenchyma cells of angiosperm trees. *Planta* **213**: 811–823.

Chaffey, N. J., Barnett, J. R., and P. W. Barlow. 1997. Endomembranes, cytoskeleton, and cell walls: aspects of the ultrastructure of the vascular cambium of tap roots of *Aesculus hippocastanum* L. (Hippocastanaceae). *Int. J. Plant Sci.* **158**: 97–109.

Chaffey, N. J., Barlow, P. W., and J. R. Barnett. 1998. A seasonal cycle of cell wall structure is accompanied by a cyclical rearrangement of cortical microtubules in fusiform cambial cells within tap roots of *Aesculus hippocastanum* (Hippocastanaceae). *New Phytol.* **139**: 623–635.

2000. Structure–function relationships during secondary phloem development in an angiosperm tree, *Aesculus hippocastanum*: microtubules and cell walls. *Tree Physiol.* **20**: 777–786.

Chaffey, H., Barlow, P., and B. Sundberg. 2002. Understanding the role of the cytoskeleton in wood formation in angiosperm trees: hybrid aspen (*Populus tremula* × *P. tremuloides*) as the model species. *Tree Physiol.* **22**: 239–249.

Cheadle, V. I. and K. Esau. 1964. Secondary phloem of *Liriodendron tulipifera*. *Univ. Calif. Publ. Bot.* **36**: 143–252.

Derr, W. F. and R. F. Evert. 1967. The cambium and seasonal development of the phloem in *Robinia pseudoacacia*. *Am. J. Bot.* **54**: 147–153.

Evert, R. F. 1960. Phloem structure in *Pyrus communis* L. and its seasonal changes. *Univ. Calif. Publ. Bot.* **32**: 127–194.

1961. Some aspects of cambial development in *Pyrus communis*. *Am. J. Bot.* **48**: 479–488.

1963. The cambium and seasonal development of the phloem in *Pyrus malus*. *Am. J. Bot.* **50**: 149–159.

Evert, R. F. and B. P. Deshpande. 1970. An ultrastructural study of cell division in the cambium. *Am. J. Bot.* **57**: 942–961.

Farrar, J. J. and R. F. Evert. 1997. Seasonal changes in the ultrastructure of the vascular cambium of *Robinia pseudoacacia*. *Trees Struct. Funct.:* **11**: 191–202.

Flanders, D. J., Rawlins, D. J., Shaw, P. J., and C. W. Lloyd. 1990. Nucleus-associated microtubules help determine the division plane of plant epidermal cells: avoidance of four-way junctions and the role of cell geometry. *J. Cell Biol.* **110**: 1111–1122.

Fukuda, H. 1997. Tracheary element differentiation. *Plant Cell* **9**: 1147–1156.

Grillos, S. J. and F. H. Smith. 1959. The secondary phloem of Douglas fir. *Forest Sci.* **5**: 377–388.

Hejnowicz, Z. 1980. Tensional stress in the cambium and its developmental significance. *Am. J. Bot.* **67**: 1–5.

Iqbal, M. 1995. Structure and behaviour of vascular cambium and the mechanism and control of cambial growth. In M. Iqbal, ed., *The Cambial Derivatives*. Berlin: Gebrüder Borntraeger, pp. 1–67.

Kidwai, P. and A. W. Robards. 1969. On the ultrastructure of resting cambium of *Fagus sylvatica* L. *Planta* **89**: 361–368.

Kitin, P., Funada, R., Sano, Y., Beeckman, H., and J. Ohtani. 1999. Variations in the length of fusiform cambial cells and vessel elements in *Kalopanax pictus*. *Ann. Bot.* **84**: 621–632.

Krabel, D., Bodson, M., and W. Eschrich. 1994. Seasonal changes in the cambium of trees. I. Sucrose content in *Thuja occidentalis*. *Bot. Acta* **107**: 54–59.

Lachaud, S., Catesson, A. M., and J. L. Bonnemain. 1999. Structure and functions of the vascular cambium. *C. R. Acad. Sci. Paris, Sciences de la vie* **322**: 633–650.

Lambert, A.-M., Vantard, M., Schmit, A.-C., and H. Stoeckel. 1991. Mitosis in plants. In C. W. Lloyd, ed., *The Cytoskeletal Basis of Plant Growth and Form*. London: Academic Press, pp. 199–208.

Larson, P. R. 1994. *The Vascular Cambium: Development and Structure*. Berlin: Springer-Verlag.

Lev-Yadun, S. and R. Aloni. 1991. Polycentric vascular rays in *Suaeda monoica* and the control of ray initiation and spacing. *Trees Struct. Funct.* **5**: 22–29.

Little, C. H. A. and J. M. Bonga. 1974. Rest in the cambium of *Abies balsamea*. *Can. J. Bot.* **52**: 1723–1730.

Little, C. H. A. and R. A. Savidge. 1987. The role of plant growth regulators in forest tree cambial growth. *Plant Growth Regul.* **6**: 137–169.

Lloyd, C. W. 1988. Actin in plants. *J. Cell Sci.* **90**: 185–192.

 1989. The plant cytoskeleton. *Curr. Opin. Cell Biol.* **1**: 30–35.

 1991. How does the cytoskeleton read the laws of geometry in aligning the division plane of plant-cells? *Development* (Suppl.) **1**: 55–65.

Rao, K. S. 1985. Seasonal ultrastructural changes in the cambium of *Aesculus hippocastanum* L. *Ann. Sci. Nat., Bot. Sér. xiii* **7**: 213–228.

Rao, K. S. and Y. S. Dave. 1983. Ultrastructure of active and dormant cambial cells in teak (*Tectona grandis* L.f.). *New Phytol.* **93**: 447–456.

Robards, A. W. and W. J. Lucas. 1990. Plasmodesmata. *Annu. Rev. Plant Physiol. Plant Mol. Biol.* **41**: 369–419.

Seagull, R. W. 1989. The plant cytoskeleton. *Crit. Rev. Plant Sci.* **8**: 131–167.

Tucker, M. C. and R. F. Evert. 1969. Seasonal development of the secondary phloem in *Acer negundo*. *Am. J. Bot.* **56**: 275–284.

Uggla, C., Moritz, T. J., Sandberg, G., and B. Sundberg. 1996. Auxin as a positional signal in pattern formation in plants. *Proc. Natl Acad. Sci. USA* **93**: 9282–9286.

Uggla, C., Mellerowicz, E. J., and B. Sundberg. 1998. Indole-3-acetic acid controls cambial growth in Scots pine by positional signaling. *Plant Physiol.* **117**: 113–121.

Van Bel, A. J. E. 1990. Xylem–phloem exchange via rays: the undervalued route of transport. *J. Exp. Bot.* **41**: 631–644.

Van Bel, A. J. E. and K. Ehlers. 2000. Symplasmic organization of the transport phloem and the implications for photosynthate transfer to the cambium. In R. A. Savidge, J. R. Barnett, and R. Napier, eds., *Cell and Molecular Biology of Wood Formation*. Oxford, UK: Bios Scientific Publishers, pp. 85–99.

Wardrop, A. B. 1954. The mechanism of surface growth involved in the differentiation of fibres and tracheids. Austral. *J. Bot.* **2**: 165–175.

Wenham, M. W. and F. Cusick. 1975. The growth of secondary wood fibres. *New Phytol.* **74**: 247–261.

Wick, S. M. 1991. The preprophase band. In C. W. Lloyd, ed., *The Cytoskeletal Basis of Plant Growth and Form*. London: Academic Press, pp. 231–244.

FURTHER READING

Bailey, I. W. 1920. The cambium and its derivative tissues. II. Size variations of cambial initials in gymnosperms and angiosperms. *Am. J. Bot.* **7**: 355–367.

   1923. The cambium and its derivative tissues. IV. The increase in girth of the cambium. *Am. J. Bot.* **10**: 499–509.

Bannan, M. W. 1955. The vascular cambium and radial growth in *Thuja occidentalis* L. *Can. J. Bot.* **33**: 113–138.

Cumbie, B. G. 1967. Developmental changes in the vascular cambium in *Leitneria floridana*. *Am. J. Bot.* **54**: 414–424.

Hejnowicz, Z. 1964. Orientation of the partition in pseudotransverse division in cambia of some conifers. *Can. J. Bot.* **42**: 1685–1691.

Iqbal, M. (ed.) 1990. *The Vascular Cambium*. Taunton, UK: Research Studies Press.

Kakimoto, T. and H. Shibaoka. 1988. Cytoskeletal ultrastructure of phragmoplast–nuclei complexes isolated from cultured tobacco cells. *Protoplasma* (suppl.) **2**: 95–103.

Koslowski, T. T. 1971. *Growth and Development of Trees*, vol. 2, *Cambial Growth, Root Growth, and Reproductive Growth*. New York: Academic Press.

Larson, P. R. 1982. The concept of cambium. In P. Baas, ed., *New Perspectives in Wood Anatomy*. The Hague: Nijhoff, pp. 85–121.

Philipson, W. R., Ward, J. M., and B. G. Butterfield. 1971. *The Vascular Cambium*. London: Chapman and Hall.

Romberger, J. A., Hejnowicz, Z., and J. F. Hill. 1993. *Plant Structure: Function and Development*. Berlin: Springer-Verlag.

Sundberg, B., Little, C. H. A., Cui, K., and G. Sandberg. 1991. Level of endogenous indole-3-acetic acid in the stem of *Pinus sylvestris* in relation to the seasonal variation of cambial activity. *Plant Cell Env.* **14**: 241–246.

Wilson, B. F. 1964. A model for cell production by the cambium of conifers. In M. H. Zimmermann, ed., *The Formation of Wood in Forest Trees*. New York. Academic Press, pp. 19–36.

   1966. Mitotic activity in the cambial zone of *Pinus strobus*. *Am. J. Bot.* **53**: 364–372.

Chapter 11

Secondary xylem

Perspective

Most of the major taxa of vascular plants produce secondary xylem derived from the vascular cambium. Pteridophytes (except some extinct taxa), most monocotyledons, and a few species of largely aquatic dicotyledons, however, produce only primary vascular tissues. In woody plants secondary xylem comprises the bulk of the tissue in the stems and roots. It is the most important supporting tissue in arborescent dicotyledons and most gymnosperms, and the major tissue for the transport of water and essential minerals in woody plants. Secondary xylem is a complex tissue that consists not only of non-living supporting and conducting cells but also of important living components (rays and axial wood parenchyma) which, with those in the secondary phloem, comprise a three-dimensional symplastic pathway through which photosynthate and other essential molecular substances are transported thoughout the secondary tissues of the plant (Chaffey and Barlow, 2001; see pp. 206–207 for more detail). Additional increments of this tissue are added during each growing season (usually annually), but in older regions of most woody species only the outer increments are functional in transport although the number of increments that remain functional varies greatly among different species. Older increments gradually become plugged by the deposition in them of waste metabolites such as resins, tannins, and in some species by the formation of tyloses (balloon-like extensions of axial or ray parenchyma cells into adjacent conducting cells). The inner non-functional secondary xylem is called heartwood, the outer functional secondary xylem, sapwood. We shall discuss the characteristics and formation of heartwood in more detail later in this chapter.

Overview of the structure of secondary xylem

Secondary xylem is made up of an **axial system** of longitudinally oriented cells, derived from the fusiform initials of the vascular cambium, and a **radial system** consisting of rays derived from the ray

Figure 11.1 Scanning electron micrograph of a block of secondary xylem of *Pinus resinosa* (red pine) showing transverse and radial surfaces. This conifer wood consists primarily of tracheids and vascular rays. The sharp distinction between growth layers results from the great difference in size and wall thickness of tracheids in the late wood (lw) and early wood (ew). Note also the resin canals (rc), the vascular ray (r) exposed on the radial surface, and the large circular bordered pits in the radial walls of tracheids. From Core *et al.* (1979). Used by permission of Syracuse University Press.

initials of the vascular cambium. In gymnosperms the axial system consists of tracheids and axial parenchyma (Fig. 11.1), and in many conifers resin ducts also develop as part of both the axial and radial systems. In angiosperms the axial system consists of tracheids, fibers, vessel members, and axial parenchyma. Secretory ducts as well as secretory cavities also occur in some angiosperms and contain, in different species, a variety of substances (see Chapter 15 on secretory structures).

The successive increments of xylem cells produced by the cambium comprise **growth layers**, often called annual rings (Fig. 11.2a, b). The cells produced in the early part of a growing season (in the inner part of the growth layers), especially the tracheary cells, are typically larger in transverse dimensions than those produced later, and may have thinner cell walls. These cells comprise the **early wood** (sometimes called spring wood). **Late wood** (sometimes called summer wood), produced later in the growing season, especially that produced just prior to cessation of activity of the cambium, is composed of smaller cells that often have much thicker walls than cells of the early wood. Consequently, there is usually a well-defined

Figure 11.2 (a) Transverse section of a two-year-old stem of *Abies*. Magnification × 7.3.
(b) Transverse section of secondary xylem of *Sequoia sempervirens* (redwood) illustrating growth layers consisting of early wood containing large, thin-walled tracheids and late wood containing thick-walled tracheids. The late wood tracheids have much narrower radial dimensions than the early wood tracheids. Note also the uniseriate and biseriate vascular rays and the axial wood parenchyma containing dark contents. Magnification × 44.

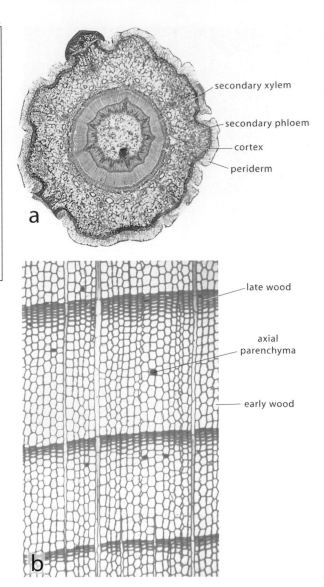

line of demarcation between successively produced growth layers (Fig. 11.2).

Whereas in temperate climatic zones growth layers are typically produced annually, in tropical regions growth layers often reflect periodicity in rainfall rather than temperature and photoperiod, and thus may not be annual increments of secondary xylem. Under some circumstances, even in temperate zones, more than one growth layer may develop during a growing season. Such false growth layers can result from environmental factors such as defoliation by insects, forest fire, or severe drought.

Rays, which comprise the radial system of the secondary xylem, function in storage as well as in lateral transport of various materials. They are composed primarily of **ray parenchyma**, but in many

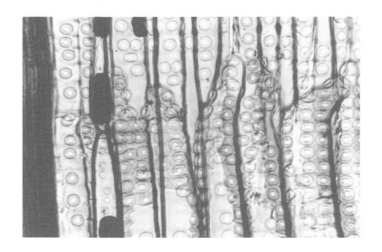

Figure 11.3 Radial section of secondary xylem of *Sequoia sempervirens* showing overlapping, blunt tracheid ends. Circular-bordered pits are especially abundant in the walls of the tracheid ends. Note the row of axial parenchyma cells on the left. Magnification × 178.

conifers, some other gymnosperms, and a few angiosperms, they also contain **ray tracheids**, non-living cells, the walls of which contain bordered pits. As we shall see later, in different species rays are variable in height, width, and arrangement of cells as well as distribution in the wood.

Secondary xylem of gymnosperms

Gymnosperm wood which consists primarily of tracheids and rays (Fig. 11.2b) is less complex than the wood of angiosperms which is characterized by a greater diversity of cell types and cell sizes. We shall now consider in some detail the secondary xylem of conifers. The conifer **tracheid** has dual functions, providing both support and conduction, and its structure is an adaptation to those functions. It may be very long, reaching 8 mm in some members of the Araucariaceae and, on average, is longer than the average length of tracheids in secondary wood of angiosperms. The tracheids of conifers have tapered to relatively blunt ends that overlap those of other tracheids and are characterized by large, usually **circular-bordered pits** that are very abundant on the overlapping ends (Fig. 11.3). The pits have highly specialized pit membranes that both facilitate the lateral transport of water and minerals from one tracheid into another, and, under certain conditions, function as valves that restrict the movement of air within the xylem.

Bordered pits occur only in the radial walls of conifer tracheids except in those in the late wood in which they may occur in the tangential walls (Fig. 11.4). They form **bordered pit-pairs** with pits of contiguous tracheids (Fig. 11.5a–c) and **half-bordered pit-pairs** with simple pits of contiguous axial or ray parenchyma cells (Fig. 11.6). Typically, in conifers, bordered pits are uniseriate or biseriate (Fig. 11.5b, c), but may also be multiseriate. It is very interesting that the Devonian plant *Archaeopteris*, a pteridophyte of arborescent stature which lived over 350 million years ago, produced secondary xylem remarkably

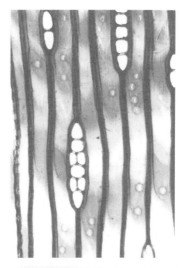

Figure 11.4 Tangential section of late wood of *Sequoia sempervirens* showing the low uniseriate and biseriate vascular rays, and the bordered pits in the tangential walls of tracheids. Magnification × 124.

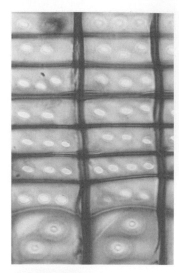

Figure 11.6 Radial section of secondary xylem of *Sequoia sempervirens* showing half-bordered pit-pairs between ray parenchyma cells and tracheids, as seen in face view. Magnification × 355.

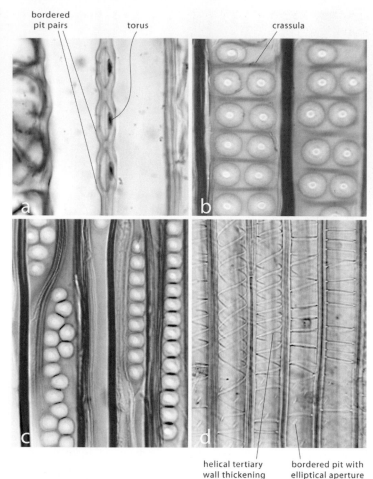

Figure 11.5 Secondary xylem in conifers. (a) Tangential section of *Pinus jeffreyi* illustrating bordered pit-pairs in sectional view. Note the aspirated pit membranes, each containing a thick, centrally located torus. Magnification × 700. (b) Radial section of *Sequoia sempervirens* showing circular-bordered pit-pairs in face view. The faint rim around the pit apertures represents the lateral extent of the torus. Radiating and anastomosing strands extend from the torus and comprise the margo, the peripheral region of the pit membrane. Compare Figs. 11.5a and b with Fig. 11.9. The dark, horizontal bands between pits are crassulae, regions of primary wall heavily impregnated with lignin that separate primary pit fields prior to the deposition of secondary wall layers. Magnification × 400. (c) Radial section showing bordered pits of angular outline in *Araucaria bidwillii*. The occurrence of pits in groups is common in this species. Magnification × 400. (d) Radial section of *Picea rubra* (spruce). Note the small, circular-bordered pits with elliptical apertures. Superimposed on the secondary wall are tertiary thickenings which form one or two helices. Magnification × 373.

similar to that of conifers (Fig. 11.7a, b). Its wood is distinctive, however, in that the circular-bordered pits occur in groups arranged in radial bands. The bordered pits of conifers are predominantly circular in shape as seen in face view, but when crowded, may be angular (Fig. 11.5c). The highly lignified, slightly thickened regions of the

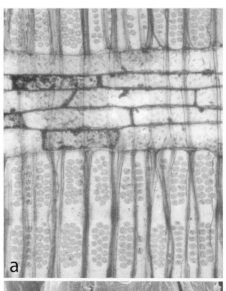

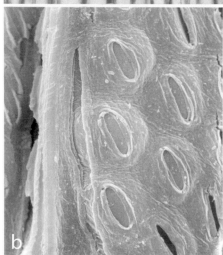

Figure 11.7 Radial sections of *Callixylon newberryi*, the secondary wood of *Archaeopteris*, an extinct Devonian plant. (a) Groups of circular-bordered pit-pairs in radial rows. Note also the vascular ray. Magnification × 136. (b) Scanning electron micrograph of circular-bordered pits showing the elliptical apertures. Magnification × 1943. From Beck *et al.* (1982).

primary walls and middle lamella between pit-pairs, often visible in radial sections as heavily stained bands (Fig. 11.5b), are **crassulae**. In older literature these are called bars of Sanio after the Italian botanist who first described them in the late nineteenth century. In some conifers helical, **tertiary wall thickenings** are synthesized between bordered pits on the inner surfaces of the secondary wall (Fig. 11.5d).

Bordered pit-pairs vary in structure depending on the thickness of the walls in which they occur. If the secondary walls are thin the pit border of each pit is usually described as having a single **aperture** and no pit canal (Fig. 11.5a). If the walls are thick, however, there will be a **pit canal**, typically in the shape of a flattened cone, with inner and outer apertures (Fig. 11.8b). The inner aperture is commonly oval to elliptical, the outer circular. In all cases, the long axis of the inner aperture is parallel to the microfibrils of the S2 layer of the secondary

Figure 11.8 (a) Bordered pit-pairs in a thick-walled cell such as a fiber, in face view. The angle of the long axis of the inner apertures reflects the orientation of the cellulose microfibrils in the S2 wall layers of the contiguous cells. (b) Sectional view of a pit-pair such as that seen in (a). The pit canal has the form of a flattened cone.

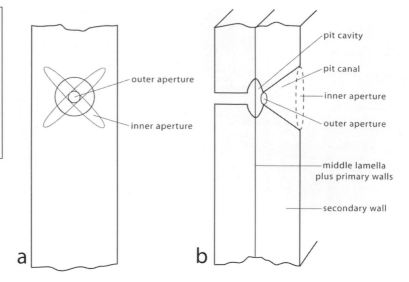

Figure 11.9 Scanning electron micrograph of the pit membrane of a bordered pit-pair of *Abies grandis* (fir). The dense central structure is the torus; the peripheral region of radiating and anastomosing strands of cellulose microfibrils is the margo. Magnification × 4192. From Preston (1974). Used by permission of Springer-Verlag New York, Inc.

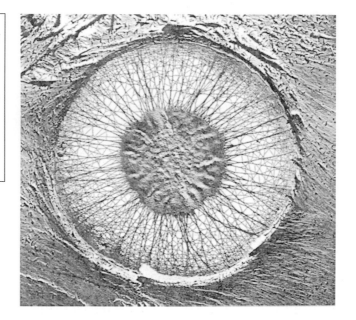

cell wall, and apertures of the opposing bordered pits of a pit-pair are always crossed (Fig. 11.8a).

In most conifers, the **pit membrane** (a highly modified region of the primary cell walls and intervening middle lamella) which separates the two **pit cavities** of a bordered pit-pair consists of a central, thickened, lignified, circular region called the **torus** and a thin surrounding region called the **margo** (Figs. 11.5a, 11.9). In the functional tracheid, the torus is dense, heavily impregnated with pectic compounds, and impermeable to the passage of water. In contrast, the margo is highly porous and consists of radiating and anastomosing bundles of lignified microfibrils (Fig. 11.9) that were exposed by hydrolysis of the encrusting pectic compounds just prior to the death

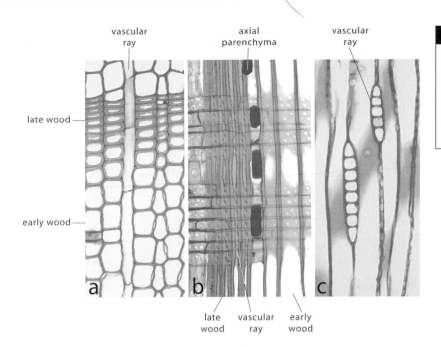

vascular ray   axial parenchyma   vascular ray

late wood

early wood

a   b   c

late wood   vascular ray   early wood

Figure 11.10 Uniseriate vascular rays in the secondary xylem of *Sequoia sempervirens*. (a) Transverse section. (b) Radial section. (c) Tangential section. Note the axial parenchyma in (b). Magnification (a), (b), and (c) × 145.

of the cell protoplast. The margo provides the major passageway for the transport of water and minerals from one tracheid to an adjacent one, and facilitates the efficient transport through the apoplast. The torus, the diameter of which exceeds that of the pit aperture, functions like a valve, closing (i.e., pressing against) a pit aperture (Fig. 11.5a) when the pressure differential between two adjacent cells is great, a condition that can occur during periods of rapid transpiration and resulting negative water pressures. During such conditions air bubbles (embolisms) may form in tracheid lumina. Closure of the pit apertures by the tori effectively confines air to limited areas which if more widespread in the wood could seriously disrupt water flow.

In addition to tracheids, axial wood parenchyma and the epithelial cells of longitudinally oriented resin ducts are also derived from the fusiform initials in conifers. **Axial wood parenchyma** (Fig. 11.10b) develops from immature derivatives of fusiform initials by transverse divisions in these derivatives resulting in longitudinal strands of axial parenchyma cells. These strands of xylem parenchyma are scattered throughout the wood (Fig. 11.2b), often in well-defined patterns. The dark contents of axial parenchyma cells of conifers were thought by early botanists to be resin. In fact, the nature of this material is unknown, but it may be tannin.

The **vascular rays** of conifers, relatively low and uniseriate to biseriate in most species (Fig. 11.10b, c), are commonly composed of two types of cells, ray parenchyma cells and ray tracheids. In species containing resin ducts the rays also contain epithelial cells lining the ducts which are large intercellular spaces.

Ray parenchyma cells are typically (but not without exception) radially elongate, that is, their radial dimension is greater than their height (Fig. 11.10a–c). Within a ray, cells are arranged like bricks in a

Figure 11.11 Radial section of secondary xylem of *Pinus strobus* showing bordered pit-pairs in the walls of ray tracheids and contiguous axial tracheids (X) (face views), and in the walls of adjacent ray tracheids (Y) (sectional views). Note also the large fenestriform (window-like) half-bordered pit-pairs in the walls of contiguous ray parenchyma cells and axial tracheids (Z) (face views). Magnification × 500.

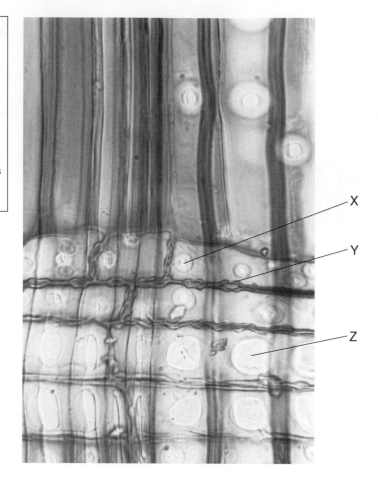

wall overlapping the ends of cells in contiguous files. Contact zones between axial tracheids and ray parenchyma cells are called **cross-fields** and are categorized by the characteristics of the half-bordered pit-pairs in these zones (Figs. 11.6, 11.11). Cross-fields are useful in determining the source of isolated secondary wood of conifers since those of different species may have different characteristics. Compare the small cross-field pits of *Sequoia sempervirens* (Fig. 11.6) with the large fenestriform (window-like) pits of *Pinus strobus* (Fig. 11.11).

Ray tracheids are non-living ray cells that occur in either **marginal** (Fig. 11.11) or marginal and **interspersed** rows. They are often more radially elongate, and their secondary walls, which contain bordered pits, thicker than those of ray parenchyma cells. As one would expect, bordered pit-pairs occur in the walls of contiguous ray tracheids as well as between ray tracheids and axial tracheids, and half-bordered pit-pairs occur in the contiguous walls of ray tracheids and ray parenchyma cells (Fig. 11.11). Ray tracheids are thought to facilitate more rapid transport of water and minerals through the rays than would be possible through ray parenchyma cells.

radial resin duct

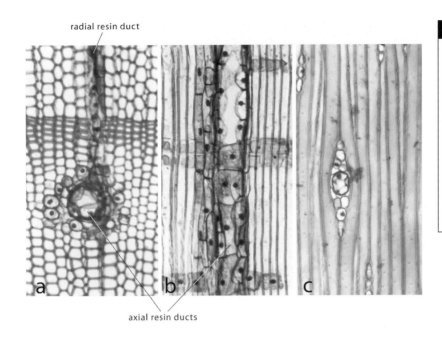

axial resin ducts

Figure 11.12 Resin ducts in the secondary xylem of *Pinus strobus*. (a) Intersection of an axial and a radial resin duct as viewed in transverse section. (b) An axial resin duct viewed in radial section. (c) A radial resin duct within a vascular ray viewed in tangential section. Note the epithelial (secretory) cells lining the ducts, and in (a) and (b) the surrounding axial parenchyma cells. Magnification (a), (b), and (c) × 174.

Resin ducts

Resin ducts (Fig. 11.12) are a distinctive feature in the secondary xylem of many conifers. They consist of long channels enclosed by **epithelial cells** which secrete resin into them granulocrinously. In length they may range up to 40 mm, and in diameter to 150 μm. They are a constant and, presumably, genetically controlled feature of *Pinus, Pseudotsuga, Larix*, and *Picea*, and have the potential to develop following wounding in *Sequoia, Cedrus*, and *Abies*. They are unknown in *Cupressus*. The frequency and distribution of resin ducts in genera in which they normally develop can be severely modified by the response to injury (see Romberger *et al.*, 1993). Resin ducts also develop in the primary xylem of roots and shoots of some conifers, and the ducts in the cortex of young shoots may extend into the mesophyll of leaves (Werker and Fahn, 1969). In the secondary xylem resin ducts are both axial, oriented longitudinally among tracheary cells (Fig. 11.12a, b), and radial, extending through vascular rays (Fig. 11.12a, c). The axial and radial ducts are connected to and continuous with each other (Fig. 11.12a), thus comprising an extensive, three-dimensional system. In some taxa the radial ducts extend into the secondary phloem, but are probably not open in the immature region across the cambial zone (see Werker and Fahn, 1969).

The epithelial cells of axial ducts (Fig. 11.12a, b) are derived from the derivatives of fusiform cambial initials whereas those of the radial ducts (Fig. 11.12c) are derived from derivatives of ray cell initials. By a series of anticlinal and periclinal divisions in fusiform cambial derivatives, columns of incipient epithelial cells are formed. As development proceeds, the epithelial cells separate from each other schizogenously, forming channels which they enclose. Radial resin ducts originate in developing uniseriate rays following anticlinal and periclinal divisions in ray cell derivatives, and schizogenous separation of incipient epithelial cells. As the resin duct develops, the ray expands and becomes multiseriate (Fig. 11.12c). The obviously complex mechanisms of the genetic control of resin duct development are unknown, but signals (probably hormonal) distinguish between the specific sites of axial and radial duct development.

The function of resin in conifers is not clearly understood, but is thought to protect the plant from invasion by fungi and insects following injury. According to Romberger *et al.* (1993) the resin in the ducts is under positive pressure, the result of turgor pressure in the epithelial cells. Upon rupture of the ducts resin flows out slowly due in part to its viscosity and in part to the local reduction in pressure. Osmotic uptake of water in adjacent epithelial cells leads to their expansion and consequent constriction of the rupture (Romberger *et al.*, 1993).

Secondary xylem of dicotyledons

The secondary xylem of dicotyledons differs from that of gymnosperms primarily in the greater complexity of the tissue (Fig. 11.13a, b) and the greater structural diversity among different taxa. The most significant difference is the functional separation in dicotyledons of cells providing mechanical support and those which provide transport of water and minerals. Whereas in gymnosperms both of these functions are served by tracheids, in dicotyledons support is provided largely by fibers and longitudinal transport by vessel members except in the most primitive angiosperms (e.g., members of the Winteraceae of the Magnoliidae) which lack vessel members. The secondary xylem of these taxa closely resembles that of gymnosperms, the axial system being composed primarily of tracheids. Another difference between some gymnosperms and angiosperms is the nature of the pit membranes in the pits of tracheary elements. Whereas in conifers the pit membranes have a central, thickened torus and a surrounding, highly porous margo of intertwined cellulosic strands, the pit membranes of tracheary elements in angiosperms is a much simpler sheet of primary wall of uniform structure (Fig. 11.14a).

Whereas one thinks of the axial system of angiosperm secondary xylem as being composed largely of fibers, vessel members, and axial parenchyma, the diversity of cell types in this wood is somewhat greater than that. The following types of tracheary cells characterize the axial system.

Tracheids
In the vessel-less angiosperms of the Winteraceae the tracheids are very long (up to 4.5 mm), relatively thin walled, and bear circular-bordered or scalariform-bordered pits. They have relatively blunt ends that overlap others in the tissue. Some taxa in more advanced families

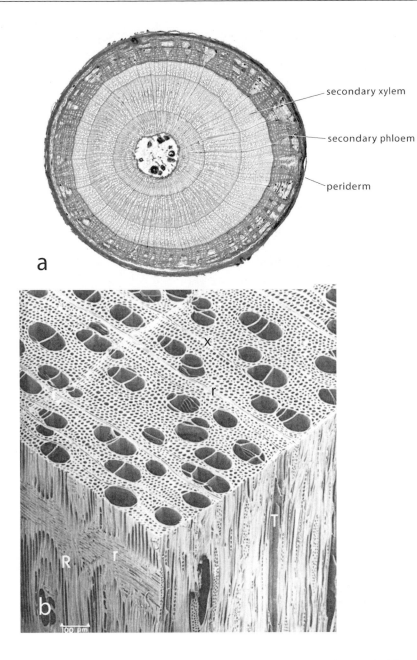

a

Figure 11.13 (a) Transverse section of a three-year old stem of *Tilia americana*. Magnification × 1.7. (b) Scanning electron micrograph of a block of secondary xylem of *Betula alleghaniensis* (yellow birch) showing transverse (X), radial (R), and tangential (T) surfaces. Note the vascular rays (r) on all surfaces. The axial system of the xylem of this dicotyledon is composed predominantly of fibers and clusters of large vessels. The fibers abruptly diminish in size at the end of a growing season, marking the boundary of the growth layers. (b) From Core *et al.* (1979). Used by permission of Syracuse University Press.

secondary xylem

secondary phloem

periderm

b

100 µm

of dicotyledons (e.g., *Quercus*, of the Fagaceae) also contain tracheids as well as fibers and vessel members. These tracheids are typically relatively short, often somewhat distorted in form, bear circular-bordered pits, and have blunt to sharply tapered ends.

Fiber-tracheids

These cells are of frequent occurrence in primitive dicotyledons, the secondary xylem of which is also characterized by long, slender vessel members (see p. 194). As their name suggests they are intermediate in structure between typical fibers and tracheids although they are usually longer than tracheids in the same wood. Their bordered pits

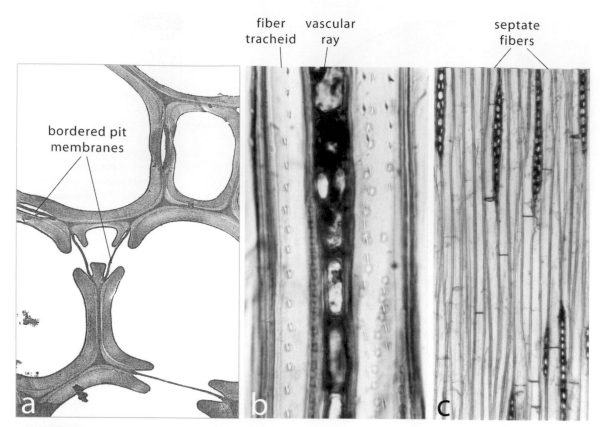

fiber tracheid   vascular ray   septate fibers

a   b   c

bordered pit membranes

Figure 11.14 (a) Transverse section of tracheids in the secondary xylem of *Quercus rubra* illustrating the homogeneous membranes of bordered pit-pairs. Magnification × 3000. (b) Fiber-tracheids in secondary xylem of *Cercidiphyllum japonicum* bearing circular-bordered pit-pairs with elliptical apertures extending to the boundary of the pit borders. Magnification × 600. (c) Septate fibers in *Albizzia saponaria*. Note that septa extend only to the inner surface of the secondary walls. Magnification × 210. (a) From Côté (1967). Used by permission of the University of Washington Press.

are relatively small with apertures that extend to or beyond the borders as in *Cercidiphyllum japonicum* (Fig. 11.14b). In some genera (e.g., *Albizzia*) of a few families they may be **septate** (Fig. 11.14c) and retain a functional protoplast containing a nucleus between each pair of septa. These living septate fiber-tracheids usually store photosynthate in the form of starch grains and may remain alive for several to many years. Thus they function in both support and storage of photosynthate. Their cell walls are typically much thinner than those of libriform fibers, and their bordered pits less reduced. Typically, the septa between protoplasts consist solely of primary wall. Since the septa extend only to the inner surface of the secondary wall of the fiber (Fig. 11.14c), it is apparent that they were deposited by the protoplasts following each nuclear division late in the development of the cell.

Libriform fibers

As the most highly specialized wood fibers, libriform fibers (Fig. 11.15) are the most abundant cell type in many woods and are the major elements of structural support. In many taxa their protoplasts are maintained for several years. They are characterized by great length, relatively small, transverse diameters, thick secondary walls (Fig. 11.15a) and sharply tapered ends (Fig. 11.15b). Typically they are several times the length of the fusiform initials from which they were derived, their greater length resulting from intrusive tip growth during development. Their pits are highly reduced with very small

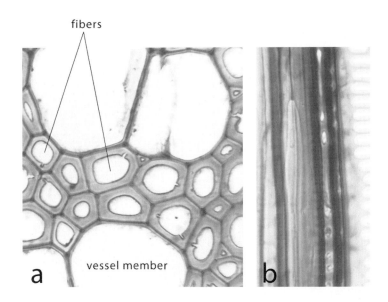

fibers

vessel member

a

b

Figure 11.15 (a) Transverse section of secondary xylem of *Liriodendron tulipifera* (tulip tree) illustrating libriform fibers. The discontinuities in the walls are cracks resulting from improper drying prior to sectioning, not simple pits. (b) A sharply tapered fiber end. Magnification (a) and (b) × 528.

borders and slit-like apertures that extend greatly beyond the pit borders. In some species the pits are so greatly reduced that they have the appearance of simple pits. It is interesting to note that the most highly specialized fibers occur in the same wood as structurally specialized vessel members which suggests that they evolved concurrently.

Gelatinous fibers

Fibers in **reaction wood** (wood of modified structure by virtue of its location in sites of tension or compression) frequently have secondary wall layers that are hygroscopic, under certain conditions absorbing water and becoming swollen. These so-called "gelatinous fibers" are common in the secondary xylem on the upper sides of tree limbs (tension wood).

Substitute fibers

In reality not tracheary elements, substitute fibers form a large part of the secondary xylem of some suffrutescent (shrubby) plants with fleshy stems such as *Pelargonium*. These cells are longitudinally elongate parenchyma cells with tapered ends that have the appearance of short tracheids or fibers, but are usually of greater diameter than fibers. They provide both support and storage functions for the plants.

Vessel members

Characteristic of all angiosperms except the most primitive taxa, vessel members function primarily in the transport of water and minerals. Unlike tracheids and fibers, vessel members have end walls, and occur in superposed columns of cells comprising **vessels** (Fig. 11.16a–c). During development, the pairs of end walls between vessel members become perforated (see pp. 195–202 for the development of tracheary elements.). Thus, vessels are elongate tubes well adapted for the movement of water and solutes in the plant. Vessel members are, typically, similar in length to the cambial fusiform

Figure 11.16 Superposed vessel members comprising vessels. (a) Tangential section of *Acer saccharum* (sugar maple). Note also the small uniseriate and very large multiseriate vascular rays. Magnification × 50. (b) Radial section of *Fraxinus americana* (ash). Axial parenchyma cells are associated with the cluster of vessels. Magnification × 150. (c) Scanning electron micrograph of a vessel of *Quercus*. Magnification × 113. (c) From Core *et al.* (1979). Used by permission of Syracuse University Press.

vessels

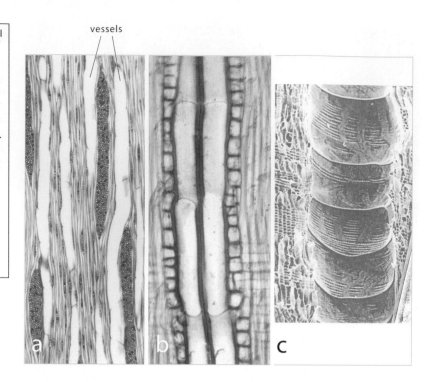

initial from which they were derived. As they develop there is little or no growth in length, but in many species there is considerable lateral growth resulting in increase in diameter. This lateral growth results in displacement of developing fibers, axial parenchyma, and rays.

Vessel members as well as libriform fibers evolved from tracheids (see Bailey, 1953, 1954, 1957). It is not surprising, therefore, that vessel members in primitive taxa (and the fusiform cambial initials from which they were derived) are quite long. Such vessel members typically have oblique, **compound perforation plates** containing several to many perforations (Figs. 11.17a–c, 11.18c). They are often angular in transverse section, of relatively small diameter, and characterized by scalariform bordered pits on their lateral walls (Fig. 11.17d). All of these characters are considered to represent primitive states. In contrast, vessel members of more highly specialized (evolutionarily advanced) taxa are short, often very broad (Fig. 11.16c), have **simple perforation plates** with one large opening in the end wall (Fig. 11.18g) and circular-bordered pits on the lateral walls. Structurally intermediate vessel members occur between these extremes (Fig. 11.18e, f). The vessel members of some taxa in diverse families are characterized by helical wall thickenings which comprise the inner part of the S3 layer as in *Tilia* (basswood) (Fig. 11.17e). Because genera are usually characterized by vessel members of a single structural type, vessel members are often used, with other evidence, to suggest their level of evolutionary specialization.

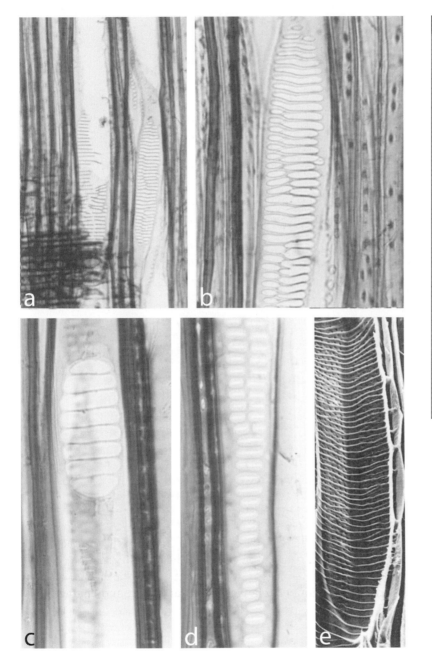

Figure 11.17 (a, b) Vessel members with very oblique scalariform to reticulate perforation plates in the secondary xylem of *Cercidiphyllum japonicum*. Note the fiber-tracheids associated with the vessel members. Magnification (a) × 159, (b) × 397. (c) Vessel member of *Liriodendron tulipifera* with a scalariform perforation plate. Note also the small, oval to circular-bordered pits and the helical wall thickening in the adjacent vessel member. Magnification × 209. (d) Scalariform-bordered pits in the wall of a vessel member in the same wood. Magnification × 412. (e) Scanning electron micrograph of a vessel member from the secondary xylem of *Tilia americana* with helical thickenings lining the secondary wall. Magnification × 408. (e) From Core *et al.* (1979). Used by permission of Syracuse University Press.

Differentiation of tracheary elements

Tracheary elements resemble the fusiform cambial initials from which they are derived, especially in length. Although there may be some intrusive growth of the cambial derivatives during differentiation of tracheids in gymnosperms, and extensive intrusive tip growth during the development of fibers in the wood of dicotyledons, there is essentially no increase in length in the derivatives that differentiate

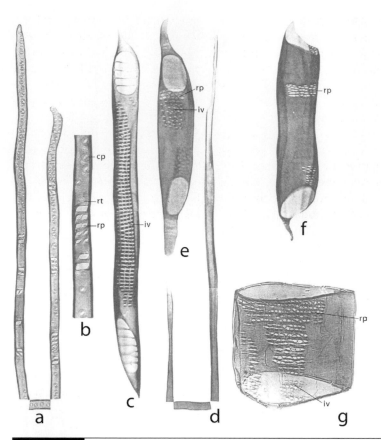

Figure 11.18 Tracheary elements from secondary xylem. (a) A tracheid from *Pinus resinosa*. Magnification × 54. (b) Enlarged segment of the tracheid shown in (a). (c) A vessel member with scalariform perforation plates from *Liriodendron tulipifera*. (d) A libriform fiber from *Populus grandidentata* (big-tooth aspen). (e) A vessel member from *P. grandidentata* with simple perforation plates. (f) A vessel member from *Salix nigra* (black willow) with simple perforation plates. (g) A vessel member from *Sassafras albidum* with simple perforation plates. All vessel members, the libriform fiber, and the enlarged part of the tracheid are shown at the same magnification. The tracheid in (a) is shown at half the magnification of the vessel members. cp, circular-bordered pits; iv, intervessel pits (bordered pits that comprise bordered pit-pairs with pits in the walls of contiguous vessel members); rp, pits of tracheids or vessel members that comprise half-bordered pit-pairs with simple pits in contiguous ray parenchyma cells; rt, small bordered pits that comprise bordered pit-pairs with pits in contiguous tracheids or vessel members. Magnification (b–g) × 109. From Carpenter and Leney (1952). Used by permission of the SUNY College of Environmental Sciences and Forestry at Syracuse University.

into the vessel members of angiosperms. This is not surprising considering the fact that vessel members become longitudinally stacked one upon another, at maturity forming long tubes (Figs. 11.16a, b). Most of the growth during differentiation of tracheids and vessel members is, therefore, diametric, resulting in lateral increase in size. As developing vessel members increase in size, diametrically, the inclination of their end walls decreases.

It is widely accepted that the growth and development of the cells in the secondary xylem are influenced by growth hormones. Research of Sachs (1981) and Aloni (1992, 2001), among many others, has convincingly demonstrated the presence of a polar flow of auxin through the cambial zone from the leaves to the roots. During periods of growth this basipetal flow of auxin plays a significant role in the continuing, generally acropetal, differentiation of vascular tissues. Differentiation of xylem occurs at high levels of auxin, and differentiation of vessel members is dependent, in addition, on cytokinin originating in root apices (Aloni, 1992). It has been suggested that cytokinin increases the sensitivity of cambial derivatives to auxin, thus stimulating them to differentiate into vessel members (Baum et al., 1991; Fukuda, 1997). High levels of auxin stimulate the rapid differentiation of fibers with thick secondary walls whereas high levels of gibberellic acid influence the development of long fibers with thin walls (Roberts et al., 1988).

Cell growth is directly related to turgor pressure in the protoplasts which is generally accepted as the motive force for cell expansion. According to Cosgrove (1993), enzymatic loosening of the cell wall results initially in a reduction of turgor pressure with the result that water is drawn into the cell which physically extends it, a viewpoint promoted earlier by Lockhart (1965) and Ray et al. (1972). Recent research indicates that the protein expansin may control wall loosening by uncoupling the molecular strands comprising the polysaccharide network of the wall (Cosgrove, 2000; Cosgrove et al., 2002). Some cells of the secondary xylem, for example, libriform fibers, become much longer during development than the fusiform initials from which they were derived. Whereas the central region of such developing cells may grow symplastically with surrounding cells in the tissue, their growth beyond the longitudinal extent of their cambial precursors results from intrusive tip growth. The primary wall of the growing tips of fibers is very thin, and as the tip is extended, new wall is synthesized. Evidence for intrusive tip growth of fibers is the presence in the growing tips of the nucleus and dense cytoplasm, and during development, the gradual addition of secondary wall toward the cell tips (see Larson, 1994). Chaffey et al. (2000) have suggested that the movement of Golgi and ER vesicles, carrying precursor compounds of wall synthesis, into the cell tip is facilitated by actin microfilaments which are thought to control cyclosis.

During **intrusive growth**, the cell tips push their way between adjacent cells which are displaced to some extent. As a result, their form is modified with the tips often becoming wedge-shaped (Wenham and Cusik, 1975). The means whereby growing cell tips intrude between adjacent cells is unclear. Early workers apparently thought that the growing tips simply pushed apart the cells and, in the process, disrupted the middle lamella. More recently, the possibility of enzymatic digestion of the middle lamella (Hejnowicz, 1980) and loosening or relaxation of the walls of cells between which intrusive growth takes place (Barnett, 1981) have been suggested as means which facilitate the intrusion of cell tips.

Figure 11.19 Bordered pit development in *Taxus* (yew). (a) Sectional view of a developing bordered pit-pair showing extension of the pit borders (arrowheads) over the pit membrane. (b) Enlargement showing a cluster of microtubules in sectional view (cluster of arrowheads) just inside the plasmalemma. These microtubules form a ring around the inner, developing edge of the pit border (PB). Microtubules also occur over the surface of the pit border, some of which are indicated by arrow heads. From Uehara and Hogetsu (1993). Used by permission of Springer-Verlag Wien.

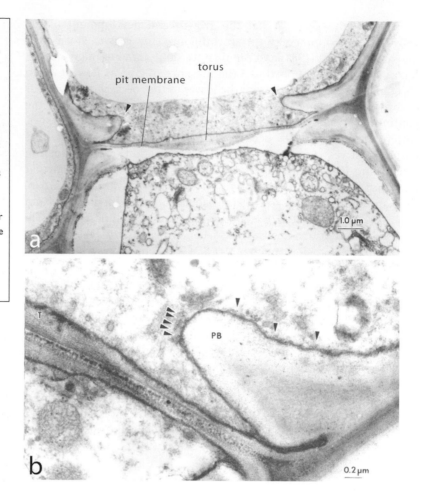

Just prior to the cessation of growth in cells of the secondary xylem, the cell wall becomes radically modified by the addition of secondary wall in various patterns. Components of the cytoskeleton, especially microtubules, are intimately associated with cellulose synthesis, the orientation of cellulose microfibrils, and with wall sculpturing (see, e.g., Baskin, 2001). During the processes associated with this phase of differentiation the nucleus enlarges, Golgi bodies and endoplasmic reticulum become associated with sites of wall synthesis, and microtubules commonly (but not always) become aligned beneath the plasmalemma in patterns that will reflect the ultimate patterns of secondary wall deposition (see, e.g., Hirakawa, 1984; Uehara and Hogetsu, 1993; Abe *et al.*, 1994). Transport of secondary wall components into the region of wall synthesis is accomplished by fusion of Golgi and/or ER vesicles with the plasmalemma, thus expelling into the sites of wall synthesis precursor compounds of cellulose and wall matrix substances. Cellulose synthase complexes (rosettes) from which cellulose microfibrils are generated (see Chapter 4 on the cell wall) are thought also to be transported to the plasmalemma in Golgi vesicles (Giddings and Staehelin, 1988).

There are many instances of the correlation between the position of microtubules and the orientation of microfibrils. For example, during the development of annular and helical secondary wall thickenings of tracheary elements in the primary xylem, helical wall thickenings in tracheids of the secondary xylem, and in developing pit borders, cellulose microfibrils are synthesized in patterns parallel to the microtubules underlying the plasmalemma (Hogetsu, 1991; Uehara and Hogetsu, 1993; Chaffey *et al.*, 1997; Funada, 2002). It is interesting to note, however, that the fusiform cambial initials from which tracheary elements and associated cells are ultimately derived typically contain randomly arranged, peripheral microtubules. As cambial derivatives begin their differentiation, the microtubules become arranged in specific patterns in relation to the ultimate orientation of cellulose microfibrils. In the conifer *Taxus* (yew), for example, very early in tracheid development regions which become free of microtubules are thought to indicate the sites in which pits will form (Funada, 2002). These incipient pits are quite large, up to 8–9 μm in diameter (Uehara and Hogetsu, 1993). They become surrounded on their inner margins by a ring of microtubules (Fig. 11.19a, b). As seen in face view in confocal immunofluorescence images they appear as a bright ring around a wide aperture (Fig. 11.20). As development proceeds and additional wall material is added to the border, the ring of microtubules decreases in diameter as the border expands over the pit cavity and the aperture becomes smaller (Fig. 11.21) (Uehara and Hogetsu, 1993; Funada, 2002). Concurrently, microtubules between adjacent pits become oriented helically and are associated with cellulose microfibril synthesis resulting in an increase in secondary wall thickness between and around the pits (Figs. 11.19b, 11.21a, b). During

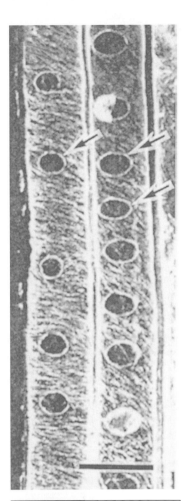

Figure 11.20 Rings of microtubules (arrows) around the inner edges of developing pit borders. Note also the helical bands of microtubules, associated with helical wall thickening, between the developing pits. Bar = 25 μm. From Funada (2002). Used by permission of Taylor and Francis Books Ltd.

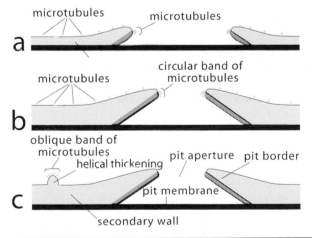

Figure 11.21 Diagram illustrating the association of microtubules with (a) the initial thickening and extension of the border over the pit cavity, (b) continued development of the border, and increase in thickness of the wall between pits, and (c) development of helical thickenings. From Uehara and Hogetsu (1993). Used by permission of Springer-Verlag Wien.

Figure 11.22 Arrangement of cellulose microfibrils in developing bordered pits in tracheary elements from the root of *Pisum sativum* (pea). (a) Initially, microfibrils are deposited in arcs above and below the pit sites. Bar = 2 μm. (b) As development continues, microfibrils are deposited around the ends of the arcs, enclosing the developing pit apertures. Bar = 2 μm. (a, b) From Hogetsu (1991). Used by permission of Springer-Verlag GmbH and Co. KG. © Springer-Verlag Berlin Heidelberg.

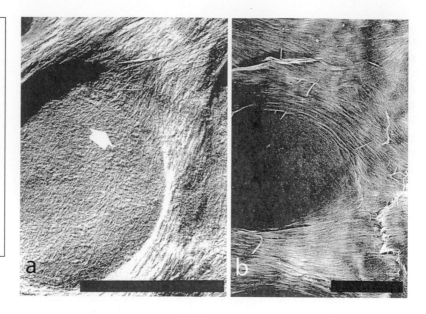

Figure 11.23 Association of microtubules with developing helical wall thickenings in *Taxus* tracheids. (a) Early stage, showing helical bands of microtubules. (b) Bands of microtubules superimposed on developing helical thickenings. (c) Helical thickenings after microtubules have disappeared. Bars (a–c) = 10 μm. (a–c) From Uehara and Hogetsu (1993). Used by permission of Springer-Verlag Wien.

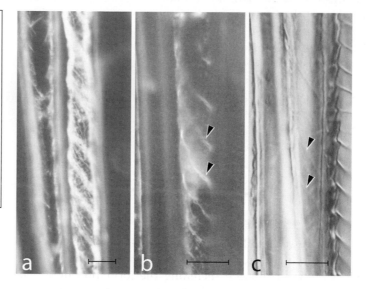

development of pit borders microfibrils are synthesized in parallel arrays that, initially, form arcs around and over the periphery of the pit sites. In face view of pit replicas from freeze-dried sections from roots of *Pisum sativum* (pea), the microfibrils appeared to Hogetsu (1991) "like a stream flowing around rocks in a river" (Fig. 11.22a). As development of the border continued, the microfibrils curved around the ends and, thus, enclosed the developing aperture (Fig. 11.22b). Following pit development, conspicuous helical bands of microtubules (Figs. 11.21c, 11.23a) appeared on the inner surface of the wall between the pits in the pattern of helical, tertiary wall thickenings that developed on the inner surface of the secondary wall (Fig. 11.23b, c) (Uehara and Hogestsu, 1993).

Similar observations have been made also on developing tracheary elements in other angiosperms. One additional example will suffice. Early in the differentiation of developing fibers and axial parenchyma cells in incipient vessel members of *Aesculus hippocastanum*, microtubules become arranged in a sort of reticulum containing microtubule-free regions (as in *Taxus*) that signal the future position of bordered pits (Chaffey *et al.*, 1999). During pit development, a ring of microtubules appears just inside the developing pit border. As the border grows and the pit opening is reduced, the diameter of the ring of microtubules also decreases (Chaffey *et al.*, 1997).

This intimate association of microtubules with cellulose microfibrils and structural modifications of the cell wall during the development of the secondary xylem in conifers and dicotyledons suggests that microtubules may play a controlling role in the processes whereby these wall modifications occur (Chaffey *et al.*, 1997; Baskin, 2001). This viewpoint is supported by experiments utilizing colchicine to destroy the microtubules prior to secondary wall formation which provides strong evidence of their influence in the orientation of wall thickenings and other structures of wall ornamentation. Lacking microtubules, wall thickenings occur in very irregular patterns. Since the early studies of Ledbetter and Porter (1963), Hepler and Newcomb (1964), Wooding and Northcote (1964) and others, the significance of this relationship has been discussed and speculated upon at length, but as yet, no widely accepted explanation of the mechanism of such control has been provided. Giddings and Staehelin (1988) hypothesized that in the alga *Closterium*, parallel microtubules attached to the plasmalemma by putative protein bridges, form "channels" in which terminal cellulose synthase complexes (rosettes) are restricted. Thus, as the cellulose microfibrils are generated the rosettes are pushed along these channels by the polymerization of cellulose (but see also the more complex hypothesis of "templated incorporation" of Baskin (2001), which accounts for microfibril alignment in association with microtubules as well as in their absence).

In developing vessel members and other developing xylem cells in the axial system, actin microfilaments also apparently play a major role in cell wall ornamentation including the formation of simple and bordered pits, helical tertiary thickenings, and the perforations in perforation plates (Chaffey and Barlow, 2002; Chaffey *et al.*, 2002). The parallel orientation of microfilaments and microtubules at all of these sites suggests interaction between these components of the microskeleton during development, but the nature of this interaction is unknown. Chaffey *et al.* (2000) have hypothesized, however, that in the development of bordered pits in vessel members F-actin, possibly linked with myosin as actomyosin, might provide the motive force for contraction of the ring of microtubules and microfilaments as the pit aperture decreases in diameter. This viewpoint is further supported and expanded by Chaffey and Barlow (2002) who hypothesize that "an acto-myosin contractile system (a 'plant muscle') is present at the cell plate, the sieve pores, the plasmodesmata, within the walls of long-lived parenchyma cells, and at the apertures of bordered pits during their development."

Microfilaments are also thought to have a significant role in the differentiation of other types of tracheary cells (Chaffey *et al.*, 2000, 2002; see also Funada *et al.*, 2000). Microfilaments are predominantly oriented longitudinally in a helix of high pitch in both fusiform cambial initials and developing fibers derived from them. They are thought to function in the delivery of secretory vesicles to the tips of the elongating cells, providing the precursor compounds for wall synthesis, possibly their primary role in the intrusive tip growth of fibers (Chaffey *et al.*, 2000).

Upon completion of secondary wall formation in tracheary elements, the protoplast may die, as in tracheids and vessel members, or it may persist for several years as in some wood fibers (Fahn, 1990; Larson, 1994). Just prior to cell death hydrolytic enzymes are believed to enter the cell protoplast from surrounding parenchyma cells, probably through and under the control of plasmodesmata (see Juniper, 1977; Ehlers and Kollmann, 2001), initiating hydrolysis of the regions of primary wall not protected by lignified secondary wall. The effect of hydrolases on the unprotected primary wall may result in the digestion of wall matrix material (see, e.g., Ohdaira *et al.*, 2002), leaving meshworks of cellulose microfibrils which comprise pit membranes. During the development of vessel members entire regions of primary wall may be removed resulting in perforations in the end walls and the formation of perforation plates. It is now well established that the first visual sign of programmed cell death, which concludes the differentiation of tracheids and vessel members, is the breakdown of the tonoplast. This releases hydrolytic enzymes sequestered in the vacuole that cause the rapid degradation of all cytoplasmic organelles and the consequent death of the protoplast (Fukuda, 1996, 2000; Groover *et al.*, 1997; Ito and Fukuda, 2002; Kuriyama and Fukuda, 2002).

Patterns of distribution of xylary elements and rays

The secondary xylem of different taxa of dicotyledons can be distinguished on the basis of the distribution in the xylem of vessels of different sizes and wall thickness, the distribution of axial parenchyma and its association with vessels, and the distribution of rays of various morphologies. For example, the secondary xylem can be characterized as ring-porous or diffuse-porous (Fig. 11.24a, b). If **ring-porous**, the early wood (that formed first during a growing season) contains one to several layers of very large, relatively thin-walled vessels and the late wood relatively uniformly distributed small, thicker-walled vessels (Fig. 11.24a). Well-known examples of taxa with ring-porous secondary xylem are *Quercus, Catalpa, Ulmus* (elm), and *Fraxinus* (ash). Among **diffuse-porous** taxa are *Betula* (birch), *Acer* (maple), and *Populus* (poplar). In these taxa, vessels of relatively uniform size are evenly distributed througout the growth layers (Fig. 11.24b). In some other diffuse-porous taxa, such as *Juglans* (walnut) and *Malus* (apple), there

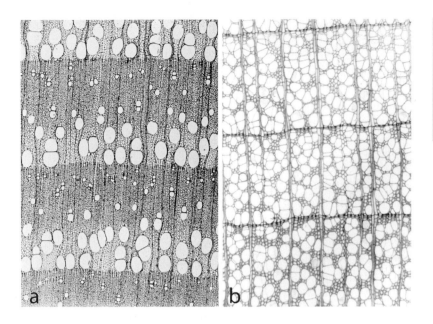

Figure 11.24 Distribution of vessels in secondary xylem as seen in transverse section.
(a) Ring-porous secondary xylem in *Fraxinus* sp. (ash).
(b) Diffuse-porous secondary xylem in *Tilia americana*.
Magnification (a) and (b) × 58.

is a gradual transition from larger to smaller vessels in the growth layers.

The distribution of axial parenchyma in the secondary xylem is relatively constant in genera and in some larger taxa. Axial parenchyma is classified as **paratracheal** when it is consistently associated with vessels, or **apotracheal** when it is not consistently associated with vessels. Three categories of apotracheal parenchyma are recognized: terminal (sometimes called boundary), diffuse, and banded (Fig. 11.25 a–c). **Terminal parenchyma** occurs in the last-formed tissue, sometimes only in the last layer of cells, in a growth layer (Figs. 11.25a, 11.26a). The term **boundary parenchyma** may appropriately be applied in instances in which parenchyma occurs on both sides of the boundary between growth layers. Examples of taxa with terminal parenchyma are *Liriodendron*, *Magnolia*, and *Salix* (willow). **Diffuse parenchyma** is scattered randomly throughout the growth layer (Figs. 11.25b, 11.26b, c), as in *Malus*, *Diospyros* (persimmon), and *Quercus*, and **banded parenchyma** occurs in narrow bands interspersed between vessels or clusters of vessels (Fig. 11.25c). Examples are *Carya* (pecan) and *Calophylum wallichianum* (Guttiferae).

Three types of paratracheal axial parenchyma are also recognized: vasicentric, aliform, and confluent (Fig. 11.25d–f). In secondary xylem with **vasicentric parenchyma**, parenchyma cells may occur in the last cell layer(s) of the growth layer, i.e., in the position of terminal parenchyma, but elsewhere in the growth layer parenchyma surrounds or contacts vessels directly or indirectly (Figs. 11.25d, 11.26d). Examples of taxa with vasicentric parenchyma are *Fraxinus* and *Catalpa*. **Aliform parenchyma** surrounds vessels and extends tangentially in wing-shaped masses (Fig. 11.25e) as in *Mangifera indica* (mango; Anacardiaceae) and *Acacia nilotica* (prickly acacia). **Confluent parenchyma** is banded, and the bands, which may branch, enclose or

Figure 11.25 Distribution of axial parenchyma in secondary xylem as viewed in transverse section. Parenchyma is indicated by fine line shading; rays by coarse line shading. (a) *Liriodendron tulipifera*. (b) *Tilia cordata*. (c) *Carya tomentosa*. (d) *Grewia mollis*. (e) *Mangifera indica*. (f) *Celtis soyauxii*. From Metcalfe and Chalk (1950). Used by permission of Oxford University Press.

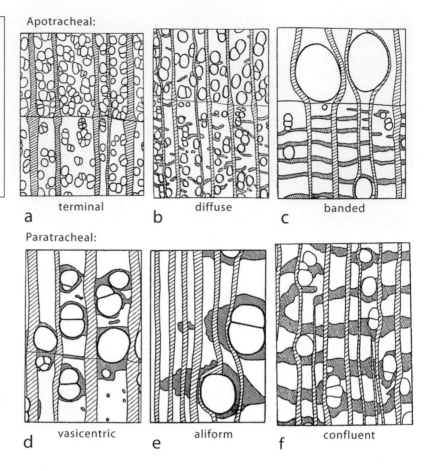

Apotracheal:

a   terminal   b   diffuse   c   banded

Paratracheal:

d   vasicentric   e   aliform   f   confluent

are in contact with vessels (Fig. 11.25f). The secondary xylem of *Celtis soyauxii* (Ulmaceae) and *Markhamia platycalyx* (Bignoniaceae) contain confluent parenchyma.

Secondary xylem is also characterized by the nature and distribution of rays. In woody dicotyledons most ray parenchyma cells are radially elongate although in some species some or all of the marginal ray cells may be longitudinally elongate. Rays in dicotyledons may be characterized as homocellular or heterocellular. **Homocellular rays** consist of ray parenchyma cells of similar size and shape which are commonly radially elongate (**procumbent**) as, for example, the rays of *Acer* (Fig. 11.27a, b). In some herbaceous plants, however, homocellular rays consist entirely of longitudinally elongate (**upright**) ray parenchyma cells. **Heterocellular rays**, in contrast, consist of cells of two shapes and sizes with the marginal cells usually differing from those of the remainder of the ray. Marginal cells may be longitudinally elongate (or sometimes with nearly equal longitudinal and radial dimensions) as in *Swietenia mahogani* (Fig. 11.27c, d) or may consist of a mixture of longitudinally and radially elongate cells. Heterocellular rays may also consist of alternating bands of marginal and internal upright cells as in *Cercidiphyllum japonicum* (Fig. 11.27e, f). The rays of conifers containing ray tracheids are also referred to as

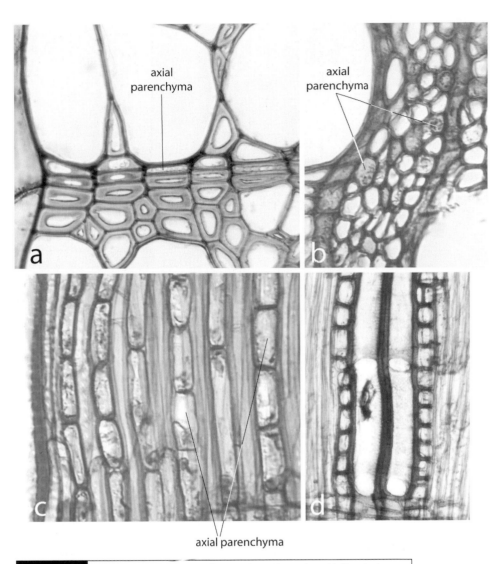

axial parenchyma

axial parenchyma

axial parenchyma

Figure 11.26 Distribution of axial parenchyma in secondary xylem. (a) Terminal parenchyma in *Liriodendron tulipifera* in transverse section. Magnification × 541. (b, c) Diffuse parenchyma in *Quercus rubra* in transverse section (b) and in radial section (c). Magnification (b) and (c) × 497. (d) Vasicentric parenchyma in *Fraxinus americana* (American ash) in radial section. Magnification × 181.

heterocellular as are those of at least two families of angiosperms, Proteaceae and Malvaceae, some species of which contain ray tracheids.

Rays vary in height from a fraction of a millimeter to 10 cm or more, and in width from one to several rows of cells. If rays are one cell wide they are referred to as **uniseriate**, if two cells wide, as **biseriate**, and if three or more cells wide, as **multiseriate**. In some species, e.g. *Alnus* (alder) and *Carpinus* (ironwood), closely spaced, narrow rays occur in groups that simulate large rays. Such groups are called **aggregate rays**. Within the secondary xylem of a single species

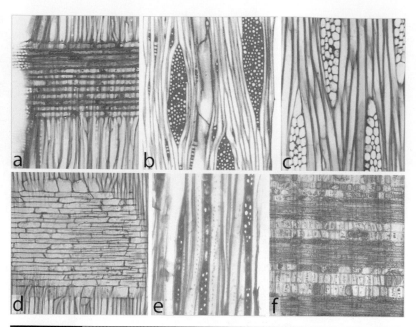

Figure 11.27 Vascular rays. (a, b) Homocellular rays from *Acer saccharum*. (a) Ray composed of radially elongate (procumbent) ray cells. Magnification × 156. (b) Tangential section containing a combination of uniseriate and multiseriate rays. Magnification × 96. (c–f) Heterocellular rays. (c) Tangential section showing multiseriate rays in *Swietenia mahogani* (mahogany) with upright marginal ray cells. Magnification × 122. (d) Radial section of the same wood. Magnification × 84. (e) Rays of *Cercidiphyllum japonicum* with marginal upright cells and internal bands of upright cells that alternate with bands of procumbent cells. Magnification × 115. (f) Radial section of the same wood. Magnification × 84.

rays may be solely uniseriate or solely multiseriate (or biseriate). In some other species uniseriate and bi- or multiseriate rays may occur together in the xylem (Fig. 11.27b). In most woods, rays are arranged so that, as viewed in tangential section, their ends extend beyond those of others. In some taxa, however, in which all rays are of similar height as, for example, in *Diospyros virginiana*, they are arranged in rows. In such cases, the rays are said to be storied.

Rays in the secondary wood of both conifers and dicotyledons have long been known as sites of storage of photosynthate in the form of starch, and pathways of lateral transport of assimilates between the xylem and the phloem. Especially in the early spring, sugars are transported to the cambium, and through pits from ray cells into adjacent vessels through which they are delivered to developing buds (Esau, 1977). Axial parenchyma cells apparently play a similar role in providing carbohydrates to growing regions of the plant. Recently, Chaffey and Barlow (2001) have proposed that axial parenchyma and rays in the xylem, which are interconnected by plasmodesmata, comprise "a **three-dimensional symplasmic continuum**" which provides pathways of radial transport between the secondary xylem and the secondary phloem (see also Van Bel, 1990; Van Bel and Ehlers, 2000). They emphasize the role of the numerous plasmodesmata in the tangential walls which would facilitate radial transport and suggest, further, that the microfilament component of the cytoskeleton might also play a role in the intracellular movement of molecules. In this system, photosynthate is transported from both the xylem and the phloem to the highly active cambial zone and its developing xylary derivatives, and growth hormones could be transported to this region from sites of synthesis in the young leaves, as well as from root and shoot apices through the interconnected system of axial and ray parenchyma. As pathways of transport to the cambial zone of proteins and nucleic acids as well as hormones and photosynthate this system apparently plays a significant role in the coordination of development.

Tyloses

Secondary xylem that functions in transport and contains living ray and axial parenchyma is called **sapwood**. Sapwood is peripherally located and encloses the **heartwood**, xylem that no longer functions in transport and contains no living cells. The volume of sapwood and heartwood varies greatly in different species, and in trees of different ages. The relative amounts may also be affected by environmental conditions. Generally, in young trees the quantity of sapwood is much greater than the quantity of heartwood whereas in old trees, heartwood typically comprises the greater volume. Not only does heartwood contain no living components, it is also characterized by the presence of large quantities of waste metabolites, primarily flavonoids and other phenolic compounds (Hillis, 1987). Plants, unlike animals, have no means of excreting waste products to the exterior. The central part of a tree trunk, therefore, becomes the repository for waste products. In some species the formation of **tyloses** (Fig. 11.28), extensions of adjacent parenchyma cells into vessel members, enhances the volume of metabolites that can be deposited in the heartwood.

Tyloses develop in a transition region between sapwood and heartwood in many angiosperms, and are common features in the heartwood of some hardwood species such as *Robinia*, *Quercus*, *Castanea* (chestnut), and *Juglans* (Fig. 11.28a–c). They are not known to occur under normal conditions in conifers although they can be stimulated to develop in response to wounding (Peters, 1974). They are uncommon in non-woody angiosperms, but do occur in some suffrutescent herbaceous taxa such as *Pelargonium* (Fig. 11.28d). In axial parenchyma and ray parenchyma cells in contact with vessel members in the region in which tyloses are forming, a thin, unlignified wall layer is deposited over the inner surface of the entire wall, including the pit membranes, adjacent to vessel members. In response to an unknown stimulus, possibly the accumulation of flavonoids to a

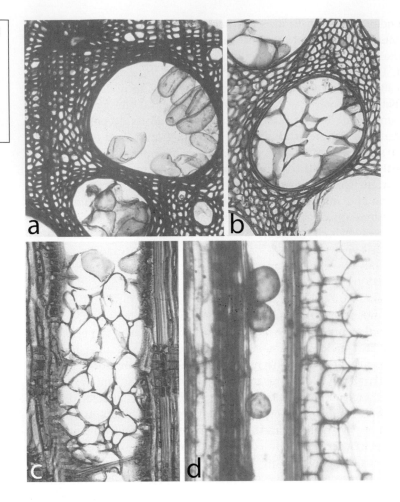

Figure 11.28 Tyloses. (a) Young tyloses in vessel members of *Quercus* sp. (b, c) Tyloses in *Quercus rubra* in transverse section (b) and radial section (c). Magnification (a), (b), and (c) × 139. (d) Tyloses in a vessel of *Pelargonium*. Magnification × 300.

level of toxicity, the pit membranes are enzymatically degraded and the unlignified wall layer lying over the pit begins to grow, extending into the lumen of the vessel member (Fig. 11.28a, d). The nucleus of the parenchyma cell often migrates into the tylosis. This expansion of the cell provides a much greater volume for the accumulation of waste metabolites prior to the death of the cell, caused, ultimately by the toxicity of these accumulations. A vessel member may be completely filled by the intrusion of many tyloses (Fig. 11.28b, c).

It is thought by some workers that vessels are no longer functional in transport at the time of tylosis development. If, on the other hand, the vessels are still functional, the formation of tyloses will greatly reduce the efficiency of transport of water and minerals, and ultimately eliminate their function in transport.

In addition to providing a site in which waste metabolites are stored, heartwood may have another adaptive value. The toxicity of the stored products in heartwood is thought to prevent or deter the invasion and decay by fungi of the central, non-living parts of tree trunks and large limbs. The presence of suberin in addition to lignin in the walls of tyloses in several tree species is believed to hinder the colonization of fungi in infected trees (Rioux *et al.*, 1995). For a detailed discussion of heartwood, and an extensive bibliography, see Hillis (1987).

Mechanism of water transport

Since the proposal of Dixon and Joly in 1894, the **Cohesion Theory** has been the most widely accepted explanation for the transport of water and minerals in the xylem. According to this theory, water loss from leaves through transpiration results in the development of tension in the water column in the tracheids and vessels that is transmitted throughout the column to the roots. As transpiration continues, water is pulled from the soil into the roots and upward through the system. Thus, as water is being lost from the plant, it is constantly being replaced. For such a mechanism to work there must be "sufficient tension to lift the water to the top of the tallest trees, and sufficient cohesive strength in the columns of water to withstand this tension" (Canny, 1995). Canny notes that for a tree 100 m tall, there must be established "a gradient in the xylem of 2 bar $(10 \text{ m})^{-1}$ and a cohesive strength of xylem sap in excess of 20 bar."

Over the years experiments using pressure chambers have led many plant physiologists to conclude that these conditions are met in nature. Some doubt was expressed very early, however, and recently has been elaborated upon by Zimmermann *et al.* (1995) who concluded, on the basis of measurements with the xylem pressure probe, that "xylem tension in the leaves of intact, transpiring plants is often much smaller than that predicted for transpiration-driven water ascent through continuous water columns." Canny (1995) states that "no one has devised a system with a known tension generated in a water column which can be put into the pressure chamber to check its reliability." Canny's experiments with the xylem pressure probe, supported in part by those of Zimmermann *et al.* (1995), suggest that "(1) The necessary high tensions in the xylem are not present; i.e., the operating tension in the xylem (both from direct measurement and from the determined thresholds or cavitation of water) is around 2 bar not 20+ bar. (2) The necessary gradient of tension with height is not present. (3) The measurements of tension with the pressure-chamber (believed to verify the Cohesion Theory) conflict with those made with the xylem-pressure probe." Consequently, he concludes that "the resolution of these conflicts demands some source of compensating pressure in the xylem to reduce the operating tension from 20 bar to <2 bar." The source of this compensating pressure, as presented in his **Compensating Pressure Theory** (Canny, 1995), is the xylem parenchyma and ray cells. He suggests that positive pressures in these cells in contact with tracheary elements "squeeze them" by "pressing onto the closed fluid spaces of the tracheary elements." He notes that under his theory, "the driving force and the transmission of the force [required in the ascent of water in tall trees] are the same as in the Cohesion Theory, but the operating pressure of the xylem is raised into a stable range by compensating tissue pressures pressing upon the tracheary elements." Thus, he concludes that whereas "the tissue pressure does not propel the transpiration stream, which is still driven by evaporation, . . . it protects the stream from cavitation."

Important new concepts and hypotheses such as those described above, need to be carefully analyzed and corroborated by many researchers before they are universally accepted or rejected. In this case, results obtained by use of the xylem pressure probe have been critized by Milburn (1996), Comstock (1999) and Stiller and Sperry (1999). Canny (1998) provides additional support for his theory, and presents (Canny, 2001) a strong rebuttal of these criticisms.

Students interested in learning in more detail about the evidence in support of the several theories of water transport in plants, the techniques used in research in this area of plant physiology, and the controversy surrounding these theories should read the papers cited above and others cited therein.

REFERENCES

Abe, H., Ohtani, J., and K. Fukazawa. 1994. A scanning electron microscopic study of changes in microtubule distributions during secondary wall formation in tracheids. *Int. Ass. Wood. Anatomists J.* **15**: 185–189.

Aloni, R. 1992. The control of vascular differentiation. *Int. J. Plant Sci.* **153**: S90–S92.

2001. Foliar and axial aspects of vascular differentiation: hypotheses and evidence. *J. Plant Growth Reg.* **20**: 22–34.

Bailey, I. W. 1953. Evolution of the tracheary tissue of land plants. *Am. J. Bot.* **40**: 4–8.

1954. *Contributions to Plant Anatomy*. Waltham, MA: Chronica Botanica.

1957. The potentialities and limitations of wood anatomy in the study of the phylogeny and classification of angiosperms. *J. Arnold Arbor.* **38**: 243–254.

Barnett, J. R. 1981. Secondary xylem cell development. In J. R. Barnett, ed., *Xylem Cell Development*. Tunbridge Wells, UK: Castle House, pp. 47–95.

Baskin, T. I. 2001. On the alignment of cellulose microfibrils by cortical microtubules: a review and a model. *Protoplasma* **215**: 150–171.

Baum, S. F., Aloni, R., and C. A. Peterson. 1991. The role of cytokinin in vessel regeneration in wounded *Coleus* internodes. *Ann. Bot.* **67**: 543–548.

Beck, C. B., Coy, K., and R. Schmid. 1982. Observations on the fine structure of *Callixylon* wood. *Am. J. Bot.* **69**: 54–76.

Canny, M. J. 1995. A new theory for the ascent of sap-cohesion supported by tissue pressure. *Ann. Bot.* **75**: 343–357.

1998. Applications of the compensating pressure theory of water transport. *Am. J. Bot.* **85**: 897–909.

2001. Contributions to the debate on water transport. *Am. J. Bot.* **88**: 43–46.

Carpenter, C. H. and L. Leney. 1952. *91 Papermaking Fibers*, Technical Publication No. 74. Syracuse, NY: College of Forestry, Syracuse University.

Chaffey, N. and P. Barlow. 2001. The cytoskeleton facilitates a three-dimensional symplasmic continuum in the long-lived ray and axial parenchyma cells of angiosperm trees. *Planta* **213**: 811–823.

2002. Myosin, microtubules and microfilaments: cooperation between cytoskeletal components during cambial cell division and secondary vascular differentiation in trees. *Planta* **214**: 526–536.

Chaffey, N. J., Barnett, J. R., and P. W. Barlow. 1997. Cortical microtubule involvement in bordered pit formation in secondary xylem vessel elements of *Aesculus hippocastanum* L. (Hippocastanaceae): a correlative study using electron microscopy and indirect immunofluorescence microscopy. *Protoplasma* **197**: 64–75.

1999. A cytoskeletal basis for wood formation in angiosperm trees: the involvement of cortical microtubules. *Planta* **208**: 19–30.

Chaffey, N., Barlow, P., and J. Barnett. 2000. A cytoskeletal basis for wood formation in angiosperm trees: the involvement of microfilaments. *Planta* **210**: 890–896.

Chaffey, N., Barlow, P., and B. Sundberg. 2002. Understanding the role of the cytoskeleton in wood formation in angiosperm trees: hybrid aspen (*Populus tremula* × *P. tremuloides*) as the model species. *Tree Physiol.* **22**: 239–249.

Comstock, J. P. 1999. Why Canny's theory doesn't hold water. *Am. J. Bot.* **86**: 1077–1081.

Core, H. A., Côté, W. A., and A. C. Day. 1979. *Wood: Structure and Identification*, 2nd edn. Syracuse, NY: Syracuse University Press.

Cosgrove, D. J. 1993. Wall extensibility: its nature, measurement and relationship to plant cell growth. *New Phytol.* **124**: 1–23.

2000. Loosening of plant cell walls by expansins. *Nature* **407**: 321–326.

Cosgrove, D. J., Li, L. C., Cho, H. T., *et al.* 2002. The growing world of expansins. *Plant Cell Physiol.* **43**: 1436–1444.

Côté, W. A. 1967. *Wood Ultrastructure: An Atlas of Electron Micrographs*. Seattle, WA: University of Washington Press.

Ehlers, K. and R. Kollmann. 2001. Primary and secondary plasmodesmata: structure, origin, and functioning. *Protoplasma* **216**: 1–30.

Esau, K. 1977. *Anatomy of Seed Plants*, 2nd edn. New York: John Wiley and Sons.

Fahn, A. 1990. *Plant Anatomy*, 4th edn. Oxford, UK: Pergamon Press.

Fukuda, H. 1996. Xylogenesis: initiation, progression, and cell death. *Annu. Rev. Plant Physiol. Plant Mol. Biol.* **47**: 299–325.

1997. Tracheary element differentiation. *Plant Cell* **9**: 1147–1156.

2000. Programmed cell death of tracheary elements as a paradigm in plants. *Plant Mol. Biol.* **44**: 245–253.

Funada, R. 2002. Immunolocalization and visualization of the cytoskeleton in gymnosperms using confocal laser scanning microscopy. In N. G. Chaffey, ed., *Wood Formation in Trees: Cell and Molecular Biology Techniques*. London: Taylor and Francis, pp. 143–257.

Funada, R., Furusawa, O., Shibagaki, M., *et al.* 2000. The role of cytoskeleton in secondary xylem differentiation in conifers. In R. A. Savidge, J. R. Barnett, and R. Napier, eds., *Cell and Molecular Biology of Wood Formation*. Oxford, UK: Bios Scientific Publishers, pp. 255–264.

Giddings, T. H., Jr. and L. A. Staehelin. 1988. Microtubule-mediated control of microfibril deposition: a re-examination of the hypothesis. In C. W. Lloyd, ed., *The Cytoskeletal Basis of Plant Growth and Form*. London: Academic Press, pp. 85–99.

Groover, A., DeWitt, N., Heidel, A., and A. Jones. 1997. Programmed cell death of plant tracheary elements differentiating *in vitro*. *Protoplasma* **196**: 197–211.

and Evert, 1991; Wimmers and Turgeon, 1991; Botha and Van Bel, 1992; Van Bel and Van Rijen, 1994) with each component of the sieve element–companion cell complex playing an integral role in the movement of photoassimilates and other compounds throughout the plant. Because during their evolution the sieve elements retained a functional plasmalemma, the evolution of a mechanism of transport through the phloem utilizing the force of osmosis, controlled by solute concentration, was possible.

Gross structure and development of the phloem

The phloem is, thus, a distinctive, important and highly complex tissue through which photosynthate is transported throughout the plant. **Primary phloem** differentiates from provascular tissue and, with primary xylem, is a major component of stem vascular bundles, leaf traces, and the vascular systems of leaves, flower parts, fruits and seeds (see Chapter 6). In roots of seed plants it usually occurs in discrete bundles that alternate with ribs or bundles of primary xylem. In pteridophytes primary phloem usually encloses either a central column or a cylinder of primary xylem. In axes with a pith, the primary xylem may be bounded on both the inside and the outside by primary phloem.

The primary phloem consists of protophloem and metaphloem. **Protophloem** differentiates earlier and nearer the apical meristem than protoxylem, and in regions that are actively elongating. Consequently (especially in stems) the conducting elements are stretched and often obliterated. **Metaphloem** differentiates later than protophloem, in regions in which growth in length has ceased (for more detail, see Chapter 6). In both structure and function the conducting elements and associated cells of the primary phloem are remarkably similar to those of the secondary phloem.

Secondary phloem, like secondary xylem, is derived from the vascular cambium, and is composed of axial and radial systems of cells (Figs. 12.1a–d, 12.2a, b). The axial system is made up of conducting cells, associated parenchyma cells, companion cells (in angiosperms), and phloem fibers. The radial system consists of phloem rays that are continuous with rays in the secondary xylem.

The conducting cells, longitudinally elongate, are collectively called **sieve elements**. There are two types of sieve elements: sieve cells characteristic of gymnosperms, pteridophytes, and other lower vascular plants, and sieve tube members, characteristic of angiosperms. These cells develop from **sieve element mother cells** which are direct descendants of fusiform cambial initials. Following periclinal, longitudinal divisions, the mother cells in gymnosperms differentiate into functional **sieve cells** (Figs. 12.1b, 12.2a), generally with little or no longitudinal (intrusive) growth. The mature sieve cells, like the fusiform cambial initials from which they, are derived, are very long with overlapping ends. By contrast, in dicotyledons, transverse, anticlinal, and/or oblique divisions in the sieve element mother cells

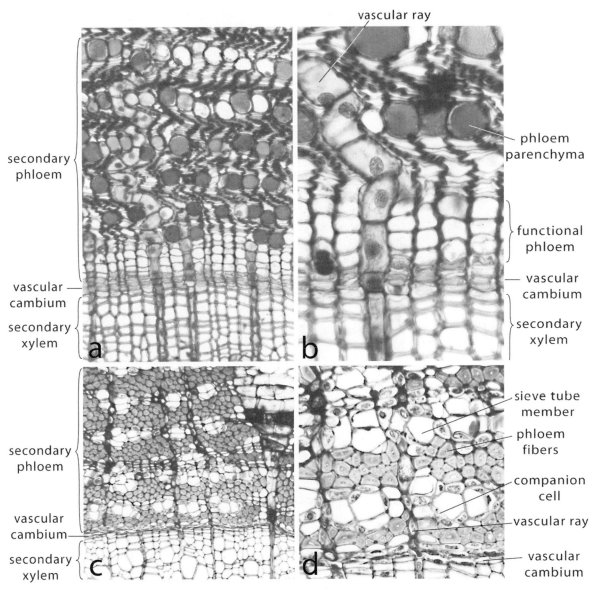

Figure 12.1 (a, b) Transverse sections of secondary phloem of *Pinus strobus*, a conifer, and (c, d) *Tilia americana*, a woody dicotyledon. The secondary phloem is derived from the vascular cambium and, like the secondary xylem, consists of axial and radial systems. Magnification (a) × 206, (b) × 520, (c) × 60, (d) × 160.

lead to columns of superposed **sieve tube members** (Fig. 12.3b). Among the distinctive features of sieve elements is the fact that they comprise longitudinal, open, but living, systems through which photosynthate is transported. At maturity, the sieve element protoplast is highly modified, lacking a nucleus and vacuolar membrane, but it retains a functional plasmalemma, and some parietal endoplasmic reticulum, mitochondria and plastids.

Another distinctive feature of sieve elements is the presence of **sieve areas** which occur on the lateral walls of sieve cells (Fig. 12.2a) and on the end walls and, in some taxa on lateral walls, of sieve tube members (Figs. 12.2b, 12.3). Sieve areas which are highly specialized and evolutionarily modified primary pit fields consist of groups of **pores** through which the protoplasts of contiguous conducting cells

Figure 12.2 (a) A longitudinal section of secondary phloem of *Pinus strobus* showing functional sieve cells characterized by lateral sieve areas. Note the columns of superposed, axial parenchyma cells in a region of the phloem in which sieve cells are no longer functional. Magnification × 555. (b) A longitudinal section of secondary phloem of *Tilia americana*, illustrating part of a sieve tube member with an oblique, compound sieve plate. Note also the lateral sieve areas, a strand of thick-walled fibers, and a column of axial parenchyma cells. Magnification × 944.

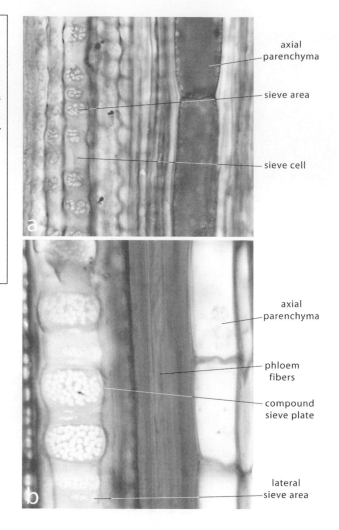

axial parenchyma

sieve area

sieve cell

axial parenchyma

phloem fibers

compound sieve plate

lateral sieve area

are connected (Fig. 12.4a, b, d, e). Pores in dicotyledons are often lined with cylinders of **callose** (the carbohydrate, β-1,3-glucan) (Fig. 12.4a, b) that enclose the plasmalemma, endoplasmic reticulum and other cytoplasmic components that traverse the pores. Callose-lined pores have been reported in the Gnetales, but only rarely have been observed in other gymnosperms. It is not clear whether the callose cylinders are a normal feature of the sieve areas of living, functioning sieve cells and sieve tube members. It is known that wounding stimulates rapid synthesis of callose, and some evidence indicates that callose cylinders in sieve areas are the result of the stimulation that results from the cutting of a segment of stem to be sectioned. As a sieve element approaches the end of its functional life, callose accumulates in large quantities, occluding the pores, and may even completely cover the surfaces of sieve areas. Callose in this state is referred to as **definitive callose** (Fig. 12.4c).

In sieve tube members sieve areas occur on end walls called **sieve plates**. A sieve plate containing a single sieve area is called a **simple sieve plate** (Figs. 12.3a, 12.4a, b) whereas one with several sieve areas

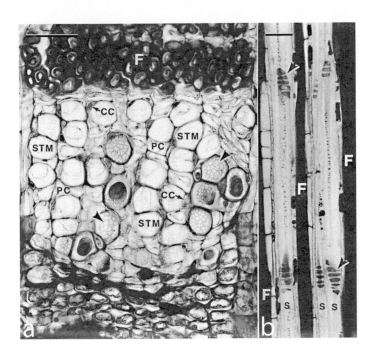

Figure 12.3 (a) Transverse section of the secondary phloem of *Robinia pseudoacacia* (black locust) illustrating sieve tube members (STM) with transversely oriented, simple sieve plates (arrowheads). Note also companion cells (CC), phloem parenchyma cells (PC) and phloem fibers (F). Bar = 50 μm. (b) Radial section of secondary phloem of *Tilia americana* showing sieve tubes consisting of sieve tube members (S) with oblique, compound sieve plates (arrowheads). Also note the small lateral sieve areas. F, fibers. Bar = 50 μm. From Evert (1990b). Used by permission of Springer-Verlag GmbH and Co. KG. © Springer-Verlag Berlin Heidelberg.

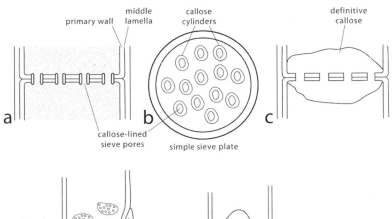

Figure 12.4 (a–c) Diagrams of simple sieve plates, each consisting of a single sieve area. (a) Sectional view of a sieve area. (b) Face view of a sieve area. (c) A sieve area enclosed in definitive callose and, consequently, non-functional. Stippling represents cytoplasmic contents continuous through sieve pores between contiguous sieve tube members. (d, e) Compound sieve plates. To simplify the diagrams callose and cytoplasmic contents are not shown. (d) Sectional view. (e) Face view.

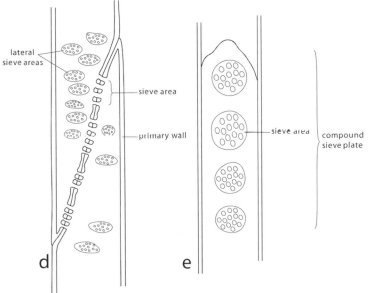

Figure 12.5 Longitudinal section of the primary phloem of *Cucurbita*. Note the companion cells associated with the sieve tube member on the left, and the sectional view of a simple sieve plate between two adjacent sieve tube members on the right. Magnification × 400.

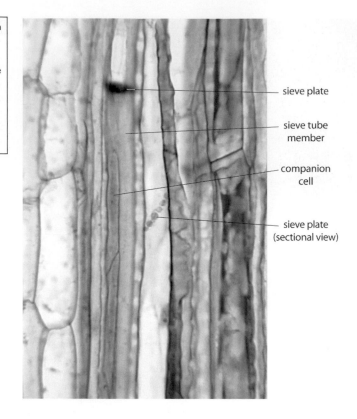

sieve plate

sieve tube member

companion cell

sieve plate (sectional view)

is a **compound sieve plate** (Figs. 12.2.b, 12.3b, 12.4d, e). Simple sieve plates usually occur on transverse end walls, and compound sieve plates on oblique end walls. The sieve tube members of some taxa also contain scattered, somewhat indistinct sieve areas in their lateral walls (Figs. 12.2b, 12.3b, 12.4d). The sieve cells of gymnosperms and lower vascular plants do not have end walls. Sieve areas occur over the entire lateral walls (Fig. 12.2a), but often occur in greater frequency in the walls of the overlapping ends of the cells.

Associated with the sieve tube members in the phloem of angiosperms are **companion cells** (Figs. 12.1d, 12.3a, 12.5), many of which, in the small veins of leaves, are transfer cells that facilitate the loading of photosynthate into sieve tube members (see pp. 230–234 for a more detailed discussion of companion cells and Strasburger cells). Although variable in the number associated with a sieve tube member (Fig. 12.6), each companion cell is derived from the same cambial initial as the sieve tube member with which it is in contact. Companion cells also accompany sieve tube members in the metaphloem of angiosperms, as well as the protophloem of some but not all species.

In conifers, (possibly other gymnosperms; see Behnke, 1990), cells similar in function to the companion cells of angiosperms differ both in origin and morphology. These cells (Fig. 12.7), called **albuminous cells** in the older literature have, in recent years been widely labeled

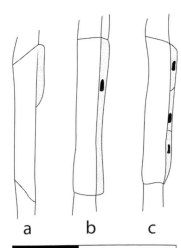

a   b   c

Figure 12.6 Diagrams of sieve tube members (details of sieve plates omitted) with associated companion cells (stippled).

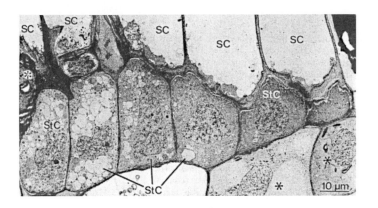

Figure 12.7 Marginal ray cells of *Abies* that have developed as Strasburger cells (StC) in contact with sieve cells (SC). Note the lobed nuclei in several Strasburger cells. Asterisks designate starch-containing cells. From Schulz (1990). Used by permission of Springer-Verlag GmbH and Co. KG. © Springer-Verlag Berlin Heidelberg.

Strasburger cells after the German botanist who first described them. They are derived largely from marginal ray cell initials as well as from some short fusiform initials (Srivastava, 1963) and generally are shorter than companion cells.

Axial parenchyma cells are usually associated with strands of conducting cells (Figs. 12.1a, b, 12.2a) and in angiosperms are often difficult to distinguish from companion cells (Fig. 12.3a). **Phloem fibers** (Figs. 12.1a, b, 12.2b) occur in longitudinal strands and appear, in transverse sections, as tangential and/or radial bands in dicotyledons. Because during cambial activity, the secondary phloem is pushed outward and compressed, only the most recently formed annual increments are functional in transport. In these increments, the living conducting cells and the associated parenchyma cells are protected from compression by the surrounding masses of phloem fibers. In increments formed earlier, however, the forces resulting from the production of secondary xylem are so great that the sieve tube members, companion cells, and phloem parenchyma cells become severely compressed and non-functional in dicotyledons. However, in some conifers (e.g., *Abies* and *Pinus*) whereas the sieve cells become non-functional, the axial parenchyma cells (Figs. 12.1b, 12.2a) resist the forces of compression, becoming repositories of phenolic compounds which provide a defense against invading insects and pathogens (see Franceschi *et al.*, 1998). This compression of the phloem explains, in part, why the annual increments (growth layers) of secondary phloem are so much less conspicuous than the growth layers of secondary xylem. Equally important is the fact that the cambium produces fewer phloem cells than xylem cells.

The nature and development of the cell wall of sieve elements

In both angiosperms and gymnosperms, with the exception of some members of the Pinaceae, the sieve elements have only primary walls although the walls may be lamellate. Consisting of cellulose and pectic compounds, the sieve element wall is of variable thickness, often

Figure 12.8 (a–f) Diagrams of stages in the development of sieve pores in a sieve plate as seen in sectional view. See the text for descriptions. Based on photographs and text descriptions in Esau *et al.* (1962) and Esau and Thorsch (1985).

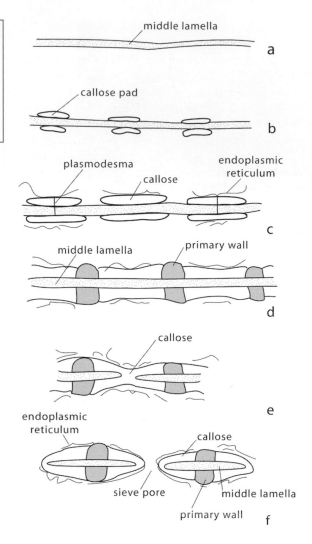

much thicker than that of associated parenchyma cells. Because of their glistening, pearly appearance in unkilled tissue, the term **nacreous wall** has been applied to walls of sieve elements.

The development of sieve areas in the walls of superposed sieve tube members has been studied extensively. For example, following mitosis of a sieve element mother cell in *Cucurbita maxima* (Esau *et al.*, 1962; Esau and Cheadle, 1965; Esau and Thorsch, 1985), the protoplasts of the incipient sieve tube members are initially separated only by the middle lamella (Fig. 12.8a). On either side of the middle lamella in the region that will develop into a sieve area, pairs of **callose platelets** (Fig. 12.8b) are synthesized synchronously. Endoplasmic reticulum, associated with the callose platelets on either side of a developing pore, is connected by the desmotubule of a single plasmodesma (Figs. 12.8c, 12.9a, b). As development of the sieve area continues, the callose platelets increase in area and thickness as primary wall material is synthesized between them (Figs. 12.8d, 12.9a). Then each pair begins to decrease in thickness centrally, presumably the result of hydrolysis of the callose under influence of the plasmodesma, and the middle lamella also begin to disappear around the plasmodesma (Fig. 12.8e). Upon completion of hydrolysis a cylindrical **pore** results (Figs. 12.8f, 12.9d). The pore may be lined with callose or, if hydrolysis has been complete, no callose will remain (Esau and Thorsch, 1985). The plasmalemmas of the two contiguous sieve elements are continuous through the pore (Figs. 12.9d, 12.10). ER cisternae which may traverse the pores are located parietally, and any P-protein in the pores is either parietally located or in a filamentous network (Figs. 12.9d, 12.10), thus the pores are unoccluded (Evert, 1990a; see also Ehlers *et al.*, 2000). The development of sieve area pores in gymnosperms is similar to that in angiosperms (see later section for details of differences), but no callose platelets have been observed in association with the developing pores. Furthermore, sieve pores in the mature sieve areas may or may not be lined with callose (see Schulz, 1990). Because it is known that callose synthesis is stimulated by injury, many researchers have suspected that the association of callose with the developing sieve pores in angiosperms is not a normal feature in the living plant, but an artifact resulting from the sectioning of the material to be studied. To test this hypothesis Walsh and Melaragno (1976) immersed an entire specimen of the diminutive, aquatic plant *Lemna minor* in the chemical fixative before sectioning. They found no callose deposition on the developing sieve areas in this specimen. In a similar specimen sectioned prior to fixation, callose was found to be associated with the sieve areas, indicating that callose formation during sieve area development in *Lemna* is an artifact. In a recent study (Ehlers *et al.*, 2000), however, in which the exposed, presumably undamaged midribs of leaves of *Vicia faba* (vetch) were fixed while attached to the living plant, callose was deposited on the developing sieve areas. Similarly, in *Lycopersicon esculentum* (tomato) vascular bundles were exposed in the stem and fixed on the living plant. Callose was also observed on the sieve areas in this plant (Fig. 12.9d). These results provide some support for the viewpoint of Esau and

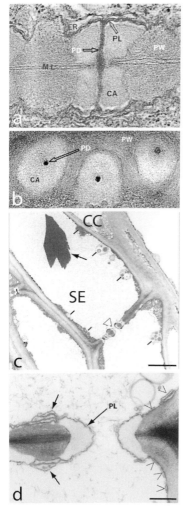

Figure 12.9 (a, b) Views of developing sieve areas in *Gossypium hirsutum* (cotton). (a) Sectional view of a developing sieve pore, showing the single plasmodesma (PD) connecting the protoplasts of the superposed sieve tube members. Note that the plasmodesma is enclosed by callose (CA). ER, endoplasmic reticulum; ML, middle lamella; PL, plasmalemma; PW, primary wall. (b) Face view of developing sieve pores. CA, callose; PD, plasmodesma; PW, primary wall. Magnification (a, b) × 32 250. (c) Part of a sieve tube member of *Vicia faba* (broad bean) illustrating a simple sieve plate in sectional view with open pores (only one shown) between superposed sieve tube members (SE). Note the peripheral position of cytoplasmic organelles. Bar = 3 μm. (d) Sectional view of an open sieve pore, lined with callose, in a mature sieve area of *Lycopersicon esculentum* (tomato). Note that the plasmalemma (PL) is continuous between the two protoplasts, covering the surface of the callose. The walls of the sieve area are covered by stacked ER cisternae (arrows). Endoplasmic reticulum also covers regions of the lateral walls of the sieve tube members (arrowheads). Bar = 0.4 μm. (a, b) From Esau and Thorsch (1985). Used by permission of the Botanical Society of America. (c, d) From Ehlers *et al.* (2000). Used by permission of Springer-Verlag Wien.

Figure 12.10 Face view of part of a simple sieve plate of *Cucurbita maxima*. The open sieve pores are lined with callose (C) which, in turn, is lined with the plasmalemma covered in some regions by endoplasmic reticulum (ER) and P-protein (unlabeled arrows). Magnification × 16 760. From Evert (1984). Used by permission of copyright-holder Ray F. Evert.

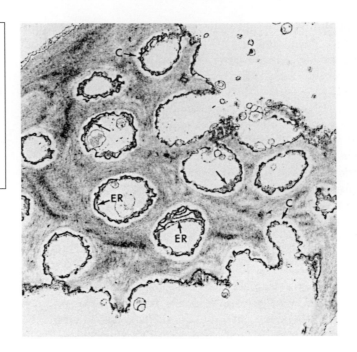

Thorsch (1985) that (the results of Walsh and Melaragno notwith-standing) "the phenomenon [production of callose during sieve area development] may be a unique feature of sieve element development in higher plants."

Role of the cytoskeleton in wall development

In the past, phloem development has been studied primarily utilizing electron microscopy. Although techniques of molecular biology such as immunolocalization of the cytoskeleton have been utilized extensively in studies of secondary xylem development, such approaches have only recently been applied to the problem of differentiation of phloem cells. Chaffey *et al.* (2000) studied the role of microtubules in the differentiation of phloem cells in the roots of *Aesculus hippocastanum*. They observed that microtubules are transversely oriented in the cambial initials and become arranged in a steep helix in all types of phloem cell derivatives as differentiation progresses. They believe that the change in orientation is correlated with an increase in synthesis and deposition of wall material. As wall thickening occurred they observed a parallel orientation of the cortical microtubules and the putative microfibrils in the developing wall, supporting the view of others that there is a functional relationship between microtubules and the synthesis of new microfibrils (see, e.g., Giddings and Staehelin, 1991; Hable *et al.*, 1998). Chaffey *et al.* (2000) observed, further, that as the walls of the developing phloem cells increased in thickness, they became laminate, but they were unable to detect any significant differences between them and the primary walls of the cambial cells from which they were derived (see also Evert, 1990b). In some phloem parenchyma cells Chaffey *et al.* (2000) observed microtubules and cellulose microfibrils arranged in rings around developing primary pit fields, a condition similar to the arrangement of these structures associated with developing bordered pits in tracheary elements described by many workers (see Chapter 11). They suggest, however, that unlike the possible function of microtubules in tracheary elements in which wall thickening occurs during the development of pit borders, the rings of microtubules in phloem parenchyma cells enclose domains in which wall thickening is restricted in the region of the developing primary pit fields. As in late stages of the development of vessel members, the peripheral microtubules in developing sieve tube members lose their helical orientation, becoming more or less transverse as the cells increase in diameter. This indicates that the microtubules are probably attached to the plasmalemma, but it is not clear "whether an increase in cell circumference is, in part, driven by a change in CMT [peripheral microtubule] arrangement, or whether CMTs are dragged to a more shallow-pitched helix as a consequence of such an increase" (Chaffey *et al.*, 2000). Although it has been assumed that microtubules disappear prior to maturity of sieve tube members (see Evert, 1990b), they were observed in possibly mature and functional sieve tube members in the secondary phloem of *Aesculus* roots. However, no evidence of any role for microtubules in sieve plate formation was found which might be related to the fact that callose rather than cellulose is synthesized during the development of sieve area pores (Chaffey *et al.*, 2000).

The nature and development of the protoplast of sieve elements

Following its formation by division of a sieve element mother cell, the protoplast of an incipient sieve element is essentially identical to that of any other immature living cell. It contains a nucleus, rough endoplasmic reticulum dispersed throughout the protoplast, mitochondria, Golgi bodies, microtubules, microfilaments, and many small vacuoles, and is enclosed by a plasmalemma. During development profound changes occur in the protoplast. With increase in size of the cell, small vacuoles fuse forming a conspicuous central vacuole, enclosed by a single vacuolar membrane, the **tonoplast**. In angiosperms, highly chromatic P-protein bodies form, and small plastids that contain starch and/or protein granules appear (Fig. 12.9c). Prior to functional maturity, the nucleus disintegrates. By the time of nuclear disintegration, Golgi bodies, microtubules, and microfilaments, associated with cell wall formation, will also usually have disappeared. However, plastids and mitochondria, peripherally located, persist and with endoplasmic reticulum which has become smooth by having lost its ribosomes, are the only organelles remaining in the functional sieve element (Fig. 12.9c). Late in development the tonoplast disintegrates, and P-protein bodies disaggregate, with P-protein becoming dispersed throughout the periphery of the protoplast.

Figure 12.11 Parietal cytoplasmic organelles of sieve tube members attached to each other and to the plasmalemma by "clamp-like" structures (at arrows). (a) Attachment of the outer membrane of a plastid (P), a mitochondrion (M), and cisternae of endoplasmic reticulum (ER) to the plasmalemma of a sieve tube member of *Lycopersicon esculentum*. A region of endoplasmic reticulum (between arrows) is also attached to the plastid. (b) A plastid of *Vicia faba* attached to the plasmalemma. Bars (a, b) = 100 nm. From Ehlers *et al.* (2000). Used by permission of Springer-Verlag Wien.

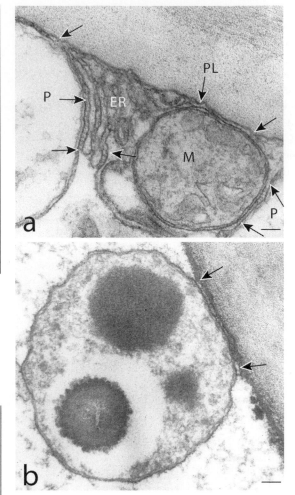

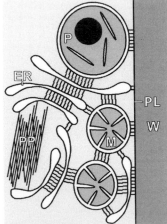

Figure 12.12 Diagram illustrating the attachment of cytoplasmic organelles in a sieve element to each other and to the plasmalemma. ER, endoplasmic reticulum; M, mitochondrion; P, plastid; PL, plasmalemma; PP, P-protein; W, cell wall. From Ehlers *et al.* (2000). Used by permission of Springer-Verlag Wien.

Recently, Ehlers *et al.* (2000) have provided evidence indicating that in *Vicia faba* and *Lycopersicon esculentum* the peripherally located sieve element organelles seem to be connected to each other and to the plasmalemma by minute "clamps" that prevent them from being moved along in the assimilate stream (Figs. 12.11, 12.12). P-protein also maintains a parietal position, but is thought to be more loosely attached than the sieve element organelles (Fig. 12.12). Since at this stage of development the sieve elements are open and any P-protein and/or ER in the pores are located parietally in the pore lumina (Fig. 12.10) the sieve elements comprise open passageways through which pressure flow of assimilates can take place with minimal impedance.

Although it is clear that, at functional maturity, sieve tube members and sieve cells are characterized by highly modified, some would say denatured, protoplasts, it has been clearly demonstrated that they possess normal properties of differential permeability and thus can exert an active influence on the movement of assimilates into and out of themselves. It is now well established, also, that the accumulation of P-protein on the surface of sieve plates commonly observed in section is not a feature of the functional cell but, rather, the result of displacement of cell contents upon the release of pressure in the cells when sections are cut. The more normal state of the protoplasts has been determined by osmotically reducing the pressure within the cells prior to preparing tissue for sectioning (see, e.g., Evert *et al.*, 1973) and by the use of other "gentle preparation" methods (Ehlers *et al.*, 2000). Recently, however, Knoblauch *et al.* (2001) have concluded that in the Fabaceae (= Leguminosae), at least, the dispersal of P-protein and its accumulation on the sieve plates may be a mechanism to control sieve tube conductivity. As sieve elements approach the end of their functional lives, callose is synthesized in large quantities (definitive callose) and deposited on the surface of sieve areas, sometimes completely covering them (in angiosperms often covering entire sieve plates), effectively restricting further transport in the system (Fig. 12.4c). Just prior to death of the protoplasts, the callose is hydrolyzed, and the open pores in the sieve areas become conspicuous.

Nature and function of P-protein

P-proteins (Figs. 12.9c, 12.12, 12.13) occur in both dicotyledons and monocotyledons but are absent from gymnosperms and lower vascular plants (see, Cronshaw and Sabnis, 1990, Evert, 1990b). As observed with the electron microscope, P-protein, called slime in the older literature, occurs in several forms (Fig. 12.13): tubular, granular, filamentous, and crystalline. Aggregation of P-protein into ellipsoidal or spheroidal **P-protein bodies** (Fig. 12.13a) occurs early in the differentiation of sieve cells. These bodies increase in size during differentiation of the cell and may fuse. Polyribosomes are associated with P-protein filaments or tubules during aggregation and are thought to have a functional role in the process. Conspicuous crystals of P-protein, commonly associated with tubular, filamentous or granular P-protein, are characteristic of many members of the Fabaceae. The small, spheroidal bodies, previously thought to be extruded nucleoli, are now considered to be a type of P-protein body.

At about the time that the nucleus begins its disintegration, components of the P-protein bodies begin to separate and to disperse within the protoplast (Fig. 12.13b). Some researchers conclude that the individual filaments or tubules form a peripheral network of P-protein whereas evidence provided by others indicates that the network extends throughout the protoplast. Cronshaw (1975) suggested that P-protein, no matter what the distribution, is relatively stationary in the mature sieve element protoplast and does not move in the assimilate stream. This viewpoint is supported by the work of Ehlers *et al.* (2000), described above, who have demonstrated that peripherally located P-protein, plastids, ER, and mitochondria are attached to

Figure 12.13 P-protein in differentiating sieve tube members. (a) A large P-protein body consisting of filamentous P-protein and a smaller body comprised of tubular P-protein in *Cucurbita maxima*. Magnification × 20 300. (b) Tubular and striated filaments in a dispersing P-protein body in *Nicotiana tabacum*. Magnification × 32 154. From Cronshaw and Sabnis (1990). Used by permission of Springer-Verlag GmbH and Co. KG. © Springer-Verlag Berlin Heidelberg.

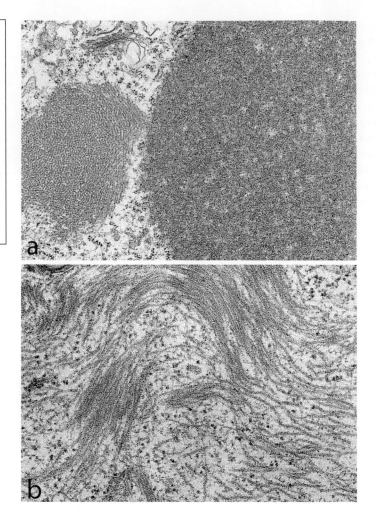

each other and to the plasmalemma by minute "clamps" (Figs. 12.11, 12.12). They emphasize further, however, that the parietal P-proteins are probably loosely attached at the cell periphery and, in the event of injury, would be quickly released in order to plug the sieve pores, thus preventing the release of assimilates. Although the function of P-protein in sieve elements is not fully understood, it has been long observed that upon the release of internal pressure, the P-protein moves suddenly in the direction of the pressure release to the ends of the cells and plugs the sieve pores. It has been shown, recently, that P-protein crystalloids in legumes can undergo rapid, reversible changes from the globular to the dispersed state in which they occlude the sieve pores (Knoblauch *et al.*, 2001). Experimentally, dispersal can be triggered by leakage of the plasmalemma caused by mechanical injury, by substances such as Ca^{2+} that increase the permeability of the plasmalemma, or by abrupt turgor changes. Reversion to the globular state can be induced by chelators, indicating that the P-protein crystalloids of legumes represent a unique class of proteinaceous structures by which conduction through sieve tubes can be controlled (Knoblauch *et al.*, 2001).

Distinctive features of the phloem of gymnosperms

Since angiosperms are believed to have diverged from gymnosperms during their evolution, it is perhaps not surprising that the sieve elements in these two groups of seed plants are similar, especially in the differentiation of the sieve element protoplast and even in the nature of the mature sieve element. One distinctive feature of angiosperms that is lacking in gymnosperms, however, is the presence of P-protein. This suggests that this substance first evolved in the sieve tube members of primitive angiosperms or the sieve elements of their immediate, extinct ancestors. It is, in fact, interesting to note that P-protein is present in primitive extant angiosperms such as *Austrobaileya*, *Degeneria*, *Drimys*, *Liriodendron*, *Magnolia*, and *Trochodendron* (see Evert, 1990b).

As noted above, the sieve cells of gymnosperms are relatively very long and have overlapping ends. Sieve areas (Fig. 12.2a) occur on the lateral walls but in greater frequency near the ends of the cells. Sieve pores tend to be smaller than those of angiosperms, but more phloem in relation to xylem is produced in conifers (possibly in other gymnosperms) than in angiosperms. This may be a compensation for the smaller sieve pores which result in a slower translocation of metabolites through overlapping sieve cells than through sieve tubes of comparable transverse area (Romberger *et al.*, 1993). In the Pinaceae, unlike other gymnosperms, the sieve cells have thick secondary walls.

In conifers as in angiosperms, endoplasmic reticulum is associated with sieve areas during their differentiation (Schulz, 1992). Tubules of ER and/or plasmodesmata mark the site of future pores, but unlike sieve area differentiation in angiosperms, no callose platelets are associated with pore formation. Pores begin their differentiation by the formation opposite each other of pore canals in contiguous cells. As the differentiating canals extend toward the middle lamella, a central cavity develops with which they merge. Upon completion of development, endoplasmic reticulum extends from one cell to the adjacent cell through the pores, connecting the protoplasts (Schulz, 1992).

As in the sieve tube members of angiosperms, the only organelles that persist in the functional sieve cells of gymnosperms are endoplasmic reticulum, mitochondria, and plastids. The plastids of all conifers contain starch grains, but plastids in most members of the Pinaceae contain, in addition, protein crystals and protein filaments. The only known exceptions are *Tsuga canadensis* (hemlock) and *Larix decidua* (larch) which contain only protein filaments.

A distinctive feature of many conifers is the presence of large, axial parenchyma cells in the secondary phloem (Figs. 12.1a, b, 12.2a).

These cells remain alive in the non-conducting part of the secondary phloem until they are incorporated into the bark (Esau, 1977). They are characterized by a dark-staining substance, recently identified in *Picea abies* (Norway spruce) as composed of phenolic compounds, and demonstrated to play an important role in defending the plant against invasive organisms such as fungi (Franceschi *et al.*, 1998). Another distinctive feature of gymnosperms is the presence of Strasburger cells (Fig. 12.7) which in these groups serve the same function as companion cells in angiosperms.

The nature and function of companion cells and Strasburger cells

Within a plant, the phloem accumulates photosynthate from the mesophyll of leaves, transports it throughout the plant, and releases it at sites where it is utilized in the growth and development of the plant. These functions are implied by the terms "source," "transport," and "sink" commonly used in the literature. Recently, Van Bel (1996) has proposed the terms **collection phloem, transport phloem**, and **release phloem** to indicate more clearly these important functions of the phloem. Each function is intimately related to the structure and activity of **companion cells** associated with the sieve elements. For example, companion cells in the collection phloem and transport phloem are relatively large in comparison with the sieve tube members (Fig. 12.14). By contrast, they are small or entirely lacking in the release phloem (Van Bel, 1996). The structure of companion cells is also directly related to the mode of **phloem-loading**, that is, the method of transport from the companion cell into the sieve tube member. If symplastic, the companion cell wall will be traversed by numerous plasmodesmata (Fig. 12.14a, b) (see Turgeon, 2000). On the other hand, if phloem-loading is apoplastic, the companion cell will have the structure of a transfer cell.

Companion cells are derived from the same mother cells as the sieve tube members with which they are in contact. In contrast to the mature sieve tube member, the companion cell is characterized by a dense protoplast (Figs. 12.14, 12.5) containing a prominent nucleus, many mitochondria, extensive rough ER and abundant ribosomes, and plastids which in a few species have been observed to contain starch grains. The density of the protoplast increases throughout the differentiation of the cell which also may become vacuolated.

Companion cells may be intimately attached to contiguous sieve tube members by symplastic connections, forming **sieve tube–companion cell complexes** (see, e.g., Van Bel and Kempers, 1990). In collection phloem companion cells function in the veins of leaves in the transfer of assimilates from photosynthetic tissues and bundle sheaths into the sieve tubes. Abundant, often branched, plasmodesmata extend through the companion cell wall and become connected

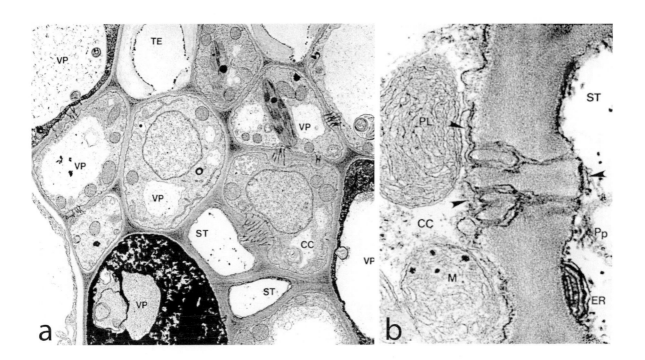

to tubules of endoplasmic reticulum in the pores of sieve areas in the wall of the sieve tube member (Fig. 12.14b) (Fisher, 1986). These symplastic connections have been designated **plasmodesmata–pore connections** by Fisher (1986; see also Van Bel and Kempers, 1997). Because of their intimate association with sieve tube members, the companion cells also play an important role in maintaining the viability and long-distance translocation system of the enucleate sieve tube members by providing them with proteins, including informational (signaling) molecules, and ribonuclear protein complexes as well as ATP (see, e.g., Schobert *et al.*, 2000; Ruiz-Medrano *et al.*, 2001).

Vascular parenchyma cells (parenchyma cells in the primary phloem) (Figs. 12.14a, 12.15) also contribute to the symplastic translocation of assimilates into sieve tube members by functioning either as intermediate providers of photosynthate to the companion cells (Fisher, 1986) or by transferring photosynthate directly into contiguous sieve tube members (see Robinson-Beers and Evert, 1991; Van Bel *et al.*, 1992; Turgeon *et al.*, 1993; Haritatos *et al.*, 2000).

Companion cells, especially those in the small veins of leaves that function as transfer cells, facilitate the apoplastic movement (through the walls) of photosynthate which then enters the sieve tube members by a process of active transport. Companion cells that function as transfer cells are characterized by extensive wall ingrowths which vastly increase the surface area of the plasmalemma. Developmentally, this formation of wall ingrowths occurs at about the same time as the leaf ceases utilization of assimilates in its own development and becomes a source of photosynthate.

Figure 12.14 (a) A transverse section of part of a minor vein in the leaf of *Populus deltoides* (poplar). Note the large companion cell (CC) associated with smaller sieve tube members (ST); also the plasmodesmatal connections between a vascular parenchyma cell (VP) and the companion cell and between the companion cell and a sieve tube member. TE, tracheary element. Magnification × 42 330. (b) Sectional view of a plasmodesmata-pore connection between a companion cell and a sieve tube member in *P. deltoides*. CC, companion cell; ER, endoplasmic reticulum (stacked cisternae); M, mitochondrion; PL, plastid; Pp, P-protein; ST, sieve tube member. Arrowheads indicate fragments of parietal endoplasmic reticulum. Magnification × 50 300. From Russin and Evert (1985). Used by permission of the Botanical Society of America.

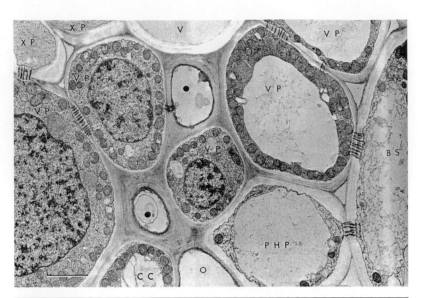

Figure 12.15 Transverse section of part of a vascular bundle in a sugar cane leaf containing two thick-walled sieve tubes (solid dots) and one thin-walled sieve tube (small circle) with associated parenchyma cells. Note the clusters of plasmodesmata connecting adjacent vascular parenchyma cells (VP), a bundle sheath cell (BS) and a vascular parenchyma cell, a bundle sheath cell and a phloem parenchyma cell (PHP), a xylem parenchyma cell (XP) and a vascular parenchyma cell, and a vascular parenchyma cell and a thick-walled sieve tube. CC, companion cell; V, vessel. Bar = 2 μm. From Robinson-Beers and Evert (1991). Used by permission of Springer-Verlag GmbH and Co. KG. © Springer-Verlag Berlin Heidelberg.

The phloem of some, perhaps most, grasses contains sieve tube–companion cell complexes as well as **thick-walled sieve tubes** which lack companion cells (Eleftheriou, 1990; Robinson-Beers and Evert, 1991; Botha, 1992; Evert *et al.*, 1996) (Fig. 12.15). Recent evidence indicates that phloem loading in grasses can follow several different pathways. For example, in sugar cane photosynthate is transferred from the mesophyll through the bundle sheath cells by way of plasmodesmata. In small and intermediate-sized veins, loading of assimilates in the thick-walled sieve tubes seems also to be largely symplastic since there are numerous plasmodesmata between the bundle sheath cells, vascular parenchyma cells, and the thick-walled sieve tube members (Fig. 12.15). However, in the sieve tube–companion cell complexes which are largely isolated symplastically (i.e., they are connected to surrounding parenchyma cells by very few or no plasmodesmata), sieve tube loading is probably apoplastic (Robinson-Beers and Evert, 1991). In contrast, Botha (1992) concluded that in two South African grasses, assimilate loading of thick-walled sieve tubes is predominantly apoplastic, but in sieve tube–companion cell complexes possibly symplastic. On the other hand, in the *Hordeum* (barley) leaf in which both sieve tube–companion cell complexes and thick-walled sieve tubes are symplastically isolated from surrounding parenchyma tissue Evert *et al.* (1996) concluded that both phloem loading and unloading were apoplastic.

It is now clear that sieve tube–companion cell complexes can be symplastically isolated from surrounding parenchyma cells by virtue of the absence or restriction of plasmodesmatal connections between them (Fisher, 1986; Van Bel and Kempers, 1990; Wimmers and Turgeon, 1991; Botha, 1992; Botha and Van Bel, 1992; Van Bel and Van Rijen, 1994; Van Bel, 1996; Botha *et al.*, 2000). The degree of isolation can vary, depending on the frequency of plasmodesmatal connections, from incomplete to almost total. Autonomy of sieve element–companion cell complexes is especially important in the transport phloem in stems where many plasmodesmata between the complexes and adjacent parenchyma cells appear to be closed (Kempers *et al.*, 1998). This presumably prevents leakage from the sieve tube–companion cell complexes into surrounding tissue, thus maintaining solute concentration sufficient to ensure pressure flow through the system and ensuring nutrition to terminal and axial sinks along the stem such as the cambial zone (Van Bel, 1996; Van Bel *et al.*, 2002). In the release phloem, where there may be no companion cells, there is a symplastic transfer of assimilates directly from the sieve tubes into the adjacent parenchyma (Van Bel *et al.*, 2002).

As in grasses and some other monocotyledons, diverse pathways characterize photosynthate transport in dicotyledons. A method of presenting graphically the differences in phloem loading involves the use of plasmodesmograms (Botha and Van Bel, 1992), diagrams that indicate the frequency of symplasmic connections between cells in phloem-loading pathways (Fig. 12.16). Because it is difficult to determine accurately the area of interface contact and the number of plasmodesmata between different types of cells, the accuracy of the assessment of plasmodesmatal frequency is fraught with uncertainty. Nevertheless, in such diagrams one can visualize the pathway of photosynthate movement from mesophyll cells to the sieve tube members, and predict the mode of phloem-loading, whether symplastic or apoplastic (Botha and Van Bel, 1992). A low frequency of plasmodesmata between cells that contact sieve tube members will indicate an apoplastic mode of phloem-loading whereas a high frequency will indicate a symplastic mode.

An interesting phylogenetic analysis by Turgeon *et al.* (2001) indicates that extensive connection by plasmodesmata between minor vein phloem and surrounding cells is a primitive character whereas limited connectivity is derived in angiosperms. This may explain the highly reduced frequency of plasmodesmata in minor vein phloem in crop plants.

Although of similar function, the Strasburger cells of gymnosperms (Fig. 12.7) differ in origin and structure from the companion cells of angiosperms. Most are highly specialized marginal ray cells in contact with sieve cells. Like the protoplast of the companion cells, that of Strasburger cells has very dense protoplasm containing abundant ribosomes, extensive rough ER, numerous mitochondria,

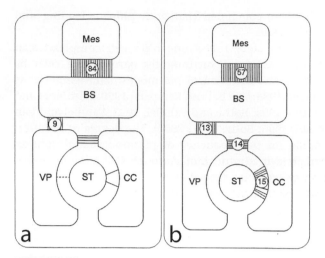

Figure 12.16 Plasmodesmograms illustrating the presumed pathways of photosynthate transport in (a) *Amaranthus retroflexus* and (b) *Cananga odorata*. Numbers in circles indicate percent plasmodesmatal frequency. Solid lines indicate plasmodesmatal frequency of greater than 1%, whereas dotted lines indicate plasmodesmatal frequency of less than 1%. BS, bundle sheath; CC, companion cell; Mes, mesophyll cell; ST, sieve tube member; VP, vascular parenchyma cell. From Botha and Van Bel (1992). Used by permission of Springer-Verlag GmbH and Co. KG. © Springer-Verlag Berlin Heidelberg.

plastids which may contain starch grains, and a prominent nucleus which may be lobed; also a large vacuole may develop. The protoplasts of Strasburger cells and adjacent sieve cells, like those of companion cells and sieve tube members, are connected through sieve area pores in the wall of the sieve cell and extensive plasmodesmata in the contiguous wall of the Strasburger cell. The Strasburger cells are also connected to ray parenchyma cells by numerous plasmodesmata, presumably facilitating the translocation of solutes from the rays into the sieve cells by way of the Strasburger cells.

The mechanism of transport in the phloem

According to the widely accepted **pressure flow hypothesis**, proposed by the German botanist Ernst Münch in 1927, the movement of solutes within the sieve tubes is essentially passive, the result of hydrostatic pressure developed osmotically along gradients established by the presence of high concentrations of sugar near the sources of photosynthate and lower concentrations in the vicinity of sinks, i.e., areas of use or storage. Photosynthate is produced largely in the leaves, and transported predominantly downward to sites of utilization and/or storage in the stem and roots, but some photosynthate and growth hormones are transported to regions of active primary growth in buds and the tips of stems distal to the sites of photosynthate production.

This hypothesis envisages an essentially open system whereby photosynthate is moved from cell to cell through the sieve pores. Because of the presence of large quantities of P-protein on the sieve plates as seen by electron microscopy, the concept of a passive flow through the sieve area pores has been severely criticized. As noted above, however, recent evidence utilizing techniques that reduce the deleterious effect of cutting on the protoplast has shown that the slime plugs are artifacts resulting from a sudden change in pressure within the system. Furthermore, visual evidence of mass flow in intact plants has been reported by Van Bel and Knoblauch (2000) utilizing confocal laser scanning microscopy. One might assume that if pressure flow were the only, or primary, means whereby photosynthate was transported through sieve tubes there would be fewer, much larger sieve pores that were free of any cell components that might tend to restrict movement of substances through them. In fact, the velocity of transport through sieve areas with many very small pores is quite efficient, reaching as high as 100 cm per hour (Romberger et al., 1993). This has led some workers to hypothesize that movement of photosynthate through the phloem might result from the active process of **electroosmosis** (see Spanner, 1974). It is known that passive flow of a solution through many small openings is less efficient than through a single large one of identical area whereas electroosmotically induced flow is more efficient through many small openings than through a single large one of comparable area. This suggests that electroosmosis, in which there is an electric potential across each pore, might be the motive force in photosynthate transport and that the electric potential powering electroosmosis might be maintained by ionic pumps in the plasmalemma lining the pores (Spanner, 1974). Romberger et al. (1993) note that pore size, on average, is smaller than would be required for the most efficient pressure flow transport, but larger than required for the most efficient electroosmotic flow. Consequently, they suggest that pressure flow and electroosmosis probably function jointly in the transport of photosynthate in the phloem.

REFERENCES

Behnke, H. D. 1990. Cycads and gnetophytes. In H. D. Behnke and R. D. Sjölund, eds., *Sieve Elements: Comparative Structure, Induction and Development*. Berlin: Springer-Verlag, pp. 89–101.

Botha, C. E. J. 1992. Plasmodesmatal distribution, structure and frequency in relation to assimilation in C_3 and C_4 grasses in southern Africa. *Planta* **187**: 348–358.

Botha, C. E. J. and A. J. E. Van Bel. 1992. Quantification of symplastic continuity as visualized by plasmodesmograms: diagnostic value for phloem-loading pathways. *Planta* **187**: 359–366.

Botha, C. E. J., Cross, R. H. M., Van Bel, A. J. E., and C. I. Peter. 2000. Phloem loading in the sucrose-export-defective (SXD-1) mutant maize is limited by callose deposition at plasmodesmata in bundle sheath–vascular parenchyma interface. *Protoplasma* **214**: 65–72.

Chaffey, N., Barlow P., and J. Barnett. 2000. Structure–function relationships during secondary phloem development in an angiosperm tree, *Aesculus hippocastanum*: microtubules and cell walls. *Tree Physiol.* **20**: 777–786.

Cronshaw, J. 1975. P-proteins. In S. Aronoff, J. H. Dainty, P. R. Gorham, L. M. Srivastava, and C. A. Swanson, eds., *Phloem Transport*. New York: Plenum Press, pp. 79–115.

Cronshaw, J. and D. B. Sabnis. 1990. Phloem proteins. In H. D. Behnke and R. D. Sjölund, eds., *Sieve Elements: Comparative Structure, Induction and Development*. Berlin: Springer-Verlag, pp. 257–283.

Ehlers, K., Knoblauch, M., and A. J. E. Van Bel. 2000. Ultrastructural features of well-preserved and injured sieve elements: minute clamps keep the phloem transport conduits free for mass flow. *Protoplasma* **214**: 80–92.

Eleftheriou, E. P. 1990. Monocotyledons. In H. D. Behnke and R. D. Sjölund, eds., *Sieve Elements: Comparative Structure, Induction and Development*. Berlin: Springer-Verlag, pp. 139–159.

Esau, K. 1977. *Anatomy of Seed Plants*, 2nd edn. New York: John Wiley and Sons.

Esau, K. and V. I. Cheadle. 1965. Cytologic studies on phloem. *Univ. Calif. Publ. Bot.* **36**: 252–343.

Esau, K. and J. Thorsch. 1985. Sieve plate pores and plasmodesmata, the communication channels of the symplast: ultrastructural aspects and developmental relations. *Am. J. Bot.* **72**: 1641–1653.

Esau, K., Cheadle, V. I., and E. B. Risley. 1962. Development of sieve-plate pores. *Bot. Gaz.* **123**: 233–243.

Evert, R. F. 1984. Comparative structure of phloem. In R. A. White and W. C. Dickinson, eds., *Contemporary Problems in Plant Anatomy*. New York: Academic Press, pp. 145–234.

1990a. Seedless vascular plants. In H.-D. Behnke and R. D. Sjölund, eds., *Sieve Elements*. Berlin: Springer-Verlag, pp. 35–62.

1990b. Dicotyledons. In H.-D. Behnke and R. D. Sjölund, eds., *Sieve Elements*. Berlin: Springer-Verlag, pp. 103–137.

Evert, R. F., Eschrich, W., and S. E. Eichhorn. 1973. P-protein distribution in mature sieve elements of *Cucurbita maxima*. *Planta* **109**: 193–210.

Evert, R. F., Russin, W. A., and C. E. F. Botha. 1996. Distribution and frequency of plasmodesmata in relation to photoassimilate pathways and phloem loading in the barley leaf. *Planta* **198**: 572–579.

Fisher, D. G. 1986. Ultrastructure, plasmodesmatal frequency, and solute concentration in green areas of variegated *Coleus blumei* Benth. leaves. *Planta* **169**: 141–152.

Franceschi, V. R., Krekling, T., Berryman, A. A., and E. Christiansen. 1998. Specialized phloem parenchyma cells in Norway spruce (Pinaceae) bark are an important site of defense reactions. *Am. J. Bot.* **85**: 601–615.

Giddings, T. H., Jr. and L. A. Staehelin. 1991. Microtubule-mediated control of microfibril deposition: a re-examination of the hypothesis. In C. W. Lloyd, ed., *The Cytoskeletal Basis of Plant Growth and Form*. London, Academic Press, pp. 85–99.

Hable, W. E., Bisgrove, S. R., and D. L. Kropf. 1998. To shape a plant cell: the cytoskeleton in plant morphogenesis. *Plant Cell* **10**: 1772–1774.

Kempers, R., Ammerlaan, A., and A. J. E. Van Bel. 1998. Symplasmic construction and ultrastructural features of the sieve element/ companion cell complex in the transport phloem of apoplasmically and symplasmically phloem-loading species. *Plant Physiol.* **116**: 271–278.

Knoblauch, M., Peters, W. S., Ehlers, K., and A. J. E. Van Bel. 2001. Reversible calcium-regulated stopcocks in legume sieve tubes. *Plant Cell* **13**: 1221–1230.

Robinson-Beers, K. and R. F. Evert. 1991. Ultrastructure of and plasmodesmatal frequency in mature leaves of sugarcane. *Planta* **184**: 291–306.

Romberger, J. A., Hejnowicz, Z., and J. B. Hill. 1993. *Plant Structure: Function and Development*. Berlin: Springer-Verlag.

Ruiz-Medrano, R., Xoconostle-Cazares, B., and W. J. Lucas. 2001. The phloem as a conduit for inter-organ communication. *Curr. Opin. Plant Biol.* **4**: 202–209.

Russin, W. A. and R. F. Evert. 1985. Studies on the leaf of *Populus deltoides* (Salicaceae): ultrastructure, plasmodesmatal frequency, and solute concentrations. *Am. J. Bot.* **72**: 1232–1247.

Schobert, C., Gottschalk, M., Kovar, D. R., *et al.* 2000. Characterization of *Ricinus communis* phloem profilin, RcPRO1. *Plant Mol. Biol.* **42**: 719–730.

Schulz, A. 1990. Conifers. In H.-D. Behnke and R. D. Sjölund, eds., *Sieve Elements*. Berlin: Springer-Verlag, pp. 63–88.

   1992. Living sieve cells of conifers as visualized by confocal, laser-scanning fluorescence microscopy. *Protoplasma* **166**: 153–164.

Spanner, D. C. 1974. The electro-osmotic theory. In S. Aronoff *et al.*, eds., *Phloem Transport*, NATO Advanced Study Institute, Series A-4. New York: Plenum Press, pp. 563–584.

Srivastava, L. M. 1963. Secondary phloem in the Pinaceae. *Univ. Calif. Publ. Bot.* **36**: 1–142.

Turgeon, R. 2000. Plasmodesmata and solute exchange in the phloem. *Austral. J. Plant Physiol.* **27**: 521–529.

Turgeon, R., Beebe, D. U., and E. Gowan. 1993. The intermediary cell: minor-vein anatomy and raffinose oligosaccharide synthesis in the Scrophulariaceae. *Planta* **191**: 446–456.

Turgeon, R., Medville, R., and K. C. Nixon. 2001. The evolution of minor vein phloem and phloem loading. *Am. J. Bot.* **88**: 1331–1339.

Van Bel, A. J. E. 1996. Interaction between sieve element and companion cell and the consequences for photoassimilate distribution: two structural hardware frames with associated physiological software packages in dicotyledons? *J. Exp. Bot.* **47**: 1129–1140.

Van Bel, A. J. E. and R. Kempers. 1990. Symplastic isolation of the sieve element–companion cell complex in the phloem of *Ricinus communis* and *Salix alba* stems. *Planta* **183**: 69–76.

   1997. The pore/plasmodesm unit: key element in the interplay between sieve element and companion cell. *Progr. Bot.* **58**: 278–291.

Van Bel, A. J. E. and M. Knoblauch. 2000. Sieve element and companion cell: the story of the comatose patient and the hyperactive nurse. *Austral. J. Plant Physiol.* **27**: 477–487.

Van Bel, A. J. E. and H. V. M. Van Rijen. 1994. Microelectrode-recorded development of the symplasmic autonomy of the sieve element/companion cell complex in the stem phloem of *Lupinus luteus* L. *Planta* **192**: 165–175.

Van Bel, A. J. E., Gamalei, Y. V., Ammerlaan, A., and L. P. M. Bik. 1992. Dissimilar phloem loading in leaves with symplasmic or apoplasmic minor-vein configurations. *Planta* **186**: 518–525.

Van Bel, A. J. E., Ehlers, K., and M. Knoblauch. 2002. Sieve elements caught in the act. *Trends Plant Sci.* **7**: 126–132.

Walsh, M. A. and J. E. Melaragno. 1976. Ultrastructural features of developing sieve elements in *Lemna minor* L.: sieve plate and lateral sieve areas. *Am. J. Bot.* **63**: 1174–1183.

Wimmers, W. E. and R. Turgeon. 1991. Transfer cells and solute uptake in minor veins of *Pisum sativum* leaves. *Planta* **186**: 2–12.

FURTHER READING

Alfieri, F. J. and R. I. Kemp. 1968. the seasonal cycle of phloem development in *Juniperus californica*. *Am. J. Bot.* **70**: 891–896.

Ayre, B. G., Blair, J. E., and R. Turgeon. 2003. Functional and phylogenetic analyses of a conserved regulatory program in the phloem of minor veins. *Plant Physiol.* **133**: 1229–1239.

Ayre, B. G., Keller, F., and R. Turgeon. 2003. Symplastic continuity between companion cells and the translocation stream: long-distance transport is controlled by retention and retrieval mechanisms in the phloem. *Plant Physiol.* **131**: 1518–1528.

Behnke, H. D. 1974. Companion cells and transfer cells. In S. Aronoff *et al.*, eds., *Phloem Transport*. New York: Plenum Press, pp. 153–175.

1995. Sieve-element plastids, phloem proteins, and the evolution of the Ranunculaceae. *Plant Syst. Evol. (Suppl.)* **9**: 25–37.

2002. Sieve-element plastids and evolution of monocotyledons, with emphasis on Melanthiaceae *sensu lato* and Aristolochiaceae–Asaroideae, a putative dicotyledon sister group. *Bot. Rev.* **68**: 524–544.

Behnke, H. D. and R. D. Sjölund (eds.) 1990. *Sieve Elements: Comparative Structure, Induction and Development*. Berlin: Springer-Verlag.

Cronshaw, J. and R. Anderson. Sieve plate pores of *Nicotiana*. *J. Ultrastruct. Res.* **27**: 134–148.

Davis, J. D. and R. F. Evert. 1970. Seasonal cycle of phloem development in woody vines. *Bot. Gaz.* **131**: 128–138.

Deshpande, B. P. 1975. Differentiation of the sieve plate of *Cucurbita*: a further view. *Ann. Bot.* **39**: 1015–1022.

Esau, K. 1965. *Vascular Differentiation in Plants*. New York: Holt, Rinehart and Winston.

1969. *Handbuch der Pflanzenanatomie*, vol. 5, part 2, *The Phloem*. Berlin: Bornträger.

Esau, K. 1973. Comparative structure of companion cells and phloem parenchyma cells in *Mimosa pudica* L. *Ann. Bot.* **37**: 625–632.

Esau, K. and V. I. Cheadle. 1958. Wall thickening in sieve elements. *Proc. Natl Acad. Sci. USA* **44**: 546–553.

1959. Size of pores and their contents in sieve elements of dicotyledons. *Proc. Natl Acad. Sci. USA* **45**: 156–162.

Eschrich, W. 1975. Sealing systems in phloem. *Encyclopedia of Plant Physiology*, 2nd Ser. **1**: 39–56.

Evert, R. F. 1963. Ontogeny and structure of the secondary phloem in *Pyrus malus*. *Am. J. Bot.* **50**: 8–37.

1977. Phloem structure and histochemistry. *Annu. Rev. Plant Physiol.* **28**: 199–222.

Evert, R. F. 1982. Sieve-tube structure in relation to function. *BioScience* **32**: 789–795.

Giaquinta, R. T. 1983. Phloem loading of sucrose. *Annu. Rev. Plant Physiol.* **34**: 347–387.

Gottwald, J. R., Krysan, P. J., Young, J. C., Evert, R. F., and M. R. Sussman. 2000. Genetic evidence for the *in planta* role of phloem-specific plasma membrane sucrose transporters. *Proc. Natl Acad. Sci. USA* **97**: 13979–13984.

Gunning, B. E. S. and J. S. Pate. 1974. Transfer cells. In A. W. Robards, ed., *Dynamic Aspects of Plant Ultrastructure.* London: McGraw-Hill, pp. 441–480.

Haritatos, E., Medville, R., and R. Turgeon. 2000. Minor vein structure and sugar transport in *Arabidopsis thaliana. Planta* **211**: 105–111.

Hoffman-Thoma, G., van Bel, A. J. E. and K. Ehlers. 2001. Ultrastructure of minor-vein phloem and assimilate export in summer and winter leaves of the symplasmically loading evergreens, *Ajuga reptans* L., *Aucuba japonica* Thunb., and *Hedera helix* L. *Planta* **2112**: 231–242.

Knoblauch, M. and A. J. E. Van Bel. 1998. Sieve tubes in action. *Plant Cell* **10**: 35–50.

Pate, J. S. and B. E. S. Gunning. 1969. Vascular transfer cells in angiosperm leaves: a taxonomic and morphological survey. *Protoplasma* **68**: 135–156.

Sjölund, R. D. 1997. The phloem sieve element: a river runs through it. *Plant Cell* **9**: 1137–1146.

Turgeon, R. and E. Gowan. 1992. Sugar synthesis and phloem loading in *Coleus blumei* leaves. *Planta* **187**: 388–394.

Van Bel, A. J. E. and M. Knoblauch. 2000. Sieve element and companion cell: the story of the comatose patient and the hyperactive nurse. *Austral. J. Plant Physiol.* **27**: 477–487.

Weatherley, P. E. 1962. The mechanism of sieve-tube translocation: observation, experiment and theory. *Adv. Sci.* **18**: 571–577.

Chapter 13

Periderm, rhytidome, and the nature of bark

Perspective

Except in the very youngest regions, the stems and roots of woody plants (specifically, gymnosperms and dicotyledons) are covered by bark consisting of the functional secondary phloem and rhytidome, a complex tissue comprised of successively formed periderms, often of overlapping shell-like morphology, between which are enclosed dead cortical and/or phloem tissues. The outer covering of stems of large monocots differs from that of woody dicotyledons and will be discussed later. The outer bark, consisting primarily of rhytidome, is a protective layer which restricts entrance of both insects and microorganisms and also protects the inner living tissues from temperature extremes. It also inhibits water loss through evaporation, but at the same time allows gaseous exchange through specialized regions in the periderm called lenticels. In addition it supplements the secondary xylem in stiffening young stems (Niklas, 1999), thus, contributing to their ability to withstand the bending forces exerted by excessive wind and/or the weight of ice.

Periderm: structure and development

Periderm consists of phellem and phelloderm, both derived from a single-layered secondary meristem, the **phellogen** (Fig. 13.1a, b). Cells of the phellogen are tabular, radially thin, somewhat elongate, and polygonal as viewed tangentially. In many plants the phellogen forms at about the same level in the stem and at about the same time as the vascular cambium. The site of its initiation is highly variable but often is an outer layer of cortical parenchyma one or two layers beneath the epidermis (Fig. 13.1b). In some other plants, however, it may form deep within the cortex, or even in the outer secondary phloem. The phellogen may become a complete cylinder relatively quickly, or it may be initiated in segmental sheets that ultimately connect, forming a cylinder.

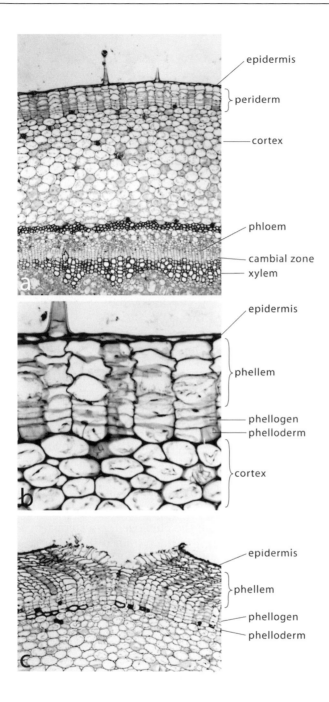

Figure 13.1 (a) Transverse section of a young stem of *Pelargonium* illustrating the periderm and its relationship to other tissue regions. Magnification × 48. (b) Enlargement of the periderm shown in (a). Note the phellogen and the proportion of phellem to phelloderm. Magnification × 113. (c) Older stem of *Pelargonium* in which diametric growth has resulted in splitting of the periderm. Note also that some cells of the phelloderm have differentiated into sclereids. Magnification × 48.

Phellogen initials arise through **dedifferentiation** (i.e., the reversion to a meristematic state) of mature parenchyma cells followed by periclinal cell divisions. Following a division, the smaller of the two daughter cells usually becomes a phellogen initial, whereas the larger differentiates into either a phellem or a phelloderm cell. During its cell divisional activity, the phellogen typically produces larger quantities of phellem cells than phelloderm cells (Fig. 13.1b, c). In many woody plants of temperate regions, the phelloderm may consist of

Figure 13.2 Transverse sections illustrating types of phellem. (a) Thick-walled phellem cells from a stem of *Quercus* sp. Magnification × 291. (b) Thin-walled phellem cells from a stem of *Abies* sp. Note also the several-layered phelloderm. Magnification × 47.

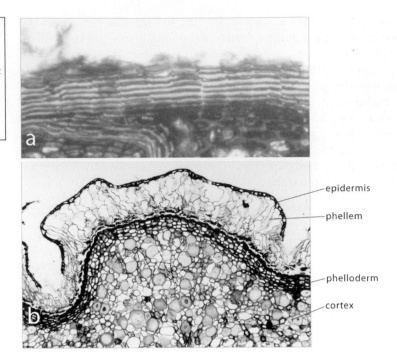

— epidermis

— phellem

— phelloderm

— cortex

only one to several layers of cells (Figs. 13.1b, 13.2) although in some tropical plants thick layers of phelloderm are produced. Since both phellem and phelloderm cells are direct descendants of phellogen initials that divide periclinally, they occur in well-defined radial files (Fig. 13.1), especially in the first-formed periderm(s). With increase in stem diameter and resulting anticlinal divisions of phellogen initials, new radial files, often of lenticular shape, are formed between the earlier formed files.

Phellem, formed to the exterior of the phellogen, is a compact tissue, consisting of cells similar in shape to the initials from which they are derived, non-living at functional maturity, and with no intercellular spaces except in lenticels. The secondary walls of phellem cells are heavily suberized, and thus relatively water impermeable. Consequently, tissues to the exterior of the phellem are destined to die.

Two distinct types of phellem cells are recognized on the basis of their morphology and cell wall structure, thick-walled and thin-walled. A single type may characterize a species, or the two types may occur in alternate layers. This latter condition characterizes some species of *Picea, Tsuga*, and *Abies*. **Thick-walled phellem cells** (Fig. 13.2a) are radially narrower than the thin-walled type and are characterized by a three-layered wall, an outer primary wall that is frequently lignified, a middle secondary wall layer that is usually heavily suberized, and an inner layer, sometimes called a tertiary wall, that often becomes impregnated with waxes. Some evidence suggests that the phellem cells are not completely impermeable to water upon suberization of the secondary wall and that at least some of the plasmodesmata are still functional. Upon impregnation of the inner, tertiary wall with waxes, the cell protoplast dies and the cell becomes fully impermeable. The lumina of thick-walled phellem cells commonly contain dark-staining resins and tannins that were translocated into the protoplasts prior to cell autolysis.

Thin-walled phellem cells (Fig. 13.2b) typically have much greater radial dimensions than the thick-walled cells, usually have a thinner secondary wall, and lack the inner secondary (or tertiary) wall layer. The protoplast is transparent in slide preparations. Phellem of this type is desirable for use in making bottle corks. At functional maturity phellem cells in some species are known to have gas-filled lumina which, with the suberized and wax-impregnated walls, contribute to their impermeability to water. Whereas typical phellem cells have suberized walls, in some species layers of cells lacking suberized walls, called **phelloid cells**, alternate with layers of suberized cells.

Phelloderm, formed to the interior of the phellogen, consists of cells that retain living protoplasts, and that resemble cortical parenchyma cells. Like the latter, phelloderm cells in the first-formed periderm(s) are photosynthetic and contain chloroplasts and starch grains. As the phelloderm ages, some of its component cells may differentiate into sclereids (Fig. 13.1c).

In many non-woody plants, following loss of the epidermis, the periderm becomes the bounding tissue and functions both in restricting the entrance of pathogens and other small organisms as well as in reducing water loss. It serves these same functions in the young stems of woody plants prior to the formation of rhytidome. Following abscission of plant parts such as branches (in some plants such as *Populus*), leaves, and flower parts, periderm usually forms just beneath the abscission site (for more detail see Chapter 16 on the leaf). Wound sites are also underlain by periderm which develops following initiation of a phellogen through dedifferentiation in parenchyma cells below the wound.

Formation of rhytidome

In most woody species in temperate climates, the initial cylindrical periderm persists for only a few years, and is followed by the formation of a succession of **internal periderm** layers. These periderms, like the initial one, may be cylindrical, completely encircling the stem, and may be formed by phellogens that differentiate in the cortex immediately below the initial cylinder of periderm, or even in a layer of phelloderm of the initial periderm. If this is the pattern, a smooth bark will result, common in many tropical trees and some temperate zone trees such as *Fagus* (beech) and *Prunus* (cherry) (Fig. 13.3a). In many trees, however, the internal periderms are lens-shaped or shell-like, partially overlapping each other (Figs. 13.4, 13.5). The internal phellogens from which they develop differentiate from parenchyma of the cortex to the interior of the initial periderm (Fig. 13.5), and ultimately in secondary phloem parenchyma. Whether

Figure 13.3 Lenticels.
(a) Transverse section showing a lenticel in sectional view from a stem of *Prunus serotina*.
(b) Drawing of an early stage in the development of a lenticel.
(c) Sectional view of a lenticel of *Prunus avium*. From Eames and MacDaniels (1925).

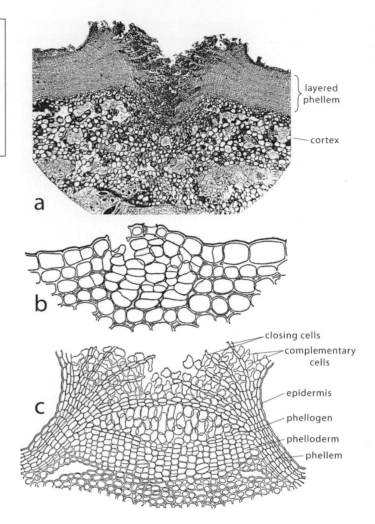

internal periderms are cylindrical or shell-like, they initially enclose regions of cortical parenchyma. Ultimately, as they progress toward the inner tissues, they enclose primary and secondary phloem which, cut off from water and solutes, will die. This complex tissue region of periderms and enclosed non-living tissue is called **rhytidome**, and in most trees comprises the bulk of the bark (Fig. 13.4e).

At the same time that rhytidome is forming the vascular cambium is producing large quantities of secondary vascular tissues that result in an outward, diametric expansion of the stem. This results not only in the compression of the outer functional phloem, but also the splitting and furrowing in the surface layers of the rhytidome (Fig. 13.4c–e). The structure of the rhytidome and the abundance and arrangement of fibers in the enclosed secondary phloem directly influence the surface morphology of the bark, and provide the unique features of particular species such as the depth and direction of furrowing and the type of exfoliation. In some species, for example, *Betula* (birch), non-suberized layers of phelloid cells in the periderm

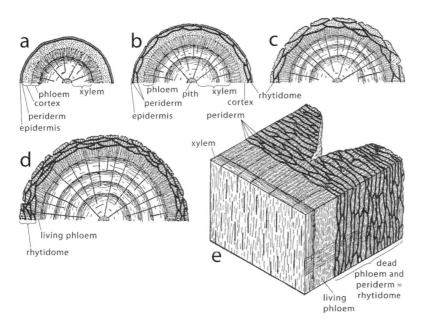

Figure 13.4 Diagrams illustrating the development and structure of rhytidome. See the text for a detailed description. From Eames and MacDaniels (1925).

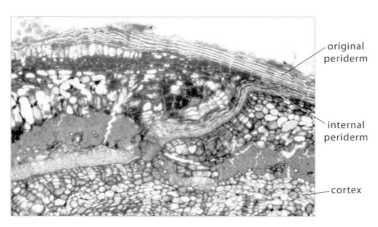

Figure 13.5 Early development of rhytidome in *Quercus* sp. oak as seen in transverse section. Magnification × 99.

function as **excision layers** that result in exfoliation of the bark. In some other species, e.g., *Platanus* (sycamore) specialized excision layers develop below layers of rhytidome. For more detail on the structure of bark, see Chattaway (1953, 1955), Chang (1954), Schneider (1955), Tomlinson (1961), Howard (1971, 1977), Borger and Kozlowski (1972), Patel (1975), Godkin *et al.* (1983), and Patel and Shand (1985).

Lenticels

Specialized regions of the periderm, called lenticels, allow for gaseous exchange between the outside atmosphere and the interior living tissues of the plant. A **lenticel** (Figs. 13.1c, 13.3a, c) is a region consisting of loosely arranged cells, derived from the phellogen. As viewed on the surface of the bark, this structure, consisting of a tissue containing

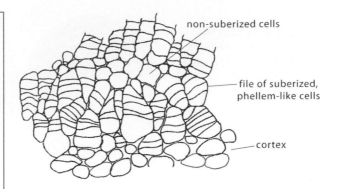

non-suberized cells

file of suberized, phellem-like cells

cortex

Figure 13.6 Drawing of the outer protective tissue of the monocotyledon *Curcuma longa* (Zingiberaceae) consisting of short, radial files of phellem-like cells. Each file is formed by a series of periclinal divisions originating in a cortical parenchyma cell. From Philipp (1923). Used by permission of E. Schweizerbart'sche Verlagsbuchhandlung. http://www.schweizerbart.de.

many large intercellular spaces, is usually lenticular in shape (thus its name) with its long axis parallel to the long axis of the stem. As the stem enlarges through diametric growth, the lenticels expand laterally, and their long axes may become oriented at right angles to the long axis of the stem. The form and distribution of lenticels on the exterior of the bark and in the internal periderms is highly variable among diverse species.

Lenticels usually originate beneath large cauline stomata (Fig. 13.3b). In the region of the developing lenticel, the phellogen functions differently than it does in other areas, in woody plants, producing parenchyma cells of two types, loosely arranged **complementary cells** and **closing cells** in tangential files interspersed among the complementary cells (Fig. 13.3a, c). As mentioned above, lenticels develop in the internal periderms as well as in the initial periderm. As a result, the system of intercellular spaces in the lenticels and the cortical parenchyma and/or phloem in the rhytidome provide conduits for essential gaseous exchange.

The outer protective layer of monocotyledons

Monocotyledons typically do not produce periderm. In the smaller, more herbaceous taxa, the epidermis may persist for the life of the plant, and the walls of outer cortical parenchyma cells may become suberized or thickened and sclerified thus forming an outer protective covering. In a few taxa characterized by secondary growth, however, a more specialized outer protective tissue develops. Within the outer cortex, parenchyma cells become meristematic and through their periclinal divisions as well as the divisions of their daughter cells produce short, radial files of cells (Fig. 13.6) that, like phellem cells, at maturity are non-living and have heavily suberized walls. In very large taxa, successive zones develop in a similar manner to the interior of the initial zone. In some taxa, non-suberized cells become trapped between zones of suberized cells (Philipp, 1923).

REFERENCES

Borger, G. A. and T. T. Kozlowski. 1972. Effects of temperature on first periderm and xylem development in *Fraxinus pensylvanica*, *Robinia pseudoacacia*, and *Ailanthus altissima*. *Canad. J. For. Res.* **2**: 198–205.

Chang, Y.-P. 1954. Bark structure of North American conifers. *U.S. Dept. Agric. Tech. Bull.* **1095**: 1–86.

Chattaway, M. M. 1953. The anatomy of bark. I. The genus *Eucalyptus*. *Austr. J. Bot.* **1**: 402–433.

Chattaway, M. M. 1955. The anatomy of bark. V. Radially elongated cells in the phelloderm of species of *Eucalyptus*. *Austral. J. Bot.* **3**: 39–47.

Eames, A. J. and L. H. MacDaniels. 1925. *An Introduction to Plant Anatomy*. New York: McGraw-Hill.

Godkin, S. E., Grozdits, G. A. and C. T. Keith. 1983. The periderms of three North American conifers. II. Fine structure. *Wood Sci. Tech.* **17**: 13–30.

Howard, E. T. 1971. Bark structure of southern pines. *Wood Sci.* **3**: 134–148.

Howard, E. T. 1977. Bark structure of southern upland oaks. *Wood Fiber* **9**: 172–183.

Niklas, K. J. 1999. The mechanical role of bark. *Am. J. Bot.* **86**: 465–469.

Patel, R. N. 1975. Bark anatomy of radiata pine, Corsican pine, and Douglas fir grown in New Zealand. *New Zeal. J. Bot.* **13**: 149–167.

Patel, R. N. and J. E. Shand. 1985. Bark anatomy of *Nothofagus* species indigenous to New Zealand. *New Zeal. J. Bot.* **23**: 511–532.

Philipp, M. 1923. Über die verkorkten Abschlussgewebe der Monokotylen. *Bibl. Bot.* **92**: 1–27.

Schneider, H. 1955. Ontogeny of lemon tree bark. *Amer. J. Bot.* **42**: 893–905.

Tomlinson, P. B. 1961. Anatomy of the Monocotyledons. II. Palmae. Clarenden Press. Oxford

FURTHER READING

Borchert, R. and J. D. McChesney. 1973. Time course and localization of DNA synthesis during wound healing of potato tuber tissue. *Devel. Biol.* **35**: 293–301.

Esau, K. 1965. *Plant Anatomy*. New York: John Wiley and Sons.
1977. *Anatomy of Seed Plants*, 2nd edn. New York: John Wiley and Sons.

Metcalfe, C. R. and L. Chalk. 1950. *Anatomy of the Dicotyledons*, 2 vols. Oxford, UK: Clarendon Press.

Mullick, D. B. and G. D. Jensen. 1973. New concepts and terminology of coniferous periderms: necrophylactic and exophylactic periderms. *Can. J. Bot.* **51**: 1459–1470.

Romberger, J. A., Hejnowicz, Z., and J. F. Hill. 1993. *Plant Structure: Function and Development*. Berlin: Springer-Verlag.

Roth, I. 1981. Structural patterns of tropical barks. In *Encyclopedia of Plant Anatomy*, 2nd edn. Berlin: Bornträger.

Sifton, H. B. 1945. Air-space tissue in plants. *Bot. Rev.* **11**: 108–143.

Solereder, H. and F. J. Meyer. 1928. *Systematische Anatomie der Monokotyledonen*, vol. 3. Berlin: Bornträger.

Srivastava, L. M. 1964. Anatomy, chemistry, and physiology of bark. *Int. Rev. For. Res.* **1**: 203–277.

Chapter 14

Unusual features of structure and development in stems and roots

Perspective

A large part of this book, thus far, has dealt with the typical condition in stems of gymnosperms and dicotyledons. This chapter will present interesting and important information about stem growth in monocotyledons as well as development and patterns of organization in some plants usually characterized as having "anomalous" structure. Unlike gymnosperms and dicotyledons, monocotyledons, even the largest taxa among the palms, do not produce a typical vascular cambium. Although most are characterized solely by primary growth, some palms, some members of the Liliaceae and Agavaceae, and a few other monocotyledons increase in size by secondary growth. The tissues derived from the secondary meristem are strikingly different from the secondary xylem and phloem of the gymnosperms and other angiosperms.

Primary peripheral thickening meristem

As in other plants, the activity of apical meristems of monocotyledons results primarily in an increase in length of the stems. The diameter of a palm stem does not vary greatly from the base to the most distal leaf-bearing region; thus considerable diametric growth must occur in the internodes just beneath the apical meristem, and this is accomplished by activity of the **primary peripheral thickening meristem**. This meristem is a rather diffuse region located in the periphery of the broad region of the stem immediately below the apical meristem. Its longitudinal extent varies in different species. Periclinal divisions in this meristem result in anticlinal files of cells that comprise cup-shaped regions of new ground tissue nearest the apical meristem and bowl-shaped to plate-shaped regions more basally, which result in the lateral expansion of the stem. Traversing this tissue are numerous provascular strands that will differentiate into axial bundles and leaf traces. The differentiation of cells in these tissues results in internodal expansion and elongation.

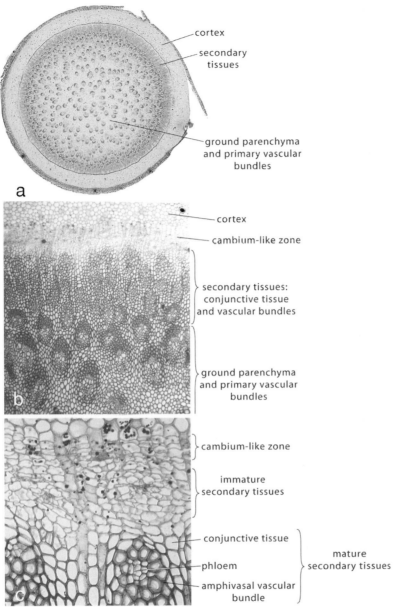

cortex

secondary tissues

ground parenchyma and primary vascular bundles

a

cortex

cambium-like zone

secondary tissues: conjunctive tissue and vascular bundles

ground parenchyma and primary vascular bundles

b

cambium-like zone

immature secondary tissues

conjunctive tissue

phloem

amphivasal vascular bundle

mature secondary tissues

c

Figure 14.1 *Dracaena.* (a) Transverse section of a stem. Magnification × 3.8. (b) Transverse section showing secondary tissue to the exterior of primary tissues. Note also the cambial zone. Magnification × 41. (c) Englargement showing detail of the cambial zone, and immature and mature secondary tissues. Magnification × 165.

(For more detail and illustrations, see Chapter 5 on meristems of the shoot.)

Secondary growth in monocotyledons

In some monocotyledons, especially in some members of the Liliaceae, the primary peripheral thickening meristem extends downward into a narrow cambium-like meristem, to the exterior of the primary vascular bundles. Activity of this meristem results in the formation of an unusual secondary tissue (Fig. 14.1a, b). As in

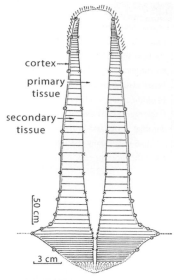

cortex

primary
tissue

secondary
tissue

50 cm

3 cm

Figure 14.2 *Cordyline australis*
(cabbage palm). Diagrammatic
representation of a young stem in
median longitudinal view showing
distribution of primary and
secondary tissues. Note the
obconical form of the primary
body. Secondary tissues are
formed in quantities that result in a
stem of approximately equal
diameter throughout its length.
Plotted from measurements on a
specimen 4 m high. The
longitudinal axis is foreshortened
16 times. The thickness of the
cortex is arbitrary. From
Tomlinson and Esler (1973). Used
by permission of the Royal Society
of New Zealand.

gymnosperms and dicotyledons, secondary tissues in monocotyledons are produced only in regions that have completed their longitudinal growth. Periclinal divisions in the cambium-like meristem result largely in secondary tissues being produced toward the inside of the stem. Only a very small amount of tissue, which will differentiate solely into parenchyma, is produced toward the outside. The cells produced to the interior of the meristem differentiate into a ground parenchyma, often called **conjunctive tissue**, the cells of which are arranged in radial files. Embedded in this tissue are numerous collateral or amphivasal vascular bundles composed of tissues similar to those of primary vascular bundles (Fig. 14.1b). Usually these bundles contain only a very small quantity of phloem (Fig. 14.1c). The xylem contains no annular or helical tracheids, consisting primarily of scalariform and pitted tracheids. Whereas there seems to be no recent, definitive supporting evidence, several workers suggest that these bundles connect with axial bundles and leaf traces of the primary body.

In taxa that feature such secondary growth (for example, *Cordyline, Dracaena, Yucca*) the primary body is obconical, and without additional support from secondary tissues the stem would be unstable. Secondary tissues are thickest at the base of such stems and thinnest near the apex. Consequently, the two tissues result in a stem of similar diameter from base to apex (Fig. 14.2). Another strategy that results in the stabilization of obconical stems is the development of prop roots as in *Pandanus* (screw pine).

Anomalous stem and root structure

An unusual distribution of primary vascular bundles, especially the occurrence of cortical bundles in dicotyledons, irregularity in the activity of the vascular cambium, and the activity of several successively formed cambia in the same stem result in unusual stem structure, often referred to as being anomalous.

Whereas dicotyledons are typically characterized by a single cylinder of stem vascular bundles, in several taxa the bundles occur in more than one cylinder, or appear somewhat scattered as seen in transverse section. *Cucurbita*, for example, is characterized by an inner cylinder of large vascular bundles and an outer cylinder of smaller cortical bundles (Fig. 14.3a). The bundles in both cylinders are bicollateral with primary phloem occupying both outer and inner parts. The genus *Piper* of the Piperaceae also has an unusual arrangement of vascular bundles. In *Piper betle*, for example, there is a somewhat irregular cylinder of vascular bundles enclosing the pith and a system of much smaller cortical bundles to the exterior of an undulating cylinder of sclerenchyma (Fig. 14.4a). In some taxa, cortical bundles are leaf traces that extend over great longitudinal distances in the stem prior to entering the leaves. *Piper excelsum* (Fig. 14.4b, c) is also characterized by two cylinders of vascular bundles. Two primary vascular bundles, often referred to as **medullary bundles**, occupy the very center of the stem. These are enclosed by an irregular inner

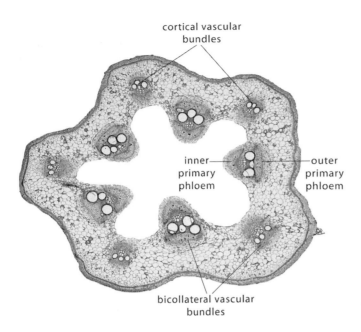

cortical vascular
bundles

inner—
primary
phloem

—outer
primary
phloem

bicollateral vascular
bundles

Figure 14.3 Transverse section of *Cucurbita*, a vine with inner and outer cylinders of primary bicollateral vascular bundles. Magnification × 6.

cylinder of similar bundles. Enclosing this cylinder is an outer cylinder of vascular bundles in which cambium develops (Fig. 14.4c). Cambial activity results in large, radially extended regions of secondary vascular tissues that are separated from each other by wide ray-like regions of primary parenchyma (often called **medullary rays**).

The structure of stems in several families (e.g., Amaranthaceae, Chenopodiaceae, Menispermaceae) results from the presence and activity of **accessory cambia**. Upon cessation of activity of an initial vascular cambium, additional cambia differentiate successively to the exterior in cortical tissue. Their activity results in a series of cylinders of secondary vascular tissues, separated by narrow cylinders of parenchyma. In *Chenopodium album* (lamb's quarters) (Fig. 14.5) a vascular cambium initially differentiates in a cylinder of vascular bundles. Cambial activity in individual bundles leads to an increase in their size by production of some secondary xylem and phloem. Upon cessation of cambial activity in these bundles a new vascular cambium, forming a continuous cylinder (Fig. 14.5a), differentiates in cortical tissue to the exterior of the vascular bundles. The activity of this cambium results in formation of a cylinder of secondary vascular tissues of unusual characteristics consisting of alternating bands of xylem, phloem, and secondary parenchyma (conjunctive tissue) (Fig. 14.5b).

Variation in the activity of the vascular cambium, either by the nature and quantity of the tissues produced by particular segments, or by variation in the frequency of cell division in different regions of the cambium, results in unusual distribution patterns of xylem and phloem, or even differently shaped stems, the exterior of which conform largely to the shape of the secondary xylem. Many tropical vines are characterized by anomalous stem structure, but such structure is not confined to them. A few examples will suffice.

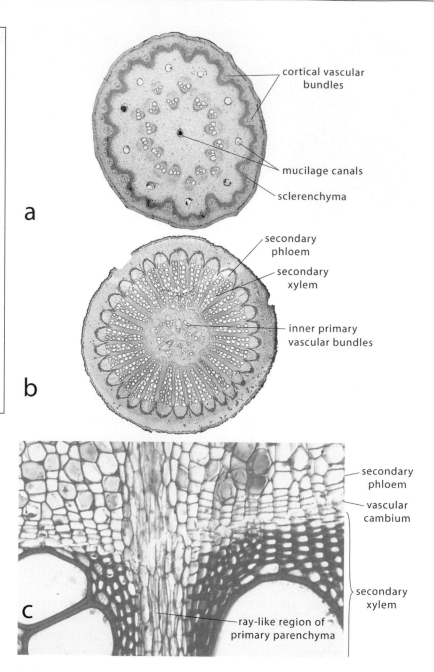

Figure 14.4 Variation in stem structure in two species of *Piper* (Piperaceae, pepper family). (a) Transverse section of a stem of *Piper betle*. The stem has an inner, irregular cylinder of primary vascular bundles, and an outer cylinder of smaller bundles to the exterior of an undulating wall of sclerenchyma. Note also the large mucilage ducts in the pith and inner cortex. Magnification × 8. (b) The stem of *Piper* sp. characterized by a central, irregularly arranged group of primary vascular bundles enclosed by a thick cylinder of secondary tissues consisting of regions of tracheary tissues capped by phloem. The secondary tracheary tissues are separated by ray-like regions of secondary parenchyma. Magnification × 3.3. (c) Detail of the vascular cambium and its derivative tissues from the stem shown in (b). Magnification × 180.

In some taxa, the vascular cambium produces clusters or bundles of phloem to the inside of itself, which become embedded within the secondary xylem. This **intraxylary secondary phloem** is characteristic of several families, including the Apocynaceae and Asclepiadaceae. In several members of the tropical family Bignoniaceae, the vascular cambium produces, in different regions, largely secondary xylem or largely secondary phloem. Consequently the xylem cylinder becomes fluted (Fig. 14.6a). By contrast, in *Aristolochia*, a vine of temperate regions, certain segments of the vascular cambium produce only

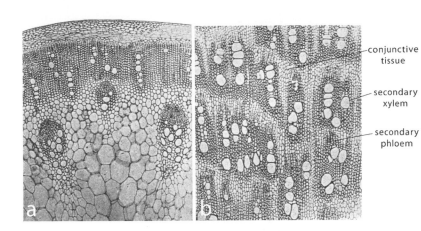

conjunctive
tissue

secondary
xylem

secondary
phloem

Figure 14.5 Transverse sections of *Chenopodium album*. (a) Part of a stem showing inner vascular bundles, consisting of both primary and secondary tissues, enclosed by a cylinder of secondary vascular tissues. (b) Part of the secondary cylinder in an older stem showing secondary xylem, conjunctive tissue, and patches of phloem. From Eames and MacDaniels (1925).

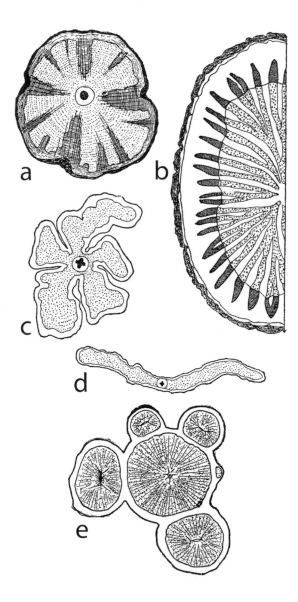

Figure 14.6 Diagrammatic representation of unusual ("anomalous") stem structure in several vines as viewed in transverse section. (a) A bignoniaceous species. (b) *Aristolochia triangularis.* (c) *Bauhinia rubiginosa.* (d) *Bauhinia* sp. (e) *Serjania ichthyoctona.* Please see the text for descriptions. From Schenk (1892–3).

Figure 14.7 Transverse sections of *Beta* (sugar beet) root. (a) Cylinders of secondary vascular tissues, produced by a succession of accessory cambia, are separated by secondary, conjunctive parenchyma. Secondary xylem and secondary phloem are composed largely of parenchyma. Magnification × 58. (b) Drawing illustrating the sparse occurrence of vessels and sieve tubes in the xylem and phloem. (a, b) From Artschwager (1926).

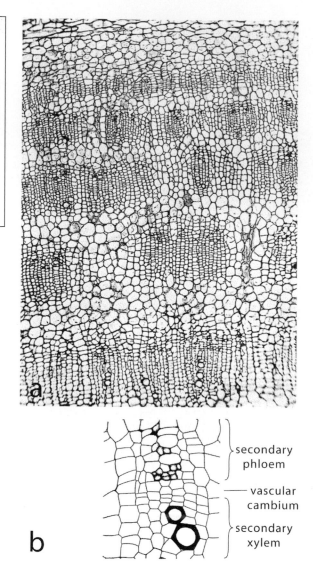

secondary phloem

vascular cambium

secondary xylem

b

secondary parenchyma whereas others produce typical proportions of xylem and phloem. Thus, the vascular cylinder consists of alternating, radially extended regions of these tissue types (Fig. 14.6b). In *Bauhinia* of the Leguminosae (sometimes included in the Caesalpiniaceae), normal cambial activity in certain areas and the restriction of activity in others results in fluted or flattened stems (Fig. 14.6c, d). In *Thinouia* (Sapindaceae) in early stages of cambial development, some regions of the cambium become convoluted, pinch off, and produce separate cylinders of secondary vascular tissue resulting in a stem of **polystelic structure** (Fig. 14.6e).

Roots as well as stems may exhibit unusual patterns of structure and development. The common root vegetable *Beta vulgaris* (beet) provides an excellent example of the formation of accessory cambia in a root (Fig. 14.7a). Following cessation of activity of the initial cambium, a succession of cambia differentiate toward the exterior in pericyclic

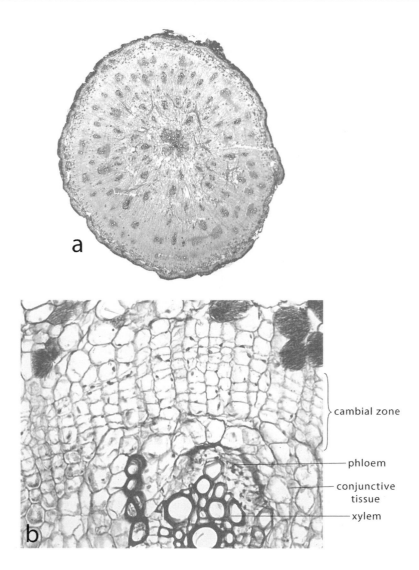

Figure 14.8 (a) A transverse section of *Boerhaavia* sp. root. The central actinostelic primary xylem column is enclosed in secondary tissue consisting of conjunctive parenchyma in which are embedded numerous vascular bundles. Magnification × 8. (b) Enlargement showing the cambial zone and part of a vascular bundle. Magnification × 216.

tissue. Each of these cambia produces, for a limited period, secondary xylem and phloem in strands of variable width, and intervening secondary parenchyma. Sieve tubes in the phloem occur in small clusters embedded in secondary phloem parenchyma. Sparsely scattered vessels occur singly or in small clusters, also enclosed in extensive secondary xylem parenchyma (Fig. 14.7b). Additional tissue is produced by cytokinesis within the pericycle to the exterior of each cylinder of secondary tissue. Consequently the beet root is composed of cylinders of highly parenchymatous secondary vascular tissues which alternate with cylinders of pericyclic parenchyma with the result that the predominant tissue of the beet root is **storage parenchyma**.

Accessory cambia also characterize both stems and roots of *Boerhaavia* of the Nyctaginaceae. Following cessation of activity in the initial cambium in roots of *Boerhaavia diffusa* (Fig. 14.8a), additional cambia differentiate to the exterior, producing small vascular bundles and intervening conjunctive tissue to the inside and secondary parenchyma to the outside. Each new cambium (Fig. 14.8b) differentiates in the secondary parenchyma produced by the previously active cambium.

For additional information about other interesting and often bizarre stem and root structure, see sections on Bignoniaceae, Apocynaceae, Amaranthaceae, Chenopodiaceae, Asclepiadaceae, Loganiaceae, Menispermaceae, Nyctaginaceae, Caesalpiniaceae, Acanthaceae, and Sapindaceae in Metcalfe and Chalk (1950).

REFERENCES

Artschwager, E. 1926. Anatomy of the vegetative organs of the sugar beet. *J. Agric. Res.* **33**: 143–176.

Eames, A. J. and L. H. MacDaniels. 1925. *An Introduction to Plant Anatomy*. New York: McGraw-Hill.

Metcalfe, C. R. and L. Chalk. 1950. *Anatomy of the Dicotyledons*, 2 vols. Oxford, UK: Clarendon Press.

Schenck, H. 1892–3. *Beiträge zur Biologie und Anatomie der Lianen in besonderen in der Brasilien einheimschen Arten*, 2 vols. Jena: G. Fischer.

Tomlinson, P. B. and A. E. Esler. 1973. Establishment growth in woody monocotyledons native to New Zealand. *N. Z. J. Bot.* **11**: 627–644.

FURTHER READING

Cheadle, V. I. 1937. Secondary growth by means of a thickening ring in certain monocotyledons. *Bot. Gaz.* **98**: 535–555.

Eckardt, T. 1941. Kritische Untersuchungen über das primäre Dickenwachstum bei Monokotylen, mit Ausblick auf dessen Verhältnis zur sekundären Verdickung. *Bot. Arch.* **42**: 289–334.

Esau, K. 1940. Developmental anatomy of the fleshy storage organ of *Daucus carota*. *Hilgardia* **13**: 175–226.

1965. *Plant Anatomy*, 2nd edn. New York: John Wiley and Sons.

1977. *Anatomy of Seed Plants*, 2nd edn. New York: John Wiley and Sons.

Esau, K. and V. I. Cheadle. 1969. Secondary growth in *Bougainvillea*. *Ann. Bot.* **33**: 807–819.

Hayward, H. E. 1938. *The Structure of Economic Plants*. New York: Macmillan.

Obaton, M. 1960. Les lianes ligneuses à structure anomale des forêts denses d'Afrique occidentale. *Ann. Sci. Nat., Bot. Ser. 12.* **1**: 1–220.

Tomlinson, P. B. 1961. *Anatomy of the Monocotyledons*, vol. 2, *Palmae*. Oxford, UK: Clarendon Press.

Chapter 15

Secretion in plants

Perspective

Waste products in animals are excreted to the exterior through the digestive system, the urinary system, and, to a lesser extent, through sweat glands. By contrast, in the plant, waste products of metabolism as well as substances that will be further utilized are stored within individual cells or transferred to regions of living or non-living tissues or into cavities and ducts within the organism. A good example is the transfer of waste metabolites into the secondary wood (with the consequent formation of heartwood) where they are isolated from the functional regions of the plant body. The transfer of metabolites from one site to another is referred to as secretion rather than excretion although some substances are transferred to the plant surface, such as precursor compounds of cutin and waxes and a variety of substances that exit the plant through glands and glandular hairs. This concept of secretion also includes the transfer of substances within single cells such as, e.g., the movement of enzymes to chloroplasts, sites of photosynthesis, and the transport in vesicles of precursors of cellulose to sites of wall synthesis. We can, thus, define **secretion in plants** as the transfer of certain intermediate or end products of metabolism from one region to another within the cell or out of the protoplast to another part of the plant body.

Substances secreted by plants

Both metabolic and non-metabolic substances are transferred within or to the exterior of the plant body. **Metabolic compounds** are those that will have a continuing function in the plant such as RNA that is transferred from the nucleus to the ribosomes, enzymes as well as precursor compounds of cellulose and/or lignin that are transferred to the region of the developing cell wall, hormones that are transferred from meristematic parenchyma into sieve tubes and through them to various parts of the organ or organism, and photosynthates transferred from mesophyll parenchyma to more interior storage parenchyma cells, or by way of companion cells into sieve tube members, among others. Whereas the movement of photosynthates fits the concept of secretion as defined, their transfer into and out of the phloem is more commonly referred to as assimilate loading and unloading (see Chapter 10).

Whereas **non-metabolic compounds** may not be further functional in the plant, they may have important adaptive value to the species. Such compounds are the essential oils, including terpenes, which provide flower fragrance and attract pollinators, alkaloids that are highly poisonous which prevent predation by herbivores (ironically *Homo sapiens* is attracted by certain alkaloids such as nicotine, caffeine, cocaine, etc.), glycosides which, on being acted upon by enzymes, release pungent odors (as in cabbage and other Cruciferae) that repel many insects, tannins and resins that may impede fungal invasion of wood, calcium oxalate crystals that make plants unpalatable to grazers, certain alcohols that accumulate in glandular hairs such as those of stinging nettle which also inhibit predation, some flavonoid pigments which, in leaves, block ultraviolet radiation that destroys nucleic acids and proteins but which admit blue–green and red light utilized in photosynthesis, and others (see, e.g., Thomas, 1991; Duke, 1994; Fahn, 2000).

Mechanisms of secretion

Two primary mechanisms of secretion have been recognized (Esau, 1977). **Granulocrinous** (or **granulocrine**) **secretion** is effected by the fusion of Golgi or ER vesicles with the plasmalemma or tonoplast resulting in the transfer of substances in the vesicles to the exterior of the plasmalemma or from them into the vacuole. A good example is the granulocrinous transfer of precursor compounds of cellulose into the region of cell wall formation. In many glands, compounds that have passed through the plasmalemma continue to the exterior through the tangential walls. They are prevented from diffusing back into the plant through the wall apoplast by the presence in radial and transverse walls of cutinized regions resembling the Casparian bands of endodermal cells. If the compounds accumulate in subepidermal or subcuticular chambers, they usually reach the exterior through epidermal pores (sometimes, modified stomata), or by disintegration of the cuticle.

Passage of small molecules directly through the plasmalemma and wall is called **eccrinous** (or **eccrine**) **secretion**. It can be either passive along concentration gradients, or active, utilizing energy in the process. Examples of eccrinous secretion are the transport of water-soluble compounds such as salts and sugars from the symplast through the cell wall apoplast, sometimes directly to the exterior as in salt glands and nectaries. In such cases the cells from which these compounds enter the cell wall are transfer cells. **Transfer cells** are widely distributed in the plant at sites of rapid flux of solutes such as between the mesophyll in leaves and sieve tube members, at transfer interfaces such as that between embryo and endosperm in seeds, and in secretory structures, especially in glands of various types. The secondary wall of transfer cells is characterized by extensive, complex ingrowths that greatly increase the wall surface area. Such wall structure is often called a **wall labyrinth**. Since the plasmalemma covers the inner surface of the wall, the area for transfer, that is, secretion through it, is greatly increased, sometimes by as much as 20 times that of cells in which there are no wall ingrowths. Substances may also be released to the exterior from glandular trichomes or other secretory structures by disintegration of the gland or secretory tissue, or its destruction by insects. Such release of compounds is called **holocrine secretion**.

Internal secretory structures

Within the plant, various substances are transported into individual cells as well as into cavities and ducts of several types. **Idioblasts** are single, isolated, and specialized secretory cells that are common within parenchyma tissue and contain distinctive substances such as mucilage, tannin (Fig. 15.1a), oils (Fig. 15.1b), and calcium oxalate (usually in crystalline form). **Mucilage cells** that occur in the secondary phloem of some plants often also contain a bundle of **raphides**, needle-like crystals composed of calcium oxalate, as in *Hydrangea paniculata*. Single large crystals (Fig. 15.1c) as well as spherical aggregates of crystals of calcium oxalate, called **druses** (Fig. 15.1d), are also common components of idioblasts in many taxa. **Lithocysts** characterize the epidermis of several genera of the Moraceae. The lithocysts of *Ficus elastica* (Fig. 15.1e) are cells that occur in the upper, multiple epidermis of the leaves. Each contains a large calcium carbonate crystalline structure called a **cystolith** attached to the cell wall by a cellulose stalk which originated as a wall ingrowth. Although lithocysts are most common in the upper epidermis of leaves they also occur in both the upper and lower epidermis in some members of the family.

Many plants contain **schizogenous ducts** or **cavities** (see, e.g., Curtis and Lersten, 1990; Turner *et al.*, 1998), formed by the separation of cells from each other during development (Fig. 15.2). The resultant cavity becomes bounded by a layer of **secretory cells** (sometimes called **epithelial cells**). These cells secrete various substances (either eccrinously or granulocrinously) into the cavities formed such as mucilage (e.g., in certain members of Malvaceae, Tiliaceae, and Sterculiaceae), oils as in Hypericaceae (Fig. 15.2a) in which the secretory cavities occur in the leaves of all genera, appearing, macroscopically, as opaque or translucent dots, and resins or gums as in many conifers as well as in several families of angiosperms (e.g., Anacardiaceae). Whereas in some species a single substance may characterize the contents in ducts and cavities, in others there may be a mixture of substances as, for example, in the Umbelliferae in which

Figure 15.1 Internal secretory cells (idioblasts). (a) Cells containing tannin in the pith of *Tilia americana*. Magnification × 300. (b) An oil idioblast in a leaf of *Annona muricata*. Magnification × 1950. (c) Axial parenchyma cells in the secondary phloem of *Tilia americana* containing large, prismatic crystals of calcium oxylate. Magnification × 482. (d) Druses, spherical aggregates of small crystals of calcium oxylate. Magnification × 165. (e) A lithocyst containing a cystolith in the multiple epidermis of *Ficus elastica* (rubber plant). The cystolith consists of a crystalline structure of calcium carbonate attached to the cell wall by a cellulose stalk that originated as an ingrowth of the wall. Magnification × 174. (b) From Bakker and Gerritsen (1990). Used by permission of Oxford University Press.

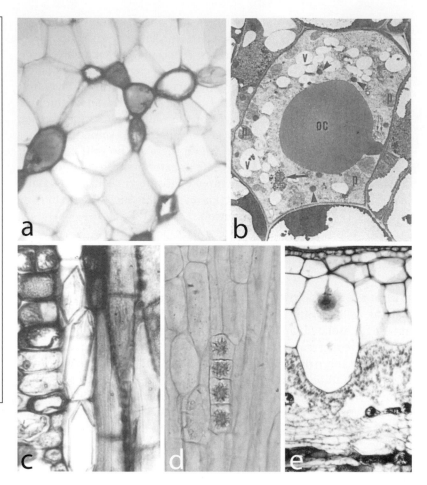

secretory canals (sometimes called resin canals) contain oils, resins, and mucilage.

Of schizogenous secretory ducts, resin ducts of conifers (Fig. 15.2b) are by far the best known. They form an extensive system in the secondary xylem, extending longitudinally as well as radially (through vascular rays). (For more detail and illustrations, see Chapter 11 on secondary xylem).

In contrast to schizogenous ducts and cavities, **lysigenous ducts and cavities** (Fig. 15.2c) are thought to develop by the lysis of cell protoplasts and dissolution of the cell walls. Consequently, the contents of the cavities and ducts thus formed are derived directly from the lysed cells. The lysigenous cavities of citrus fruits, which contain essential oils among other substances, provide a good example. Although the concept of lysigeny has been widely accepted since early in the last century, Turner *et al.* (1998) have recently concluded that supposed lysigeny in the development of oil glands in *Citrus limon* (lemon) results from an artifact of fixation. They believe that the swelling of epithelial cells in hypotonic fixatives has led workers to interpret this as an indication of "early senescence" and, thus, lysigeny in *Citrus* glands.

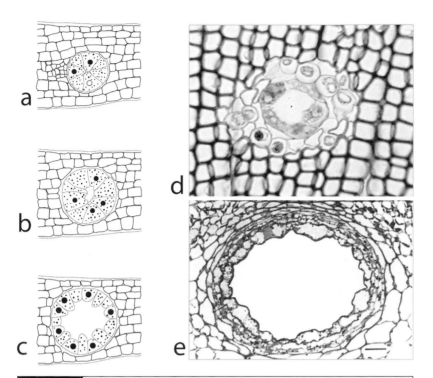

Figure 15.2 Schizogenous ducts and cavities. (a–c) Three stages in the development of a schizogenous oil cavity in the leaf of *Hypericum* sp. Note the beginning of separation of cells in (b). Cell divisions in the incipient secretory cells in (b) resulted in a more extensive single-layered epithelium in the nearly mature cavity in (c). (d) Transverse section of a schizogenous resin duct in the secondary xylem of *Pinus strobus*. Note the single, inner layer of epithelial cells and the thicker-walled parenchyma cells enclosing the duct. Magnification × 246. (e) An oil cavity in *Citrus limon* (lemon) often considered to have lysigenous development, i.e., to be formed by autolysis of the secretory cells which, as a result, release their contents into the cavity that forms. Recent research (Turner *et al.*, 1998) indicates that development of the oil cavities of *Citrus* as well as those of other taxa may actually be schizogenous having been incorrectly interpreted in the past. See the text for more detail. Bar = 20 μm. (e) From Turner *et al.* (1998). Used by permission of the University of Chicago Press. © 1998 The University of Chicago. All rights reserved.

Laticifers, which contain latex, differ greatly in development from both schizogenous and lysigenous ducts. Two types, articulated and non-articulated laticifers, are recognized on the basis of their development. **Non-articulated laticifers** (Fig. 15.3a, b) are found in many members of the Euphorbiaceae, Asclepiadaceae, and Apocynaceae among other families. They originate in the embryo from single-celled laticifer primordia and, as development of the plant proceeds, the primordia grow longitudinally and symplastically with some apical intrusion and, often, extensive branching (Mahlberg, 1961). Nuclei divide repeatedly but without accompanying cytokinesis and cell wall formation (Mahlberg and Sabharwal, 1967). The result is a coenocytic system that extends throughout the primary body of the plant, and in some cases into the secondary body as well.

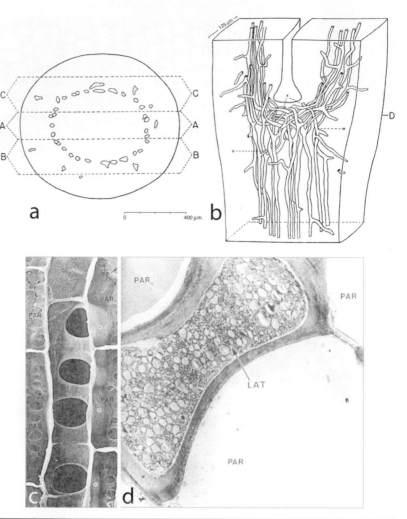

Figure 15.3 (a, b) Non-articulated laticifers in an embryo of *Nerium oleander* (Apocynaceae). (a) Drawing of a transverse section at the cotyledonary node indicated in (b) as D. (b) Three-dimensional drawing of a thick, longitudinal section indicated in (a) as B-B showing the complex system of laticifers. (c, d) Articulated laticifers in *Hevea brasiliensis* (rubber tree). (c) Scanning electron micrograph of an articulated laticifer and associated axial parenchyma cells (PAR) in the secondary phloem as seen in longitudinal section. PF, primary pit field. Note the large perforations between this and a contiguous laticifer. The contents of this laticifer have been removed. Magnification × 504. (d) Electron micrograph of a transverse section showing the thick, non-lignified primary cell wall and the cell contents, including numerous vesicles (lutoids) which contain polyterpenes and other compounds. LAT, laticifer. Magnification × 2584. (a, b) From Mahlberg (1961). Used by permission of the Botanical Society of America. (c, d) From de Faÿ et al. (1989). Used by permission of Springer-Verlag Wien.

Articulated laticifers (Fig. 15.3c, d) characterize members of the Euphorbiaceae (e.g., *Hevea*, rubber tree); Compositae (e.g., dandelion); Papaveraceae (e.g., poppy), etc. They originate in the apical meristem as single cells. As growth proceeds they increase in length acropetally by the addition of new cells without intrusive growth. As cells mature

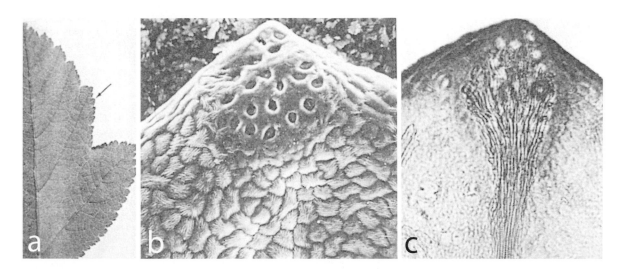

in the developing articulated laticifer, the end walls between longitudinally superposed cells are absorbed resulting in a coenocyte which may be much branched (see Esau, 1975). Thus, mature non-articulated and articulated laticifers have a similar appearance, and comprise a system of slender, branched or unbranched, tubes although they differ greatly in ontogeny.

Laticifers have non-lignified, primary walls, and contain latex, the composition of which may vary greatly in different species. The color of the latex varies from clear to white, brown, yellow, and orange. Polyterpenes are common constituents of latex which, with other components, occur in small vesicles. In rubber-producing plants (e.g., *Hevea brasiliensis* and *Ficus elastica*) these vesicles are called **lutoids** (Fig. 15.3d). The vesicles may ultimately fuse forming a large vacuole. In addition to polyterpenes, laticifers may also contain alkaloids (as, e.g. in *Papaver somniferum*, opium poppy), sugars, proteins, enzymes, starch grains, etc.

Figure 15.4 Hydathodes from the marginal teeth of a leaf of *Physocarpus opulifolius*. (a) Leaf showing location of the hydathode in (b) at arrow. Life size. (b) Scanning electron micrograph illustrating numerous water pores on the adaxial surface of a hydathode. Magnification × 200. (c) Hydathode from a cleared leaf showing the relationship between the water pores (light circular areas) and the tracheary tissue. Magnification × 140. From Lersten and Curtis (1982). Used by permission of the National Research Council of Canada.

External secretory structures

External secretory structures occur in many forms, ranging from simple hydathodes and glandular hairs to complex salt glands and nectaries, among others. **Hydathodes** are modified stomata and associated tissues through which excess water is released from the plant through a process called **guttation**. This occurs when transpiration is low and root pressure is high. Although somewhat variable in structure, a typical hydathode (Fig. 15.4) which is located at the leaf margin (Fig. 15.4a), consists of parenchyma tissue, lacking chlorophyll, termed **epithem**, between the end of a vein and the external pore (or pores) (Fig. 15.4b, c). The parenchyma (epithem), in contact with the substomatal chamber, unlike that associated with the substomatal chambers of typical stomata, is wettable, thus facilitating the passage of water to the exterior. The movement of water through the epithem may either be passive, the result of root pressure, or

Figure 15.5 Glandular trichomes. (a) Peltate (p) and capitate (c) trichomes from the abaxial surface of a leaf of *Mentha* (mint). Bar = 10 μm. (b–d) Capitate trichomes from *Plectranthus ornatus*. (b) Long-stalked capitate trichomes. Arrow indicates short-stalked capitate trichome; stars indicate long-stalked capitate trichomes. (c) A conoidal trichome consisting of a large conical, glandular cell and a bicellular stalk. Arrow indicates apical pore. (d) Digitiform trichomes of three to four cells. Bars (b–d) = 25 μm. (a) From Colson *et al.* (1993). Used by permission of the National Research Council of Canada. (b–d) From Ascensão *et al.* (1999). Used by permission of Oxford University Press.

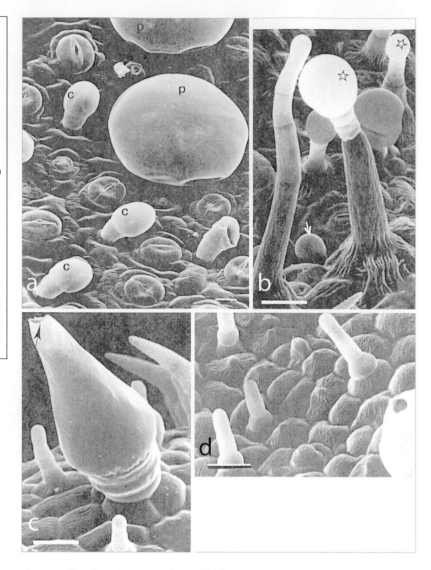

the result of active transfer, utilizing energy. In the latter case the epithem cells may function as, and have the structure of, transfer cells. The number of pores (Fig. 15.4b) in a hydathode varies from one to many. Some hydathodes, especially those located at the tips of teeth on leaf margins, may consist simply of a single modified stoma without associated epithem.

It has been estimated that 20–30% of vascular plant species contain **glandular trichomes** on their aerial surfaces (Fahn, 1988; Duke, 1994). Although these trichomes are of diverse morphology (Fig. 15.5a–d), the glands, with few exceptions, are characterized by a globular appearance resulting from the separation of the cuticle from the walls of the secretory cells and filling of the subcuticular space by chemical compounds produced by these cells (Duke and Paul, 1993; Duke, 1994; Bourett *et al.*, 1994). Perhaps the most common type is the **peltate trichome** (Fig. 15.6a–d), consisting of a stalk one or more cells long bearing an expanded gland of several cells. Among the

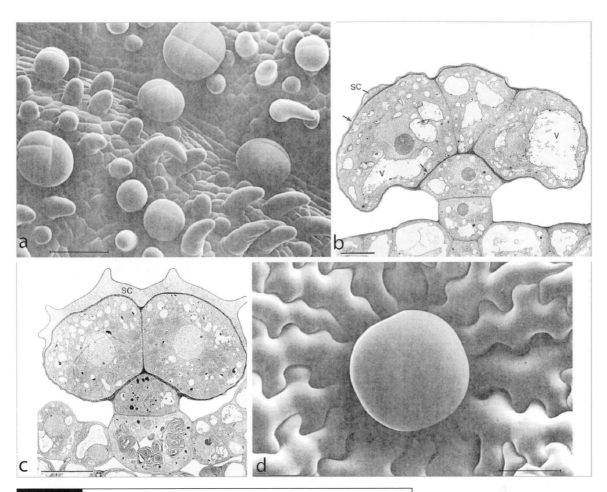

Figure 15.6 Peltate glandular trichomes of *Nepeta racemosa* (catmint). (a) Young peltate glands with two, three, or four secretory cells representing stages in development. These glandular trichomes are scattered among immature, non-glandular hairs. Bar = 25 μm. (b) Transmission electron micrograph of a sectional view of a peltate gland. The cuticle is pulled slightly away from the cell wall but there is very little material within the subcuticular cavity (sc). Secretory cells contain large vacuoles (v). Arrows indicate electron-opaque cell wall. Bar = 5 μm. (c) A sectional view of a peltate gland at a more mature stage with a somewhat distended cuticle and a large quantity of material in the subcuticular cavity. Bar = 10 μm. (d) A mature peltate gland with a fully extended cuticle which obscures the sutures between secretory cells. Bar = 25 μm. From Bourett *et al.* (1994). Used by permission of the University of Chicago Press. © 1994 The University of Chicago. All rights reserved.

compounds produced by plants, terpenoids and their derivatives are among the most common in glandular trichomes. Some are highly toxic whereas others provide the flavors of spices such as oregano and of essential oils such as mint, lavender, and sweet basil. Others form the basis for fragrances in perfumes (see, e.g., Werker *et al.*, 1993; Bourett *et al.*, 1994; Ascensão *et al.*, 1999).

Glandular trichomes are abundant on the surfaces of young leaf primordia and develop very early (Werker *et al.*, 1993; Bourett *et al.*,

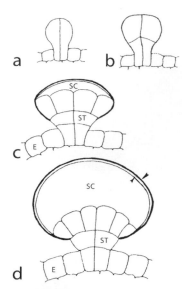

Figure 15.7 Diagrams illustrating the development of the glandular trichome in *Cannabis sativa*. (a) A cell of the protoderm expands and divides anticlinally followed by (b) a periclinal division in each cell forming an upper and a lower tier of two cells each. (c) By a series of anticlinal divisions in the upper tier a discoid layer of secretory cells (SC) is formed. Periclinal divisions in the lower tier result in the formation of the stipe (ST) and the basal cells which comprise a part of the epidermis (E). (c, d) The secretory cavity develops, as the cuticle (large arrowhead) expands, and cell wall substance, possibly derived from the outer walls of the secretory cells, forms a subcuticular wall (small arrowhead). See the text for more detail. From Kim and Mahlberg (1991). Used by permission of the Botanical Society of America.

1994). By leaking from the glands, their contents are apparently spread over the surface of the primordia, protecting them from insect damage by various biological effects, among which are insecticidal, repellent, and behavior-modifying effects as well as feeding deterrents (Duke, 1994). In addition, the young regions of some plants become coated with resinous materials derived from glandular hairs in which insects may become entrapped. Such coatings are also thought to reduce solar heating or water loss through evaporation (Duke, 1994). Glandular trichomes that produce resinous and other sticky compounds are called **colleters**. They are common on the young leaves and flower parts of many species of woody dicotyledons in a large number of families. Some are of complex structure, even containing laticifers and vascular tissue (Thomas, 1991).

Glandular hairs may also contain potent insecticidal phytotoxins. For example, artemisinin, a sesquiterpenoid compound, produced by *Artemisia annua*, is highly toxic to the plant that produces it as well as to other plant species. Because it is sequestered in the subcuticular space in the glands, however, *Artemisia* is protected from its toxic effects. This and other toxic substances produced in glandular trichomes probably function as a defense against insects and microbial pathogens (Duke, 1994). It is interesting to note, that artemisinin is also an important antimalarial drug, effective against *Plasmodium falciparum* strains that have become resistant to more commonly used drugs.

An especially interesting and important role of chemical substances produced in glandular trichomes is **allelopathy**, the inhibition of one plant species by chemicals produced by another. The phytotoxin 1,8-cineole (a monoterpene) produced by species of *Salvia* (Kelsey *et al.*, 1984), is thought to inhibit competitors by volatilization from the ground litter (Muller and Muller, 1964; Duke, 1994).

We shall consider in detail the structure of several types of glandular hairs including those of *Nepeta racemosa* (catmint), *Cannabis sativa* (hemp, marijuana), and the stinging hairs of *Urtica dioica* (stinging nettle).

The peltate oil glands of *Nepeta* (Fig. 15.6a–d), typically borne on the abaxial surface of the leaves, consist of a short stalk, composed of a basal cell located within the epidermis and a single stalk cell upon which rests a cluster of four secretory cells, covered by a cuticle (Fig. 15.6b) (Bourett *et al.*, 1994). The aromatic oil is synthesized within these cells and transferred granulocrinously to the exterior, expanding the cuticle (Fig. 15.6c). Upon complete expansion of the cuticle in mature glands, the sutures between the four secretory cells disappear (Fig. 15.6d).

The glandular trichome of *Cannabis sativa* is very similar in external appearance to that of the peltate glands of other taxa, but differs strikingly in that the secretory cavity of the mature gland is bounded by both a cuticle and a subcuticular wall (Fig. 15.7a–d) (Kim and Mahlberg, 1991; Mahlberg and Kim, 1991). During early stages of secretory cavity development, the outer walls of the secretory cells become "loosened." Precursor materials of cutin and, presumably, the active ingredient in the drug marijuana (tetrahydrocannabinol), are thought to form in hyaline areas in the walls, become membrane-enclosed, and move in these vesicles into the developing secretory cavity, nearly filling it. Wall material, referred to by Kim and Mahlberg (1991) as "wall matrix," derived from the outer walls of the secretory cells, moves through the developing cavity in a hydrophilic phase between the vesicles, and is deposited under the cuticle, forming the subcuticular wall. The substructure of this wall has not been well defined, and it is not clear whether or not it contains cellulose microfibrils. However, Kim and Mahlberg (1995) have described the presence in the wall of elongate fibrils in parallel pairs. Vesicles in the secretory cavity enlarge and some migrate to the subcuticular wall where the vesicular membrane disintegrates. Precursor compounds of cutin apparently move through the wall, are sythesized into cutin and, thus, contribute to the expanding and thickening cuticle (Mahlberg and Kim, 1991; Kim and Mahlberg, 1995). This unusual development wherein cutin precursors contained in vesicles and non-vesiculate subcuticular wall material are transported through a non-living space characterizes both *Cannabis* and *Humulus* (hops) (Kim and Mahlberg, 2000).

The stinging hairs of *Urtica dioica* (Fig. 15.8) consist of a cluster of basal cells, both epidermal and subepidermal in origin, in which rests the base of a long needle-like cell with a rigid, siliceous wall and a small bulbous tip. Upon contact with an animal, the tip breaks off releasing the toxic substances in the cell as it penetrates the skin. The composition of the toxic material has not been definitively determined but has been reported to contain a histamine and an acetylcholine (see Esau, 1965).

Salt glands (Fig. 15.9) provide a mechanism for the plant to rid itself of excess salt absorbed from the environment. This is an essential function of plants living in saltmarshes and salt-infiltrated soils in coastal marine areas. Salt glands of several types are modified

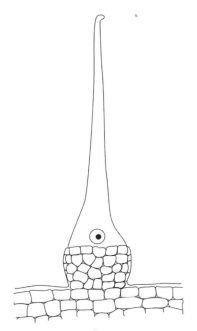

Figure 15.8 Glandular hair of *Urtica dioica* (stinging nettle).

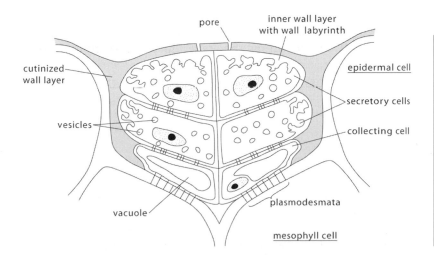

Figure 15.9 Salt gland of *Tamarix aphylla* (tamarisk). Note that the gland, which consists of two basal collecting cells and four secretory cells, is sunken within the epidermis. Salt is transferred symplastically from the leaf mesophyll into collecting cells and secretory cells, and apoplastically through the wall labyrinth to the exterior. See the text for more detail. Redrawn from Thomson *et al.* (1969). Used by permission of Dr. W. W. Thomson.

Figure 15.10 Nectaries. (a) The floral nectary of *Cornus sanguinea* (dogwood) surrounds the style at the base of the corolla. (b, c) Extrafloral nectaries. (b) Flattened, scale-like nectaries on the stem of *Bignonia capreolata*. (c) Nectary of *Gossypium hirsutum*, embedded in the midvein on the abaxial surface of the leaf. Note the numerous, raised secretory cells in the cavity. (a) From Jaeger (1961). Used by permission of Chambers Harrap Publishers, Ltd. (b, c) From Elias (1983). Used by permission of Columbia University Press. © 1983 Columbia University Press.

multicellular trichomes borne on stems and leaves. Among the best known are those of members of the Tamaricaceae and Chenopodiaceae. In many species of *Atriplex* the gland consists of a large, bladder-like cell attached terminally to a uniseriate stalk consisting of one to several cells. Salt moves apoplastically from the xylem of a nearby vein to parenchyma cells subtending the base of the gland and symplastically through plasmodesmata in the walls of the stalk cells and the bladder where it is deposited within its vacuole (see Osmond *et al.*, 1969). Upon disintegration of the gland the salt is released, forming a white residue on the surface of the leaf or stem. The movement of salt ions into the gland against a concentration gradient requires an expenditure of energy. Salt is prevented from moving back into the leaf or stem through the cell wall apoplast by the presence of cutinized transverse and radial walls of cells subtending the gland.

A very different, more complex type of salt gland characterizes *Tamarix aphylla* (tamarisk) (Fig. 15.9). It is also a modified, multicellular trichome, but one that is sunken within the epidermis. Salt in solution is transported symplastically by way of plasmodesmata through two basal collecting cells into the several secretory cells that comprise the bulk of the gland. The secretory cells are transfer cells with extensive wall ingrowths and heavily cutinized walls except in the region of contact with the two basal cells. The protoplasts of the transfer cells contain numerous small vesicles into which the salt accumulates. Fusion of the vesicles with the plasmalemma empties the salt into the wall where it then moves through the inner, uncutinized layer toward the surface of the gland and is released to the exterior through small pores. Movement of salt back into the leaf through the wall along the concentration gradient is prevented by the cutinized layer enclosing the gland (see Thomson *et al.*, 1969).

Unlike salt glands that function to eliminate a substance potentially harmful to the plant, **nectaries** benefit plants in very different ways. By secreting a sugar solution (nectar) that attracts bird and insect pollinators, floral nectaries (Fig. 15.10a) play a significant role in plant reproduction. Whereas extrafloral nectaries (Fig. 15.10b, c) which occur on the stem, leaves, leaf petioles, midribs, stipules and flower pedicels play no role in reproduction, they attract insects, the presence of which may prevent predation by animals or other insects.

Nectaries are highly variable in morphology, some being no more than a small cluster of secretory cells, others being raised mounds of secretory tissue of various forms (circular or kidney-shaped), whereas some are conspicuous outgrowths containing vascular tissue and both secretory and non-secretory parenchyma. Floral nectaries in many flowers take the form of a continuous ring around the base of the gynoecium (Fig. 15.10a; see Chapter 18 for another illustration). The movement of sugar solution derived from the phloem into the nectaries as well as from them to the exterior of the protoplasts can be either eccrinous (i.e., apoplastic) or granulocrinous. If eccrinous, the secretory cells have the structure of, and function as, transfer cells. An interesting example of granulocrinous transfer occurs in *Lonicera japonica* (honeysuckle). Sugar secretion occurs through short

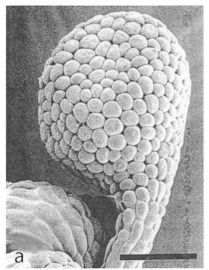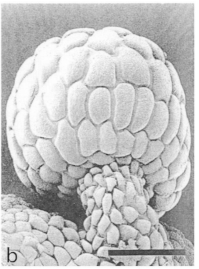

Figure 15.11 Secretory glands on anther connectives. (a) Anther gland of *Stryphnodendron* sp. Bar = 50 μm. (b) Anther gland of *Adenanthera pavonina*. Bar = 120 μm. From Luckow and Grimes (1997). Used by permission of the Botanical Society of America.

epidermal hairs, each consisting of a single large bulbous cell. Prior to secretion, many small ER and/or Golgi vesicles containing a sugar solution fuse with the extensive plasmalemma lining the wall ingrowths, thus extruding their contents into the wall and ultimately into an expanding nectar-filled cavity between the wall and the cuticle. The nectar is released by rupture of the cuticle. In some members of the Myrtaceae, e.g., *Chamelaucium uncinatum*, sugar is transferred granulocrinously from the protoplasts, and through modified stomata into subcuticular cavities. The nectar moves to the exterior through openings in the cuticle that reflect the positions of the subtending stomata (O'Brien *et al.*, 1996). Less well known than nectaries, and uncommon among plants (occurring in the Violaceae and some mimosoid Leguminosae), are secretory glands on anther connectives (Fig. 15.11a, b). They are thought to provide a food reward for pollinators. In some species they secrete a sticky material by which pollen is attached to the body of insects. In others they may function as **osmophores**, structures that emit a perfume which becomes a reward for pollinators such as euglossine bees (see Beardsell *et al.*, 1989; Luckow and Grimes, 1997).

REFERENCES

Ascensao, L., Mota, L., and M. De Castro. 1999. Glandular trichomes on the leaves and flowers of *Plectranthus ornatus*: morphology, distribution and histochemistry. *Ann. Bot.* **84**: 437–447.

Bakker, M. E. and A. F. Gerritsen. 1990. Ultrastructure and development of oil idioblasts in *Annona muricata* L. *Ann. Bot.* **66**: 673–686.

Beardsell, D. V., Williams, E. G., and R. B. Knox. 1989. The structure and histochemistry of the nectary and anther secretory tissue of the flowers of *Thryptomene calycina* (Lindl.) Stapf (Myrtaceae). *Austral. J. Bot.* **37**: 65–80.

Bourett, T. M., Howard, R. J., O'Keefe, D. P., and D. L. Hallahan. 1994. Gland development on leaf surfaces of *Nepeta racemosa*. *Int. J. Plant Sci.* **155**: 623–632.

Colson, M., Pupier, R., and A. Perrin. 1993. Etude biomathématique du nombre de glandes peltées des feuilles de *Mentha* × *piperita*. *Can. J. Bot.* **71**: 1202–1211.

Curtis, J. D. and N. R. Lersten. 1990. Internal secretory structures in *Hypericum* (Clusiaceae): *H. perforatum* L. and *H. balearicum* L. *New Phytol.* **114**: 571–580.

Duke, S. O. 1994. Glandular trichomes: a focal point of chemical and structural interactions. *Int. J. Plant Sci.* **155**: 617–620.

Duke, S. O. and R. N. Paul. 1993. Development and fine structure of the glandular trichomes of *Artemisia annua* L. *Int. J. Plant Sci.* **154**: 107–118.

Elias, T. S. 1983. Extrafloral nectaries: their structure and distribution. In B. Bentley and T. Elias, eds., *The Biology of Nectaries*. New York: Columbia University Press, pp. 174–203.

Esau, K. 1965. *Plant Anatomy*, 2nd edn. New York: John Wiley and Sons.

1975. Laticifers in *Nelumbo nucifera* Gaertn.: distribution and structure. *Ann. Bot.* **39**: 713–719.

1977. *Anatomy of Seed Plants*, 2nd edn. New York: John Wiley and Sons.

Fahn, A. 1988. Secretory tissues in vascular plants. *New Phytol.* **108**: 229–257.

2000. Structure and function of secretory cells. *Adv. Bot. Res.* **31**: 37–75.

Faÿ, E. de, de Sanier, C., and C. Hebant. 1989. The distribution of plasmodesmata in the phloem of *Hevea brasiliensis* in relation to laticifer loading. *Protoplasma* **149**: 155–162.

Jaeger, P. 1961. *The Wonderful Life of Flowers* (Engl. translation). Edinburgh, UK: Chambers Harrap.

Kelsey, R. G., Reynolds, G. W., and E. Rodriguez. 1984. The chemistry of biologically active consituents secreted and stored in plant glandular trichomes. In E. Rodriguez, P. L. Healey, and I. Mehta, eds., *Biology and Chemistry of Plant Trichomes*. New York: Plenum Press, pp. 187–241.

Kim, E.-S. and P. G. Mahlberg. 1991. Secretory cavity development in glandular trichomes of *Cannabis sativa* L. (Cannabaceae). *Am. J. Bot.* **78**: 220–229.

1995. Glandular cuticle formation in *Cannabis* (Cannabaceae). *Am. J. Bot.* **82**: 1207–1214.

2000. Early development of the secretory cavity of peltate glands in *Humulus lupulus* L. (Cannabaceae). *Mol. Cells* **10**: 487–492.

Lersten, N. R. and J. D. Curtis. 1982. Hydathodes in *Physocarpus* (Rosaceae: Spiraeoideae). *Can. J. Bot.* **60**: 850–855.

Luckow, M. and J. Grimes. 1997. A survey of anther glands in the mimosoid legume tribes Parkieae and Mimoseae. *Am. J. Bot.* **84**: 285–297.

Mahlberg, P. G. 1961. Embryogeny and histogenesis in *Nerium oleander*. II. Origin and development of the non-articulated laticifer. *Am. J. Bot.* **48**: 90–99.

Mahlberg, P. G. and E.-S. Kim. 1991. Cuticle development on glandular trichomes of *Cannabis sativa* (Cannabaceae). *Am. J. Bot.* **78**: 1113–1122.

Mahlberg, P. G. and P. S. Sabharwal. 1967. Mitosis in the non-articulated laticifer of *Euphorbia marginata*. *Am. J. Bot.* **54**: 465–472.

Muller, W. H. and C. H. Muller, 1964. Volatile growth inhibitors produced by *Salvia* species. *Bull. Torrey Bot. Club* **91**: 327–330.

O'Brien, S. P., Loveys, B. R., and W. J. R. Grant. 1996. Ultrastructure and function of floral nectaries of *Chamelaucium uncinatum* (Myrtaceae). *Ann. Bot.* **78**: 189–196.

Osmond, C. B., Lüttige, U., West, K. R., Pallaghy, C. K., and B. Shacher-Hill. 1969. Ion absorption in *Atriplex* leaf tissue. II. Secretion of ions to epidermal bladders. *Austral. J. Biol. Sci.* **22**: 797–814.

Thomas, V. 1991. Structural, functional and phylogenetic aspects of the colleter. *Ann Bot.* **68**: 287–305.

Thomson, W. W., Berry, W. L., and L. L. Liu. 1969. Localization and secretion of salt by the salt glands of *Tamarix aphylla*. *Proc. Natl Acad. Sci.* USA **63**: 310–317.

Turner, G. W., Berry, A. M., and E. M. Gifford. 1998. Schizogenous secretory cavities of *Citrus limon* (L.) Burm. f. and a re-evaluation of the lysigenous gland concept. *Int. J. Plant Sci.* **159**: 75–88.

Werker, E., Putievsky, E., Ravid, U., Dudai, N., and I. Katzir. 1993. Glandular hairs and essential oil in developing leaves of *Ocimum basilicum* L. (Lamiaceae). *Ann. Bot.* **71**: 43–50.

FURTHER READING

Boughton, V. H. 1981. Extrafloral nectaries of some Australian phyllodineous acacias. *Austral. J. Bot.* **29**: 653–664.

Curtis, J. D. and N. R. Lersten. 1974. Morphology, seasonal variation, and function of resin glands on buds and leaves of *Populus deltoides* (Salicaceae). *Am. J. Bot.* **61**: 835–845.

1994. Developmental anatomy of internal cavities of epidermal origin in leaves of *Polygonum* (Polygonaceae). *New Phytol.* **127**: 761–770.

Fahn, A. 1979. *Secretory Tissues in Plants*. London: Academic Press.

Hillis, W. E. 1987. *Heartwood and Tree Exudates*. Berlin: Springer-Verlag.

Keeler, K. H. 1977. The extrafloral nectaries of *Ipomoea carnea* (Convolvulaceae). *Am. J. Bot.* **64**: 1182–1188.

Lersten, N. R. and J. D. Curtis. 2001. Idioblasts and other unusual internal foliar secretory structures in Scrophulariaceae. *Plant Syst. Evol.* **227**: 63–73.

Lersten, N. R. and H. T. Horner. 2000. Calcium oxalate crystal types and trends in their distribution patterns in leaves of *Prunus* (Rosaceae: Prunoideae). *Plant Syst. Evol.* **224**: 83–96.

Mahlberg, P. G. and E.-S. Kim. 1992. Secretory vesicle formation in glandular trichomes of *Cannabis sativa* (Cannabaceae). *Am. J. Bot.* **79**: 166–173.

Metcalfe, C. R. 1967. Distribution of latex in the plant kingdom. *Econ. Bot.* **21**: 115–127.

Pate, J. S. and B. E. S. Gunning. 1972. Transfer cells. *Annu. Rev. Plant Physiol.* **23**: 173–196.

Scheres, B. 2002. Plant patterning: try to inhibit your neighbors. *Curr. Biol.* **12**: R804–R806.

Schnepf, E. 1974. Gland cells. In A. W. Robards, ed., *Dynamic Aspects of Plant Ultrastructure*. London: McGraw-Hill.

Swain, T. 1977. Secondary compounds as protective agents. *Annu. Rev. Plant Physiol.* **28**: 479–501.

Thurston, E. L. 1974. Morphology, fine structure, and ontogeny of the stinging emergence of *Urtica dioica*. *Am. J. Bot.* **61**: 809–817.

Werker, E. and A. Fahn. 1969. Resin ducts of *Pinus halepensis* Mill.: their structure, development and pattern of arrangement. *Bot. J. Linn. Soc.* **62**: 379–411.

Wooding, F. B. P. and D. H. Northcote. 1965. The fine structure of mature resin canal cells in *Pinus pinea*. *J. Ultrastruct. Res.* **13**: 233–244.

Chapter 16

The root

Perspective: evolution of the root

The anatomy of the root reflects its origin, its subterranean environment, and its function. The first vascular plants (Rhyniophyta) lacked roots, and absorption of water and nutrients was facilitated by rhizoids. Roots evolved in the seed plant clade (rhyniophytes, trimerophytes, progymnosperms, seed plants) as well as in lycophytes, sphenophytes, and ferns in response to the pressures of a land environment, enhanced by increasing plant size. During their evolution important functions such as anchorage, absorption and transport of minerals and water, and storage of photosynthate were established. In some ways, however, roots changed relatively little through time. This is the result of the subterranean environment in which they evolved, and the fact that roots were, thus, not exposed to the same intense selection pressures as stems.

The seed plant root (Fig. 16.1a, b) is considered by most researchers to be an evolutionarily modified stem although it has also been suggested that it might be an entirely new organ that evolved independently of the stem. The predominant view is supported by the fact that the structure of the root of extant plants is remarkably similar to the anatomy of the stem of their ancestors. Even in many plants with stems that feature specialized siphonostelic or eustelic structure, the roots are protostelic (Fig. 16.1b), also a feature of the stems of very primitive plants. Roots with central piths have an alternate arrangement of xylem and phloem that may reflect a protostelic origin.

The origin of the root in the lycophyte clade was apparently quite different from that in the seed plant clade. The zosterophyllophytes, from which the lycophytes are thought to have evolved, lacked roots. In the most highly specialized lycophytes, the Lepidodendrales, roots were borne on stigmarian appendages which were similar to stems in anatomy. The roots were helically arranged and are considered by many paleobotanists to be leaf homologs. It is hypothesized, therefore, that in this clade roots and leaves evolved from similar structures (see Stewart and Rothwell, 1993). The roots of all lycophytes, herbaceous and arborescent, are of remarkably similar anatomy, thus supporting

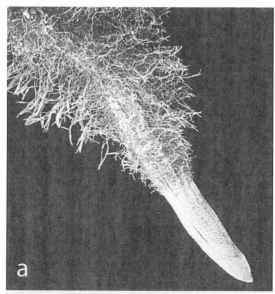

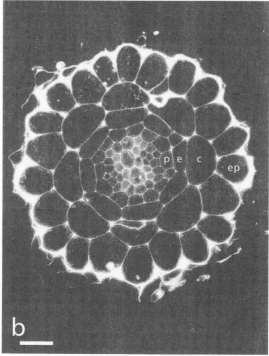

Figure 16.1 (a) Part of the root of a seedling of *Rhaphanus sativus* (radish) showing the root tip and a zone of numerous root hairs. Magnification × 20. (b) Transverse section of a root of *Arabidopsis thaliana* showing the central column (a protostele) of primary xylem and primary phloem enclosed by the pericycle (p), the endodermis (e), a single-layered cortex of eight large cells (c), and the epidermis (ep) Bar = 10 μm. (a) From Troughton and Donaldson (1972). Used by permission of the New Zealand Ministry of Research, Science and Technology. (b) From Bowman (1994). Used by permission of Springer-Verlag Heidelberg. ©️ Springer-Verlag Berlin Heidelberg.

the concept of a single origin of roots in this group. Within vascular plants, however, it seems likely that roots are polyphyletic in origin, having evolved independently in seed plants and lycophytes (see Knoll and Niklas, 1987). The origin of roots in sphenophytes and ferns is unclear.

The major structural adaptations of roots are directly related to their functions. As plants increased in size during their evolution, the problem of anchorage became progressively more important, and this is reflected in root structure. Most large plants have tap root systems which develop directly from the radicle in the embryo whereas smaller plants are often characterized by fibrous root systems consisting of many small roots that spread out laterally from the base of the stem. There are, however, many variations in the morphology of root systems, often related to the habitat of the plant.

As adaptations to the function of absorption, roots are characterized by the production of absorptive roots hairs in a region just proximal to the root apex. In addition, suberized Casparian bands in the walls of endodermal cells prevent back-flow of nutrients through the apoplast and out of the region of vascular tissue. A subepidermal exodermis, similar in structure to the endodermis, has the important function of restricting water loss from the root during periods of drought. In some taxa, absorbing roots are greater in diameter and have proportionately more vascular tissue containing broader tracheary and sieve elements than anchoring roots (Wilder and Johansen, 1992). The symbiotic relationship of most seed plants with fungi results in the formation of mycorrhizae that greatly facilitate the absorption of nutrients from the surrounding soil environment. Cluster roots enhance the absorption of phosphorous in mineral-deficient regions. In line with their function as absorbing organs, the roots of most plants penetrate the soil substrate. The apical meristem is covered by the root cap, a conical structure that protects and lubricates the root tip as it grows through the soil.

The structure of roots is also adapted to the function of storage. In roots that have primary structure only, the cortex is usually large, providing an expansive area for storage of photosynthate, often in the form of starch. In woody roots the phellogen differentiates in the pericle which typically results in an early elimination of the cortex. Consequently, the secondary xylem which is highly parenchymatous becomes the major storage tissue. We shall discuss later, in more detail, each of these aspects of root structure and function.

Gross morphology

Gymnosperms and dicotyledons are characterized, typically, by tap root systems, whereas monocotyledons most commonly have fibrous root systems. The roots of extant ferns, sphenophytes, and lycophytes are usually adventitious roots that originate from stems. A **tap root** develops from a meristem at the lower end of the hypocotyl of the embryo. Once developmentof the root has begun in gymnosperms and dicotyledons its growth continues throughout the life of the plant. The main axis of a tap root system may extend to great depths in the soil, and often becomes woody, especially in trees, providing effective anchorage for these large plants. Lateral roots which develop from the tap root commonly branch, and extend outward for great distances. These lateral roots may also become woody and, with the main axis of the tap root system, are effective in maintaining the vertical orientation of the plant under adverse environmental conditions such as flooding and/or high wind velocities. Their location just beneath the soil surface is also highly beneficial to the nutrition of the plant since they occupy nutrient-rich surface areas (Coutts and Nicoll, 1991). Although the biological and/or physical factors that result in the shallow subsurface orientation of lateral roots are not clearly understood, Coutts and Nicoll (1991) observe that some of the lateral roots of a tree are **plagiogravitropic**, that is, they grow obliquely upward, but just prior to reaching the soil surface they begin a downward deflection, thus maintaining their subsurface orientation. The biological and/or physical signals that result in the downward deflection of the roots are unknown.

In monocotyledons an initial tap root usually aborts early in development, and the root system becomes composed of numerous **adventitious roots**, lacking secondary growth, which develop from the base of the stem. **Fibrous root systems** thus formed are effective in anchorage as well as in nutrient absorption and transport.

Contractile roots and other highly specialized root systems

Although in seed plants, tap root and fibrous root systems are very common, highly specialized root systems may characterize plants growing in specific habitats or in which a particular function is enhanced. For example, mangroves, which grow in water, develop large prop roots that extend from the stem into the sub-aquatic substrate. The roots of plants growing in wet environments are often characterized by extensive development of cortical aerenchyma. The roots of some plants, such as carrot, rutabaga, beet, radish, etc. are fleshy storage organs, consisting almost entirely of parenchyma. **Cluster roots** are common in many families of dicotyledons growing in regions deficient in phosphorous, and provide an important mechanism for the acquisition of this element, essential in plant development and function. Cluster roots consist of a group of small, determinate roots that are initiated in the pericycle opposite protoxylem strands (Skene, 2000). Organic acids and phosphatases which are released from these roots into the substrate function in phosphorous acquisition and, possibly, the absorption of other soil nutrients (Skene, 1998; Watt and Evans, 1999).

Contractile roots are common in many taxa of both monocotyledons and dicotyledons. They are especially common in plants characterized by bulbs or corms such as crocus and hyacinth as well as some that are not such as dandelion. By contracting, the roots pull the plant closer to the soil surface, or deeper within the soil. This adaptation assures that the bulbs or corms of such plants (called **geophytes**) become established in the soil at physiologically and ecologically effective depths (Pütz, 1991) where they or their developing aerial shoots will have access to light and adequate soil moisture (among other environmental factors), and will be protected from low winter and high summer temperatures.

The mechanism of root contraction in geophytes has been of considerable interest for many years. Recent histological studies have shown conclusively that root contraction in many geophytes is caused by a change in dimensions and volume of cells in the inner and middle cortex. For example, in both *Chlorogalum pomeridianum* and *Hyacinthus orientalis*, members of the Liliaceae, the inner and middle cortical cells increase in radial dimension and decrease in longitudinal dimension. There is also a substantial increase in the volume of cells that comprise the outer part of the middle cortex. These changes result in shortening the root and the concomitant collapse of the outer cortical cells and epidermis (Jernstedt, 1984a, b). Although the mechanism of change in cell shape is not well understood, there is a correlation between the change in cell shape and changes in cell wall structure, differential wall extensibility, and deposition of new cell wall material (see Wilson and Honey, 1966; Mosiniak *et al.*, 1995). As the process of contraction continues over several periods of growth, the outer, collapsed layers of root tissue form conspicuous and characteristic ridges and folds.

Induction of the development of contractile roots in geophytes is apparently related to the depth of the planted bulb or corm. It was known as early as the late nineteenth century that deeply planted geophytes do not produce contractile roots. Research by Halevy (1986) demonstrated that contractile roots developed in *Gladiolus grandiflorus* only on small offshoot corms located close to the soil surface. He observed, further, that light and temperature fluctuations were essential for the development of contractile roots in this taxon. As growth of the aerial shoot began, two basal, sheathing leaves of specific lengths were produced. The upper sheathing leaf that extended above the soil level functioned as a light receptor which controlled depth. If the plant was deprived of either light or fluctuating temperatures, contractile roots failed to develop.

Whereas the corms of some taxa are pulled downward through the soil by contractile roots and must overcome the soil resistance, those of others, especially small, offshoot corms, move through channels in the soil made by a contractile root produced by the offshoot. In the latter situation, if the diameter of the corm and root are similar, the corm is pulled through the channel made by the contracting root with little or no soil resistance (Pütz, 1991). For more detail about the interesting subject of contractile roots, please refer to the references cited above.

Apical meristems

Several major histological patterns characterize the apical meristems of roots. In many pteridophytes there is a **single apical initial** from which all other tissues in the root are ultimately derived whereas

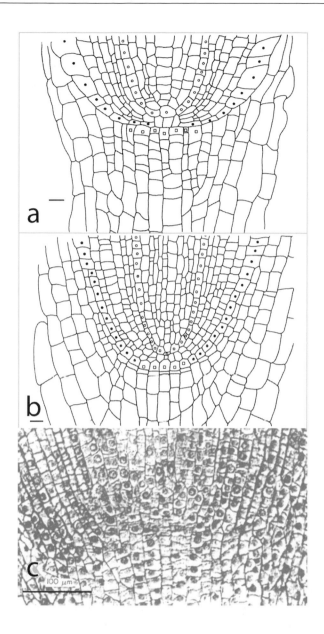

Figure 16.2 (a, b) Drawings of median longitudinal sections of the apical meristems of roots of a monocotyledon and a dicotyledon. (a) The apical meristem of *Triticum aestivum*, a monocotyledon, is characterized by three tiers of initials. The provascular column, the outer layer of which is indicated by circles, can be traced back to the basal tier, the ground tissue (unmarked) and protoderm (black dots) to the middle tier, and the root cap to the apical tier (calyptrogen) (squares). (b) The apical meristem of *Sinapis alba*, a dicotyledon, is also characterized by three tiers of initials. The provascular column (circles) can be traced back to the basal tier (as in the monocotyledon), the ground tissue (unmarked) to the middle tier, and the protoderm (black dots) and root cap to the apical tier (squares). Bar (a, b) = 20 μm. (c) Photograph of a median longitudinal section of the apical meristem of a root of *Vicia faba* consisting of a single zone of initials from which all tissue regions are derived, including a columella centrally located in the root cap. (a, b) From Clowes (1994). Used by permission of Blackwell Publishing Ltd. (c) From Clowes (1976). Used by permission of Elsevier.

lycophytes are characterized by a small **cluster of apical initials**. In most monocotyledons **three tiers of initials** have been recognized (Fig. 16.2a). By tracing cell lineages, the provascular column seems to arise from the basal tier, the ground meristem and the protoderm from the middle tier, and the root cap (or calyptra, a term rarely used today) from the apical tier, often referred to as the **calyptrogen**. Three tiers have also been recognized in many dicotyledons (Fig. 16.2b). As in the monocoyledons, the provascular tissue can be traced through cell lineages to the basal tier, the ground meristem, alone, to the middle tier, and both the protoderm and the root cap to the apical tier. In some taxa, including many tree species, the root apex consists of a single zone of initials (Fig. 16.2c) from which all regions of the root

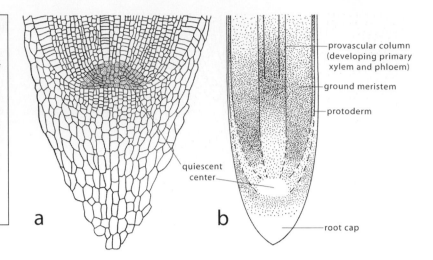

Figure 16.3 Median longitudinal views of root apices showing the quiescent center. (a) *Zea mays*. (b) *Allium cepa* showing distribution of cell divisions. Frequency of cell division is indicated by density of stippling. Light shading indicates low frequency, dark shading, high frequency. (a) From Clowes (1959). Used by permission of the Cambridge Philosophical Society. (b) From Jensen and Kavaljian (1958). Used by permission of the Botanical Society of America.

provascular column (developing primary xylem and phloem)

ground meristem

protoderm

quiescent center

root cap

a

b

appear to arise, including a **columella** centrally located in the root cap.

In roots of both monocotyledons and dicotyledons in which the various mature regions can be traced to distinct meristematic tiers, the apical meristems are considered to be **closed** (Fig. 16.2a, b). In trees and other woody, as well as herbaceous, species of dicotyledons characterized by a single apical group of initials from which root cap, epidermis, cortex, and primary vascular system are all derived, the apical meristem is considered to be **open** (Fig. 16.2c). For exceptions and variations see Clowes (1994). Those who accept the concept that root apical meristems are composed of tiers of initials believe that the ultimate fate of the files of cells derived from the initials is restricted by cell lineages since these cells can be traced back to the initials. However, in many taxa a very low frequency of cell division has been observed in the region of the tiers of initials. This has led to the establishment of the concept of the **quiescent center** (see Clowes, 1959; 1961), a region of minimal cell division enclosed by a region of active cell division (Fig. 16.3a, b). According to Clowes, the quiescent center and the active initials which surround it comprise the **promeristem** in roots. Recent research has shown that the quiescent center plays an important role in development which we shall consider further in a later section.

The nineteenth-century German worker Hanstein proposed that the major tissue regions of the root were derived from a large meristem comprising the root tip. He recognized three distinct regions, or **histogens**, in this meristem, which he believed were derived from tiers of initials in the root tip. He proposed that the central histogen, called the **plerome**, gave rise to the vascular column, that a surrounding histogen, called the **periblem**, gave rise to the cortex, and that an outermost histogen, called the **dermatogen**, gave rise to the epidermis. If one excludes the apical tiers of initials in the

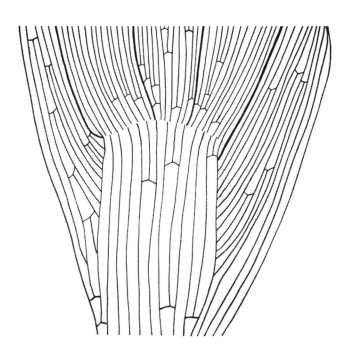

Figure 16.4 Arrangement of tissue domains in the root apex of *Vicia faba* as based on the Körper–Kappe concept. At levels at which new cell rows are formed there is a T-configuration. In the Körper (body) domain the T configuration is inverted whereas in the Kappe (cap) domain the T configuration in normal. This pattern is not, however, consistent as illustrated in *Vicia* in which the Körper includes not only the provascular tissue and ground meristem but also the columella (the central part of the root). Only the peripheral part of this root cap would be included in the Kappe domain. Note that the protoderm is also a part of the Kappe domain. From Clowes (1959). Used by permission of the Cambridge Philosophical Society.

concept, Hanstein's histogens become approximately equivalent to the provascular column, ground meristem, and protoderm located between the apical initials and the mature tissue regions. However, today we recognize that these are not permanent meristems or components of a larger meristem but, rather, transitional regions between the apical initials and mature tissue regions in which cell division, growth, and differentiation occur.

Another way to look at root tip structure is the **Körper–Kappe** (body–cap) concept (Fig. 16.4) proposed by Schuepp in 1917. He noted that, during growth in the central region of the root tip at the level at which new files of cells form, there is an inverted T configuration, whereas in the outer region and in the root cap of some roots, the beginning of new files of cells is marked by a normal T configuration. The region in which the cross bar of the T faces the root apex is designated **Körper**, or body. The region in which the cross bar faces the base of the root is designated **Kappe**, or cap although in some taxa an inverted T configuration extends into the **columella**, the central part of the root cap (Fig. 16.4). In some cases Kappe configurations coincide solely with the root cap, and Körper configurations with the root tip proper (i.e., provascular column, ground meristem, and protoderm). In others Kappe is composed of root cap and protoderm, and may even include some of the outer ground meristem. When the Kappe of a particular root consists of root cap only we can conclude that there is a separate root cap initial. If the Kappe consists of root cap and protoderm, we know that the root cap and protoderm have a common initiating layer, or tier of initials. If the Kappe, in addition, includes some ground meristem, we can conclude that there is a cluster of apical initials, not distinctly tiered.

The quiescent center and its role in development

Although, as noted above, there has been great emphasis in the literature on the presence of tiers of cells that function as initiating layers in the roots of seed plants, work during the 1950s using radioactive isotopes demonstrated that the region in which these initials occur in mature roots is, in fact, a zone of minimal cell-divisional activity. Consequently, this zone was labeled the **quiescent center** (Fig. 16.3) by the British botanist Clowes (see, e.g., Clowes, 1959, 1961). At about the same time, American botanists Jensen and Kavaljian, and the French botanist Buvat, arrived at similar conclusions in studies of the frequency of mitoses in root tips (Fig. 16.3b). Subsequently, the quiescent center has been documented in the root tips of many seed plants. Whereas early in development, the original tissue pattern in many species is apparently derived from tiers of meristematic initials in the root apex, with the establishment of the quiescent center the active zone of cell division is shifted to its periphery. Experiments in which auxin or gibberellin production was restricted by the use of chemical inhibitors (Barlow, 1992; Kerk and Feldman, 1994) and which led to increased cytokinesis in the quiescent center support the viewpoint that the low rate of cell division as well as the maintenance of the quiescent center are controlled, at least in part, by hormones produced in the root cap or root cap initials.

Studies of the embryogenesis of *Arabidopsis thaliana* have shown that the quiescent center and the columella of the root cap originate from direct derivatives of the hypophyseal cell of the developing embryo whereas the more proximal initials which enclose the quiescent center are derived from the apical cell of the globular embryo. The quiescent center and enclosing initials comprise the promeristem, described by Scheres *et al.* (1996) as the "functionally integrated root meristem." Genetic studies of root meristem development in *Arabidopsis* demonstrate that specimens which have the mutant *HOBBIT* (*hbt*) gene lack a quiescent center. In addition the calyptrogen (root cap meristem) develops abnormally resulting in a root cap lacking a columella, demonstrating that the *hbt* gene is required for normal root meristem development (Willemsen *et al.*, 1998). Other studies, by these workers, of mutant embryos in which major tissue regions fail to develop suggest that regional identity of tissue domains is established early in embryo development. During normal embryogenesis in *Arabidopsis*, protoderm, ground meristem, and provascular tissue become defined very early as clonal domains that seem to be derived from specific initials. This supports the viewpoint that these tissue regions are derived, initially, as cell lineages from specific initials or groups (tiers) of initials as described above (see Scheres *et al.*, 1996). This research group has shown, further, however, that following ablation (removal by killing) of initials from which a particular tissue type is directly derived, the adjacent cells that take over the positions of the ablated cells produce cells of the tissue type of the ablated initials. For example, if root cap columella cells were normally derived from the ablated initials, their replacements, which had originally been programmed to produce cells of vascular tissue "switch fate" and begin producing columella cells (see also van den Berg et al., 1997), thus supporting a conclusion that, in some species, root meristem cells and their derivatives develop according to position rather than as cell lineages, the predominant past viewpoint. Additional evidence of the function of the quiescent center has been provided recently by Ponce et al. (2000), who demonstrate that following excision of the root cap in *Zea mays* genes that code for cell wall proteins or enzymes are expressed in the regenerating root cap only after the quiescent center has been re-established.

These studies provide strong evidence for communication between the quiescent center and the associated apical initials, the operation of a system of positional information, and some control of development by the quiescent center in both *Arabidopsis* and *Zea* (see Scheres et al., 1996; Schiefelbein et al., 1997; Ponce et al., 2000).

Ponce et al. (2000) describe the quiescent center "as an architectural template in the root apical meristem of all angiosperm and gymnosperm root tips . . . [which with surrounding initials] may regulate the positional and structural expression of . . . genes [which control the differentiation of tissue regions in roots]." Evidence that the quiescent center has a significant role in development is one of the exciting conclusions to come from recent research in plant molecular biology.

Primary tissues and tissue regions

The arrangement of primary tissues in the root parallels that in the stem, yet differs in several distinctive ways. In roots of many plants the central region consists of a solid column of primary xylem (Figs. 16.1b, 16.5) whereas in others there is a central pith. The column of primary xylem, whether with or without a pith, is often fluted, appearing ribbed when observed in transverse section. The primary phloem commonly comprises small longitudinal bundles between the ribs of xylem (Fig. 16.5). In some dicotyledons and many monocotyledons in which a pith is present, the xylem and phloem occur in alternating bundles (Fig. 16.6b, d). The primary xylem and phloem (and the pith, when present) plus the pericycle (Figs. 16.1b, 16.5) comprise the **central column**.

The primary xylem is commonly two-ribbed, three-ribbed, four-ribbed, or five-ribbed, and is referred to as being diarch, triarch, tetrarch, or pentarch; but it may also be polyarch (Fig. 16.7a–e). Order of maturation of primary xylem is always exarch with protoxylem located in the tips of the ribs, or if considered in three dimensions,

Figure 16.5 Transverse section of a root of *Ranunculus* sp. (buttercup) illustrating the ribbed protostele with strands of primary phloem between the ribs. Magnification × 260.

Figure 16.6 Roots of monocotyledons. (a) Transverse section of a root of *Smilax* containing a central column consisting of a compact cylinder of radiating primary vascular strands enclosing a well-defined pith. Magnification ×15. (b) Enlargement of a part of the central column in (a) illustrating histological details. Note the lateral root and the bundles of primary phloem that alternate with strands of primary xylem. Magnification × 275. (c) Transverse section of a root of *Zingiber officinate* (ginger) with a central column consisting of a pith and numerous alternating strands of primary xylem and primary phloem embedded in parenchyma. A lateral root extends through the cortex. Magnification × 12. (d) Enlargement of part of the central column of the root shown in (c). Magnification × 180.

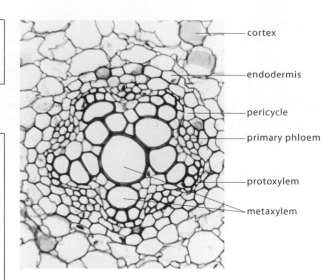

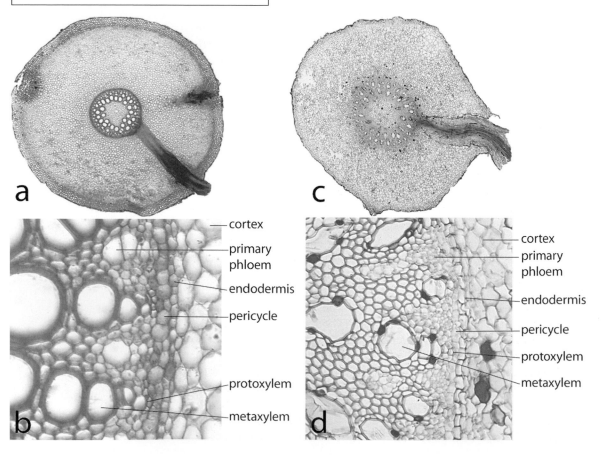

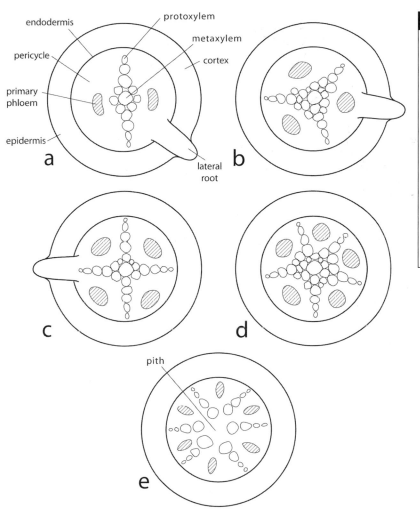

comprising the outermost edges of the ribs. The number of xylem ribs or bundles (where there is a pith) is a constant character in some taxa, but variable in others. The histological characteristics of both primary xylem and phloem in roots are similar to those in stems. It is, however, more difficult to distinguish accurately protoxylem from metaxylem and protophloem from metaphloem in roots because during development there is commonly considerably less elongation of tissues in the region of first differentiation of these tissues than in stems. Consequently, there are often no protoxylem tracheids or vessel members with annular or helical secondary wall thickenings that would allow for stretching, and there is less obliteration of protophloem than is typical in stems.

The **pericycle** (Figs. 16.5, 16.6b, d) is a narrow layer of potentially meristematic parenchyma between the vascular tissues and the endodermis from which lateral roots develop (see pp. 295–298 for a detailed discussion of lateral root development). In roots with secondary growth, the phellogen also develops within the pericycle.

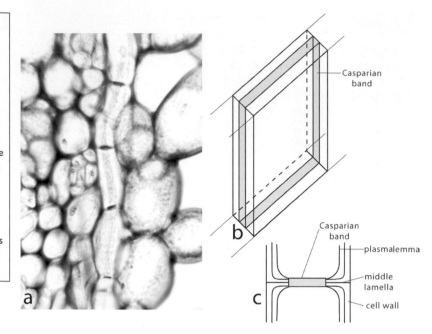

Figure 16.8 Endodermis illustrating Casparian bands. (a) Endodermis in the stem of *Acorus calamus*. Note Casparian bands in surface and sectional views. (b) Three-dimensional diagram of an endodermal cell showing the Casparian band in radial and transverse walls. (c) Sectional view showing that the suberized Casparian band extends through the walls of contiguous cells. Note that the plasmalemma of each cell is tightly attached to the Casparian band. Upon plasmolysis, the plasmalemma pulls away from the walls except in the region of the Casparian band.

The **endodermis** (Figs. 16.1b, 16.5, 16.6b, 16.8a) is a highly specialized, single layer of cells that comprises the innermost layer of the cortex. Although it is characteristic of roots where it serves very important functions (see below), it also occurs in leaves, stems, or rhizomes of some angiosperms, the leaves of many pteridophytes, and the leaves of some conifers (Lersten, 1997).

The endodermis consists of tabular cells, the transverse and radial walls of which are characterized by the presence of a continuous suberized and often lignified region in the primary wall called the **Casparian band** (Fig. 16.8a, b, c) (see Schreiber, 1996), a prominent feature of seed plants, but which also occurs in many species of pteridophytes (Damus *et al.*, 1997). In addition to suberin and lignin, the Casparian bands may also contain cell wall proteins and carbohydrates (Schreiber *et al.*, 1999). In roots, the Casparian bands in the endodermis as well as that in the exodermis of many plants (see below) restrict, and possibly regulate, the apoplastic uptake and loss of water and solutes through the radial and transverse walls (Hose *et al.*, 2001). The Casparian band extends through the walls and middle lamella of contiguous cells, and is in contact with the plasmalemmas of these cells which adhere tightly to it. Upon plasmolysis of the protoplasts, the plasmalemmas are pulled away from the cell walls except in the region of the bands (Fig. 16.8c), often forming taut sheets resulting in so-called **band plasmolysis**. In addition to the Casparian band, the inner layer of the primary walls in many taxa become impregnated with suberin and is known as the **suberin lamella**. In many plants the suberin lamellae are covered by a cellulosic layer of variable thickness which may become lignified (see Seago *et al.*, 1999; Ma and Peterson, 2001a). Solutes that move into the root are prevented by the Casparian bands from moving through the transverse and radial walls and by the suberin lamellae from moving through the tangential walls of endodermal cells. Consequently, transport across the endodermis must occur via plasmodesmata (at least in many angiosperms) which are not severed by development of the suberin lamellae (see Verdaguer and Molinas, 1997; Ma and Peterson, 2001a, b), or apoplastically through the tangential walls of **passage cells** (thin-walled endodermal cells that lack suberin lamellae) (Clarkson, 1991; Peterson and Enstone, 1996). Thus, as the concentration of ions in solution increases to the inside of the endodermis, the Casparian bands, impermeable to the passage of small molecules, prevent the diffusion of ions out of the apoplast within the vascular column through the transverse and radial walls of the endodermal cells (see Clarkson and Robards, 1975; Clarkson, 1991; Peterson *et al.*, 1993). In many monocotyledons which lack secondary growth and herbaceous dicotyledons that produce only small quantities of secondary tissue the endodermis may become very thick-walled and persist for many years or for the life of the plant. The presence of passage cells may be especially important in such plants. Another possible function of passage cells in the endodermis is the transfer of calcium and magnesium into the transpiration stream. When the epidermis and central cortex die they are the only cells exposed to the soil solution that are capable of ion uptake (Peterson and Enstone, 1996). No plasmodesmata have been observed in the endodermal cell walls in the roots of conifers so it appears likely that movement of solutes through the tangential walls of these cells is apoplastic (see Verdaguer and Molinas, 1997).

The endodermis begins its development very close to the root tip, and is functional in the region of root hair development. In the primary (tap) root of *Quercus suber* (cork oak), for example, the developing endodermis was observed within about 1 mm of the junction of the root cap and the root proper, and at about the same level as the first evidence of maturation of protophloem elements (Verdaguer and Molinas, 1997). The Casparian band began to develop at 20–25 mm from the root tip opposite the protophloem, and plasmodesmata were apparent, except in the region of the Casparian band. By 70–80 mm from the root tip, the primary cell walls were covered on their inner surfaces by a uniformly thick suberin lamella except in the region of plasmodesmata and the Casparian band where it was very thin or absent. At maturity, the suberin lamellae were covered by a thin, cellulosic layer (Verdaguer and Molinas, 1997). Throughout development, the protoplast contained conspicuous ER and Golgi bodies and numerous associated vesicles which, presumably, transported, granulocrinously, precursor compounds of suberin, lignin, cellulose, and other components of the cell walls. The endodermal cell walls of all species previously reported, from pteridophytes to gymnosperms and angiosperms, are similar in ultrastructure, differing only in the thickness of the suberin lamellae (see references cited in Verdaguer and Molinas, 1997).

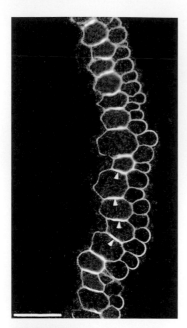

Figure 16.9 Casparian bands as seen in sectional views (at arrow heads) in the exodermis of the root of *Allium*. The layer of cells on the right is the epidermis. Bar = 50 μm. From Peterson (1988). Used by permission of Blackwell Publishing Ltd.

To the exterior of the endodermis, the **cortex** consists predominantly of parenchyma, but may contain extensive regions of collenchyma and/or sclerenchyma, especially near the periphery. The parenchyma tissue in the cortex typically contains a complex system of **intercellular air channels** which facilitates aeration, i.e., entry of oxygen and release of carbon dioxide. Such tissue, often referred to as **aerenchyma**, is especially well developed in roots that grow in aquatic or highly moist environments (see Peterson, 1992; Seago *et al.*, 2000a, b).

A major function of the cortex in most roots is storage of photosynthate, usually in the form of starch. Entire tap roots of carrot, beet, rutabaga, etc. are highly specialized as storage organs. Even the secondary xylem in such roots is highly parenchymatous, and functions primarily in storage.

The roots of many herbaceous plants and the apical regions of roots of some woody plants are characterized by a specialized tissue immediately below the epidermis called the exodermis (Fig. 16.9). The **exodermis** is similar in both structure and function to the endodermis, and like the latter, functions in controlling the passage of water and solutes into and out of the root (Ma and Peterson, 2001a, b). Its major function is thought to be the restriction of apoplastic water loss from roots during conditions of low water potential resulting from drought (Enstone and Peterson, 1997; Taleisnik *et al.*, 1999) or their presence in saline substrates (Taleisnik *et al.*, 1999). The exodermis consists of one to several layers of cells characterized by suberin lamellae which comprise the inner wall layer. Casparian bands (Fig. 16.9) extend through the radial and transverse walls of contiguous cells. Band plasmolysis occurs in the exodermis, but is less common than in the endodermis (Enstone and Peterson, 1997). The presence of Casparian bands and suberin lamellae in exodermal cells prevents ions in the soil solution from moving through the apoplast of these cells into the cortex. Consequently, with the exception of passage cells that lack suberin lamellae, movement of ions through the exodermal cells must be symplastic, i.e., through plasmodesmata (Peterson, 1988; Peterson and Enstone, 1996).

The effectiveness of the exodermis in restricting water loss from roots is directly related to the degree of maturity of the tissue. During root development, cells of the exodermis have been observed to differentiate asynchronously in some species (e.g., *Zea mays*), with mature cells occurring in groups or "patches" (Enstone and Peterson, 1997) whereas others have reported a completely developed exodermis close to the root tip (Wang *et al.*, 1995). Consequently, restriction by the exodermis of apoplastic water loss can be variable. Although Taleisnik *et al.* (1999) observed that water retention in root segments containing an exodermis was "significantly higher" than that in non-exodermal segments, they concluded that **rhizosheaths** (see Wang *et al.*, 1991), sheaths of soil that form around roots in very dry soil, may be more effective than the exodermis in restricting water loss.

protoderm

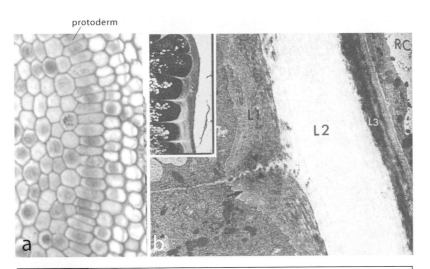

Figure 16.10 (a) Protoderm in the root tip of *Allium* with numerous nuclear division figures indicating cell divisional activity. Note also, the numerous recently formed anticlinal walls. Magnification × 343. (b) Highly magnified region of a transverse section of the root tip of *Zea mays* showing the primary wall (L1) of protodermal cells and outer L2 and L3 layers which comprise the pellicle, a protective structure. Magnification × 4800. Inset, a series of epidermal cells. Magnification × 490. (b) From Abeysekera and McCully (1993a). Used by permission of Blackwell Publishing Ltd.

Prior to the formation of secondary tissues and in roots in which little or no secondary tissues develop, the **epidermis** (Figs. 16.1b, 16.9) functions primarily in absorption, enhanced by the presence of specialized root hairs (see later section). Since, however, the root tip is a site of active cell division and differentiation, the developing epidermis (protoderm) provides, in addition, a very important protective function. The nature of the surface of the protoderm is of great importance since it must not only serve in protecting the root from insects and pathogens and the detrimental effects of moving through the soil, but also must accommodate the multiple developmental changes occurring in both the protoderm itself as well as more internal cells (Abeysekera and McCully, 1993a, b, 1994). Among these changes are rapid cell divisions (Fig. 16.10a) and the consequent expansion of the protodermal surface as well as cell division and extension growth (elongation) of cells of the ground meristem and provascular column, among others. Abeysekera and McCully describe a specialized outer surface of the protodermal cells in *Zea mays* which they describe as a distinct structure consisting of three layers (Fig. 16.10b). They consider the primary cell wall of the protodermal cells, characterized by a helicoidal arrangement of cellulose microfibrils, to be the inner layer (L1) of this surface structure. The outer layers, L2 and L3, comprise the **pellicle**. These layers differ in both staining characteristics and structure from the underlying cell walls. L2 is thick with an amorphous texture and a smooth outer boundary which does not follow the contours of the protodermal cells. Fine microfibrils, difficult to

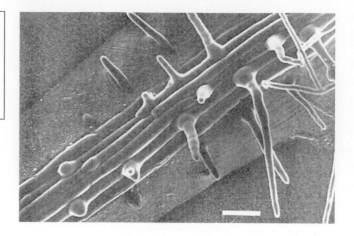

Figure 16.11 Developing root hairs in the root hair zone of *Arabidopsis thaliana*. Bar = 50 μm. From Schiefelbein and Somerville (1990). Used by permission of the American Society of Plant Biologists.

distinguish, lie parallel to the long axis of the root. L3, the outer, very thin layer "is coarsely and irregularly fibrillar." The pellicle is a strong, elastic cover over the root tip of *Zea mays* that protects it during early development stages.

Development of the pellicle begins near the interface between the root proper and the root cap. It attains maximum thickness over the meristematic region and extends into the root hair zone where the L3 layer disappears and the L2 layer thins conspicuously, remaining only as "collars" around the base of root hairs that have pushed through it and the primary cell wall during their development. Whereas it seems likely that the development of a pellicle is a common feature of root tips, there is, as yet, little evidence to support this view since it has been observed primarily in maize and a few other grasses (Abeysekera and McCully, 1993a).

Root hairs develop in a region that extends from just proximal to the apical meristem for one to several millimeters toward the base of the plant (Figs. 16.1, 16.11). The formation of root hairs greatly expands the area of absorption. As development of the root proceeds, the **root hair zone** maintains its size and its position relative to the apical meristem with new root hairs being developed distally as older hairs die proximally. As might be expected, the root epidermis has no cuticle or only a very thin cuticle. In plants, usually herbaceous perennials, that live for years without secondary growth, the epidermis proximal to the absorbing zone may produce a thick cuticle and function as a protective layer. In others it disintegrates and is replaced by development of an exodermis (see above), the walls of which become lignified and suberized. Whereas cells of the exodermis often have the appearance of sclerenchyma, they typically retain a living protoplast.

Some aerial roots of tropical plants such as orchids have a **multiple epidermis**, that is, one consisting of several cell layers. This structure, the **velamen**, consists of non-living cells. Although it has often been considered to be a water-storage region, it is now thought to function primarily in preventing water loss.

Lateral transport of water and minerals in the young root

Perhaps the most important function of the young root is the absorption of water and minerals from the soil and their transfer, laterally, into the stelar region from which they move longitudinally through the primary xylem into the shoot. During their course across the root, water and ions in solution must traverse a complex of structures and tissue regions. In general, water moves passively from the soil into the stelar region as it is being lost from the plant through transpiration. Nutrient ions, on the other hand, are actively absorbed across the plasmalemmas of living cells with the expenditure of energy and then are transferred from cell to cell through plasmodesmata. As we now know, the movement of both water and ions into the region of the xylem is influenced by the presence of Casparian bands and suberin lamellae in the endodermis and in some plants by comparable structures in an exodermis. Whereas these structures, often impregnated with both suberin and lignin, increase the resistance to water flow, they do not entirely restrict it. Furthermore, it is generally believed that water moves across the plasmalemmas of the cortical parenchyma cells relatively freely (see, e.g., Steudle and Peterson, 1998). It should be noted, however, that water not only crosses cell membranes by diffusion through the lipid bilayer, but also through specialized water channel proteins in the plasmalemma called **aquaporins**. Roots have the ability to alter their water permeability in response to environmental signals, and such changes are thought to be related to changes in cell membrane permeability mediated by aquaporins (Javot and Maurel, 2002). Steudle and Peterson conclude that as water reaches the primary xylem, it moves through the lignified cell walls of vessel members primarily through the thin, highly porous pit membranes. In the young roots of *Zea mays*, pit membranes occupy only about 14% of the total surface area of metaxylem vessel members (Steudle and Peterson, 1998). They calculate that the vessel walls provide 10–30% of the resistance to radial water flow, but note that resistance would be variable depending on the area of wall surface comprised of primary wall. For example, in primary xylem vessel members with helical secondary wall thickenings and large areas of exposed primary wall, the resistance to water flow would be considerably less than in pitted vessel members.

Three **radial pathways** are generally recognized for the movement of water in young roots (Fig. 16.12), an apoplastic path, a symplastic path through plasmodesmata, and a transcellular path, across cell walls, protoplasts, and plasmalemmas. These parallel

Figure 16.12 Diagrams illustrating the paths of water movement from the exterior to the interior of roots. (a) Apoplastic path: water moves through the cell walls (light shading) around the protoplasts (dark shading). (b) Symplastic path: water moves through the protoplasts of adjacent cells and across the walls between protoplasts through plasmodesmata. (c) Transcellular path: water moves across the walls, plasmalemmas, and protoplasts. From Steudle and Peterson (1998). Used by permission of Oxford University Press.

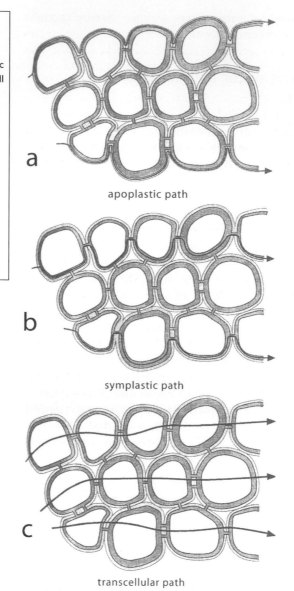

a

apoplastic path

b

symplastic path

c

transcellular path

pathways are characterized by different degrees of resistance to flow, and all are probably utilized in variable degrees during the radial transport of water across the young root (Steudle and Peterson, 1998).

Nutrient ions, upon accumulation in high concentration in the stelar region of roots, would tend to diffuse outward through the apoplast along the concentration gradient were it not for the Casparian bands in the endodermal walls (see Peterson et al., 1993; Steudle and Peterson, 1998). Consequently, unlike the passive flow of water across the tissues of the root, nutrient ions are actively pulled through the root tissues against concentration gradients through plasmodesmata or across cell membranes. Ma and Peterson (2001b) concluded that in *Allium cepa* there was a high degree of ion transport across cell membranes between epidermis and exodermis, a high degree of plasmodesmatal transport of ions through the cortex, across the endodermis and into the pericycle, and a high degree of membrane transport from pericyle into stelar parenchyma from which ions were transported directly into protoxylem vessel members. It seems likely that a similar pattern characterizes other taxa. (For more detail about the lateral transport of water and nutrient ions across the root, please see Steudle and Peterson (1998) and Ma and Peterson (2001b).)

Development of primary tissues

New cells produced by root apical meristems grow and differentiate. Often emphasis has been placed on cell elongation as the first stage in development. Recent studies, however, demonstrate, at least in *Zea mays*, that cell growth is initially isodiametric (Baluska *et al.*, 1990, 1993). During this phase of growth, cells increase essentially equally in length and width, and peripheral microtubules, which direct the orientation of cellulose microfibrils, become arranged transversely to the long axes of the cells in preparation for active cell elongation. Gibberellic acids are thought to regulate this change in orientation of microtubules and, thus, control the polarity of cell growth during the elongation phase (Baluska *et al.*, 1993). Whereas the zone of isodiametric growth is conspicuous in maize, it is relatively inconspicuous in some other species (see, e.g., Bell and McCully, 1970).

Growth and differentiation immediately proximal to the root meristem occur in provascular tissue, ground meristem, and protoderm (Fig. 16.13a). Cell growth and increase in vacuolar volume (Fig. 16.13b) in these regions result from uptake of water and resulting turgor pressure. Maintenance of turgor pressure requires wall loosening resulting in increase in wall area which, with adequate solute concentration, results in water uptake and continued cell growth (Cosgrove, 1993; Pritchard, 1994). In addition to increase in cell size by turgor pressure, cell membranes and organelles as well as cell wall components are synthesized during growth. As in other parts of the plant, communication between cells or cell domains in roots (i.e., signaling and molecular trafficking) commonly takes place, symplastically, by way of plasmodesmata, although the transfer of small molecules may occur apoplastically. In a study of distribution of plasmodesmata in root tips of *Arabidopsis*, Zhu *et al.* (1998) found that primary plasmodesmata were abundant in transverse walls of all transitional tissue regions (i.e., protoderm, ground meristem, and provascular tissue). Although some secondary plasmodesmata occurred in transverse walls they were more frequent in longitudinal walls between cell files of adjacent tissues, and between

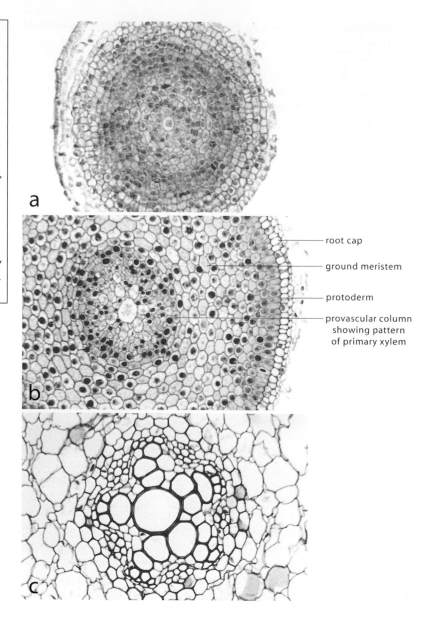

Figure 16.13 Transverse sections illustrating three stages in the development of the root of *Allium*. (a) Transverse section just proximal to the root cap in which tissue regions are immature and indistinct. (b) A more mature region showing well-defined protoderm, ground meristem, and provascular column. At this level the cells are more highly vacuolate, and the pattern of the primary xylem is apparent. (c) A mature region in which the tissues of the central column are distinct. Note the pentarch primary xylem column and the bundles of primary phloem between the ribs of xylem. Magnification (a–c) × 183.

root cap

ground meristem

protoderm

provascular column showing pattern of primary xylem

different tiers of apical initials. They concluded, further, that plasmodesmatal distribution was tissue specific. Signals (e.g., hormones) that control gene expression in particular cells or tissue domains could be restricted by secondary plasmodesmata in the longitudinal walls between cells or tissues, and longitudinal flow of information could be facilitated by the abundant primary plasmodesmata in the transverse walls. However, Zhu *et al.* (1998) emphasized that during development, as cells elongated and differentiated (i.e., approached maturity) in the proximal regions of these tissue regions, primary plasmodesmata disappeared from the transverse walls, but the density of secondary plasmodesmata in the longitudinal walls remained constant.

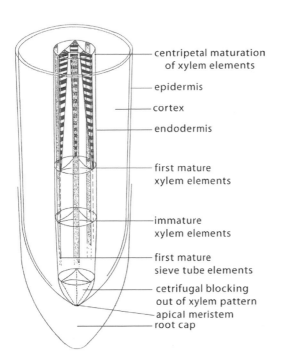

- centripetal maturation of xylem elements
- epidermis
- cortex
- endodermis
- first mature xylem elements
- immature xylem elements
- first mature sieve tube elements
- cetrifugal blocking out of xylem pattern
- apical meristem
- root cap

Figure 16.14 Diagram of a root of *Pisum* (pea) showing the spatial relationship between primary tissue regions and their relative levels of differentiation. Note that the primary phloem differentiates closer to the apical meristem than the primary xylem. From Torrey (1953). Used by permission of the Botanical Society of America.

Differentiation of mature tissues (Fig. 16.13c) from derivatives of provascular tissue, ground meristem, and protoderm is acropetal, that is, maturation proceeds from more proximal to more distal regions. In many taxa, the pericycle is the first tissue region to differentiate visually from provascular tissue. The primary phloem begins its differentiation earlier than the primary xylem and, thus, functional protophloem occurs closer to the apical meristem than functional protoxylem (Fig. 16.14). In a protostelic root, the first incipient primary xylem cells to become apparent are large, metaxylem cells in the center of the xylem column (Fig. 16.13a, b). The first primary xylem cells that acquire secondary walls and become functional, however, occur at its periphery. As development proceeds in dicotyledons and gymnosperms the ribbed pattern of the primary xylem column becomes apparent (Fig. 16.13a). Maturation of the primary xylem is centripetal, that is, the first functional primary xylem cells, the protoxylem, develop first at the periphery of the xylem column and maturation proceeds toward the center through the metaxylem (Fig. 16.14). The order of maturation is, therefore, exarch (Figs. 16.13c, 16.14). Maturation in the bundles of incipient primary phloem is also centripetal in that the protophloem matures first at the outer edge followed by maturation of metaphloem toward the center.

Controversy surrounds the nature of the tissue from which the pith differentiates in some roots. Some botanists consider this tissue to be ground meristem, others provascular tissue. The latter view is supported by the fact that plants that have a pith in their most basal regions are often characterized in more apical regions by its absence. In other words, in such plants, as the roots become older, the central meristematic tissue differentiates into primary xylem.

Concurrently with development of primary vascular tissues, cortical tissues develop through growth and differentiation of the ground meristem and its derivatives. Close to the root apex, periclinal divisions result in files of cells that increase the diameter of the root, followed by anticlinal divisions which, with cell elongation, result in increase in its length. Early in the process of differentiation, and close to the apical meristem, as cortical cells grow, the system of intercellular channels develops. The innermost layer of cells in the cortical region develops as the endodermis (Fig. 16.13c) at the same level as the root hair zone in the epidermis. As in the vascular column, differentiation of cortical tissue is acropetal.

The protoderm (Fig. 16.13b) and the epidermis which differentiates from it have somewhat different origins in different major groups. In monocotyledons, the epidermis and the cortex have a common origin from the central tier of apical initials. By contrast, in many dicotyledons the epidermis and root cap are both derived from the basal tier of initials. However, in many dicotyledons (including many tree species) that lack tiers of initials, epidermis, cortex, vascular column, and root cap all originate from a single cluster of initials in the root apex (for more detail, see the section above on apical meristems). In some taxa, in the incipient root hair zone, certain small cells from which root hairs will develop are formed by unequal divisions. These cells, called **trichoblasts**, differ from adjacent cells by having densely staining cytoplasm and larger nuclei. In other taxa, cells from which root hairs develop exhibit no apparent structural differences from adjacent cells. New root hairs develop at about the same level as the first mature protoxylem.

For many years, epidermal cells that produce root hairs have been used for the study of cell growth, especially cell elongation. Recent analyses of root hair development in *Arabidopsis thaliana* have provided significant information on the role of genes in development (see, e.g., Schiefelbein and Somerville, 1990; Dolan *et al.*, 1994; Galway *et al.*, 1994; Masucci and Schiefelbein, 1996; Berger *et al.*, 1998; Schiefelbein, 2000). *Arabidopsis* and other members of the Brassicaceae produce root hairs "in a position-dependent pattern of cells types" (Schiefelbein, 2000). Protodermal cells that will differentiate into root hair cells (trichoblasts) are located over the anticlinal walls of subjacent cells and those that will not produce root hairs are located next to them in contact with the outer periclinal walls of subjacent cells. Genetic studies have demonstrated the presence of genes in protodermal cells that specify those that will differentiate into either root hair cells or non-root hair cells (see Berger *et al.*, 1998). Furthermore, ablation experiments in which adjacent cells moved into the space vacated by the killed cells clearly demonstrate that development of these cells is position dependent. Root hair cells that become located in the former positions of non-root hair cells, and non-root hair cells that move into the space vacated by root hair cells, take on the characteristics of the ablated cells (Berger *et al.*, 1998). It is believed that root hair initiation is controlled by the plant hormones auxin and/or ethylene (Schiefelbein, 2000). Following initiation, wall loosening, controlled by expansin genes (see Baluska *et al.*, 2000; Cho and Cosgrove, 2002), leads to the formation of an outward bulge in the periclinal wall of the trichoblasts followed by tip growth and elongation of the root hair. During root hair growth, the cytoskeleton is involved in the movement of vesicles containing the precursor compounds necessary for cell wall synthesis (see Chapter 5 for more detail on cell elongation).

Auxin and tissue patterning

The various histological patterns described above develop in tissues derived from the apical meristem. That pattern formation is under genetic control and mediated by the hormone auxin is widely accepted, but the mechanism of this control of cell and tissue differentiation is only beginning to be understood (see also Chapter 5). Recent studies of patterning in the root tip of *Arabidopsis thaliana* indicate that histological pattern formation is influenced by auxin concentration gradients and the positioning of PIN proteins that function as **auxin efflux transporters** (proteins that transfer auxin from cells in one region to another) (Sabatini *et al.*, 1999; Friml and Palme, 2002; Friml *et al.*, 2002; Benkova *et al.*, 2003).

Auxin moves from apical regions of the shoot through the living cells of undifferentiated vascular tissues (provascular strands and the vascular cambial zone) to other parts of the plant (**polar auxin transport**). When it reaches the apical meristem of the root it is redistributed through the ground meristem and protoderm to the root cap and the more proximal regions of the root apex. Auxin can only leave cells by **PIN efflux transporters**. The genes that encode the auxin reflux transporters are called PIN genes. Under the influence of specific PIN genes, mediated by auxin efflux transporters, auxin can become concentrated in specific areas. Through the establishment of auxin concentration gradients these genes can control tissue differentiation and patterning.

Lateral root development

Initiation of lateral roots is **endogenous**, occurring, primarily, in the pericycle in seed plants and in the endodermis or cortex in pteridophytes. In many taxa there is a specific and constant position of lateral root initiation in relation to the ribs of the primary xylem. In pteridophytes lateral root initiation is always directly opposite the ends of the xylem ribs whereas in seed plants the site of initiation is variable, sometimes opposite the ends of xylem ribs, sometimes opposite the strands of primary phloem, sometimes between the ends of ribs and the adjacent strands of primary phloem (Fig. 16.7a–c).

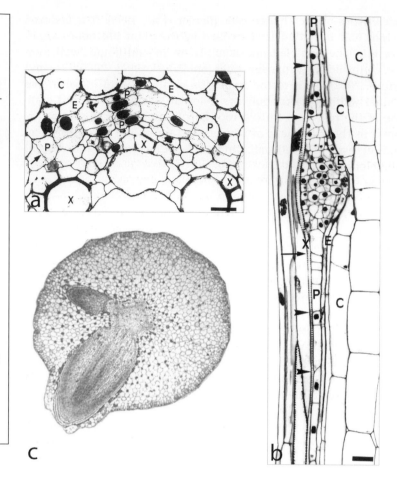

Figure 16.15 Lateral root development in *Allium*.
(a) Transverse section showing short files of cells formed by periclinal divisions in the pericycle at the site of lateral root initiation. C, cortex; E, endodermis; P, pericycle; X, xylem. Arrow indicates an anticlinal wall in a cell in the pericycle. Bar = 50 μm.
(b) Radial longitudinal section illustrating the young lateral root. Bar = 50 μm. Arrowheads indicates sites of proliferation of pericycle cells around the developing lateral root primordium; arrows delimit the root primordium. (c) Transverse section of a root of *Salix nigra* (black willow) with lateral roots. The larger lateral root has grown through the cortex, rupturing the epidermis. At this stage primary tissue regions are apparent behind the apical meristem. (a, b) From Casero *et al.* (1996). Used by permission of Springer-Verlag Wien. (c) From Eames and MacDaniels (1925).

In gymnosperms and angiosperms lateral **root primordia** are initiated in the pericycle proximal to the zone of elongation by periclinal divisions (Fig. 16.15a) followed by anticlinal divisions and cell growth (Fig. 16.15b). Prior to primordium formation in *Arabidopsis*, auxin accumulates at the sites of primordium initiation (Benkova *et al.*, 2003). Subsequently, under the control of PIN efflux transporter proteins, auxin is transferred into the primordia and auxin concentration gradients are established with their maxima occurring at the tips of the primordia. Under the influence of PIN genes, the various tissue regions differentiate.

In general, following primordium formation, and as development of the lateral root continues, anticlinal divisions often occur in the endodermis opposite the initial site of cell division in the pericycle, and in some species the endodermis becomes biseriate locally. Periclinal divisions in the endodermis result in the formation of a sheath over the developing lateral root as it pushes its way through the cortex, and a root cap and transitional tissue regions – provascular tissue, ground meristem, and protoderm – differentiate in the apical region of the young lateral root (Fig. 16.15c).

Connections between the tissues of the lateral and the parent root occur through differentiation within parenchyma tissue of pericyclic origin (see, e.g., Casero *et al.*, 1996) in the intervening regions. It is especially important to understand the structure of the connection of vascular tissues between parent and lateral roots because lateral roots play a significant role in absorbing and transporting water and minerals from the soil solution into the main root. In both maize (McCully and Mallet, 1993; Shane *et al.*, 2000) and barley (Luxova, 1990) the vascular systems of main and lateral roots are connected by an extensive vascular plexus (Fig. 16.16). This vascular plexus, the components of which differentiate from parenchyma in the pericycle, consists of a mixture of numerous small, short tracheary elements (some of which are vessel members) and xylem parenchyma bounded by sieve tube members, all described by McCully and Mallett (1993) as primarily cuboidal (Fig. 16.16b, c). The tracheary elements which are connected to each other by bordered pits apparently provide an effective apoplastic conduit for the movement of water and solutes from lateral roots into the main root. The pits in the contiguous walls between vessel members of the primary xylem of the main root and the tracheary elements of the lateral roots, termed **boundary pits**, are very large and have conspicuous pit membranes (Fig. 16.16a). Because the boundary pit membranes have very small pores and can filter out particles with mean diameters as small as 4.9 ± 0.7 nm, Shane *et al.* (2000) concluded, with other evidence, that boundary pit membranes are efficient filters for microbes and particulates.

The primary phloem which is largely peripheral to the tracheary tissue in the vascular plexus connects directly to that of the main root. McCully and Mallett (1993) also observed sites of direct contact between sieve tube members and tracheary elements (Fig. 16.16a), and conclude that, as in the leaves of a few grasses (see Eleftheriou, 1990) and ferns (Evert, 1990), these might play a role in nutrient recycling between the xylem and the phloem.

Adventitious roots

Roots that arise from stems, other plant parts (e.g., the fleshy leaves of plants such as *Sedum* and *Begonia*), and regions of main roots other than the pericycle proximal to the zone of elongation are called **adventitious roots**. Such roots may originate from callus tissue associated with wounds, or endogenously from pericyclic parenchyma, from interfascicular parenchyma, and in old roots, from vascular rays or the axial parenchyma in secondary phloem. As noted earlier in this chapter, the root system of most monocotyledons is composed of adventitious roots that develop at the base of the stem, but in some grasses they also develop from axillary shoots. The root systems of pteridophytes consist of adventitious roots that develop in internodal regions or at nodes in association with leaves. In some extinct ferns, e.g., *Botryopteris*, adventitious roots were even borne on petioles.

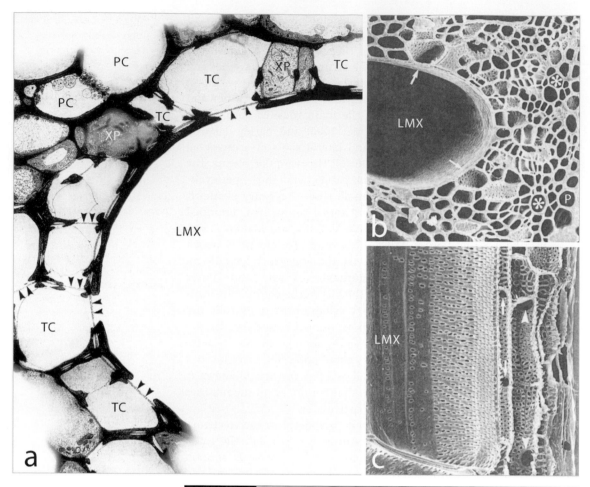

Figure 16.16 Vascular connection between the main root and a lateral root of *Zea mays*. (a) Transmission electron micrograph of a transverse section of part of a late metaxylem vessel member (LMX) of the main root. The wall of this vessel member, facing the exterior of the root, is bounded by numerous tracheary connector elements (TC) of the lateral root. The contiguous walls between the connector elements, and between connector elements and the vessel member of the main root, contain bordered pit-pairs with wide apertures and uniformly thick pit membranes (arrowheads). XP, xylem parenchyma; PC, connector sieve elements. Magnification × 2750. (b) Scanning electron micrograph of a transverse section of a late metaxylem vessel member and associated tracheary connector elements of a lateral root. Asterisks indicate sieve tube members; arrows indicate tracheary connector elements. P, pericyle of the main root. Magnification × 194. (c) Scanning electron micrograph of a longitudinal section through a region similar to that shown in (b). Some of the tracheary connector elements are vessel members. Note the vessel member end wall perforations (arrows and arrowheads). Magnification × 262. (a) From McCully and Mallet (1993). Used by permission of Oxford University Press. (b, c) From Shane et al. (2000). Used by permission of Oxford University Press.

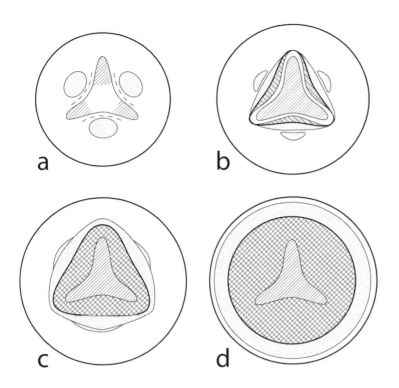

Figure 16.17 Diagrams of transverse sections illustrating the origin of secondary tissues in roots. Primary xylem is lined; secondary xylem is cross-hatched; primary phloem is unshaded; secondary phloem is stippled. (a) Differentiation of primary xylem is incomplete. The vascular cambium is indicated by dashed lines. (b) As the cambium produces secondary xylem and phloem, the primary phloem is compressed. The cambium is indicated by a heavy line. (c, d) With continued activity, the vascular cambium attains a circular outline in transverse view, and subsequent increments of secondary xylem and phloem will appear as cylinders.

Secondary growth

As in the stem, the production of secondary vascular tissues in roots is initiated by activity of the vascular cambium. The cambium differentiates first from provascular tissue in the bays between ribs of primary xylem and to the inside of the strands of primary phloem (Fig. 16.17a). These regions of the cambium commonly become active in producing secondary xylem and phloem before the cambium has become continuous around the outer edges of the primary xylem ribs (Fig. 16.17b), differentiating in these latter regions from pericycle. Because of this developmental pattern, the cambium only gradually attains a circular form as viewed in transverse section (Figs. 16.17b–d, 16.18a). In many plants, especially woody plants, as secondary phloem and xylem are produced, the primary phloem is compressed and obliterated. In some herbaceous perennials in which secondary xylem is produced in limited quantities, little or no secondary phloem may be produced and the metaphloem functions as the conduit for transfer of photosynthate throughout the life of the plant (Fig. 16.18c, d). In such plants protophloem is usually compressed and obliterated.

If no secondary vascular tissues, or only small amounts are produced, a periderm does not develop and the cortex is retained for the life of the plant (Fig. 16.18c). In such plants the epidermis may develop a thick cuticle, thus becoming the protective outer boundary of the root system, or an exodermis may develop in the outermost cortex, comprising a protective layer. The part of the outer cortex in contact with and to the interior of the exodermis may consist of one to several

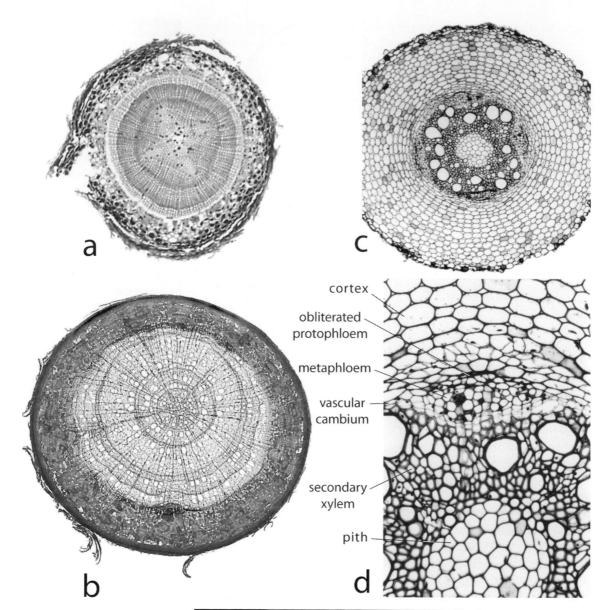

cortex

obliterated
protophloem

metaphloem

vascular
cambium

secondary
xylem

pith

a

b

c

d

Figure 16.18 Transverse sections of roots with secondary tissues. (a) A root of *Cedrus deodara* with tetrarch primary xylem. Note that the vascular cambium was not continuous around the ends of the primary xylem ribs until near the end of the first growing season. Periderm is in contact with the secondary phloem. The cortex and epidermis have sloughed off. Magnification × 43. (b) A root of *Tilia americana*. The secondary xylem contains numerous rays and very large vessels. Note the periderm in contact with the secondary phloem. Magnification × 24. (c, d) Root of *Pelargonium*. Note the pith, the cortex, and the large vessels in the secondary xylem. (d) The vascular cambium produces little or no secondary phloem, and the metaphloem functions as the major conduit for the transport of hormones and photosynthate for the life of the plant. Note the compressed and obliterated protophloem. Magnification (c) × 50, (d) × 185.

layers of sclerenchyma. This region and the exodermis comprise the **hypodermis**. Root structure of this type is characteristic not only of many herbaceous taxa of both dicotyledons and monocotyledons, but also of some large plants such as the palms.

The tracheids and vessel members in the secondary xylem of roots are commonly large in transverse dimensions and thin-walled (Fig. 16.18b–d). Both secondary phloem and secondary xylem in roots are characterized by large amounts of parenchyma, including that of abundant, large rays. These adaptations are directly related to the storage function of roots.

In roots that produce large amounts of secondary vascular tissues, a phellogen differentiates within the outer pericyclic parenchyma shortly after the beginning of cambial activity. As secondary growth leads to diametric expansion of the root, the continuity of the phellogen is maintained by numerous anticlinal divisions that result in an increase in its circumference. The single-layered phellogen produces phellem to the exterior and phelloderm to the interior, producing a periderm that is in contact with the secondary phloem (Fig. 16.18a, b). Cut off from a source of water and photosynthate, the cortex and epidermis die and slough off. In old roots, rhytidome similar to that in stems develops, but as it develops the outer part decays. Consequently the rhytidome remains relatively thin and the root surface smooth.

The root cap: its function and role in gravitropism

The **root cap** covers the apical meristem of the root and traditionally has been thought of, primarily, as a protective structure. We now know that it has other very important functions. As new root cap cells are continually produced by the root cap meristem, they gradually move to the periphery and ultimately are sloughed off. Initially they function as statocytes (perceptors of gravity), but as they near the periphery they become secretory cells that produce and secrete polysaccharide mucilage (Barlow, 1975, 2002), thought to function in lubricating the root as it pushes its way through the soil. The secretion of mucilage is granulocrinous (see Schnepf, 1993). Golgi vesicles containing mucilage fuse with the plasmalemma of cells in the outer part of the root cap, and the mucilage is thereby released into the cell walls through which it migrates to the exterior, coating the root tip. This coat of mucilage is also thought to protect the root tip from desiccation, prevent or decrease the probability of entry into the root of toxic substances, and facilitate the exchange of ions.

The degree to which roots are **gravitropic**, that is, responsive to gravity, is highly variable. Tap roots are strongly gravitropic whereas lateral roots vary from being only slightly to being not at all responsive to gravity. In strongly gravitropic roots, amyloplasts (starch grains) in the cells of the root cap and, in some plants (e.g., *Equisetum*: Ridge and Sack, 1992) in cortical cells in the elongating region, seem to function as sensors of gravity. The amyloplasts are known as **statoliths** whereas the cells in which they are contained are called **statocytes**. When the root is vertical, the statoliths are located near the distal transverse walls of the statocytes (Fig. 16.19a, b). However, when the root is horizontal, the statoliths sink toward the lateral walls (Fig. 16.19c), and cells in the elongation zone of the root slowly begin to grow faster on the upper surface than on the lower which results in its downward curvature and a return of the starch grains to their original position near the distal transverse walls of statocytes in the root cap. The mechanism of this gravitropic response and the precise role of the statoliths is unclear. It is surmised that a chemical signal such as a growth hormone which is produced in the statocyte is transferred to the region of elongation in the root resulting in a higher concentration of the hormone in the lower part than in its upper part. Since auxin, the hormone long believed to control root curvature, inhibits growth in high concentrations, cell elongation occurs in the upper part of the elongation zone and is restricted in the lower part, resulting in a downward curvature of the root. This concept of differential growth related to differential distribution of auxin, and resulting in positive geotropism in roots, was proposed by Cholodny and Went in the early part of the twentieth century. The **Cholodny–Went hypothesis** has been widely accepted for many years and recently has been strongly supported by the work of Ottenschläger *et al.* (2003). On the other hand, it also has been challenged on several bases. The most serious challenge comes from experiments that have shown, conclusively, in several species that when auxin synthesis is inhibited, the gravitropic curvature response still occurs (see Evans, 1991; Konings, 1995; Evans and Ishikawa, 1997). It has been further demonstrated that root curvature can occur in the absence of amyloplasts (statoliths), and it has been surmised that sedimentation of cell organelles might serve the same function. In this regard, it is interesting to note that in *Equisetum*, sedimentation of the cell nuclei occurs simultaneously with that of the amyloplasts, usually following them and coming to rest just above them (Fig. 16.19b, c).

Although research in many laboratories has resulted in advances in our knowledge of the factors related to gravitropism in both stems and roots, the role of statoliths, the routes of transport of the mediating signal, the mechanism and control of cell wall relaxation (wall loosening) required during cell elongation, the role of the cytoskeleton, if any, even the nature of the signal which elicits differential cell elongation in the gravitropic response are still not clearly understood.

Mycorrhizae

The root tip with its root hairs is an efficient region of water and mineral absorption from the soil under optimal conditions of water and mineral availability. The area through which absorption occurs, however, in a single root is relatively small. Furthermore, under conditions of drought, the volume of soil that provides the source of water and minerals for a root tip can soon be depleted. Consequently, as an

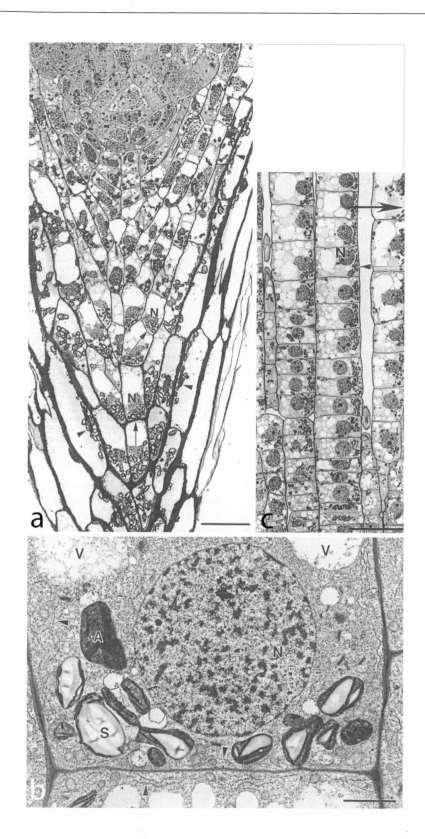

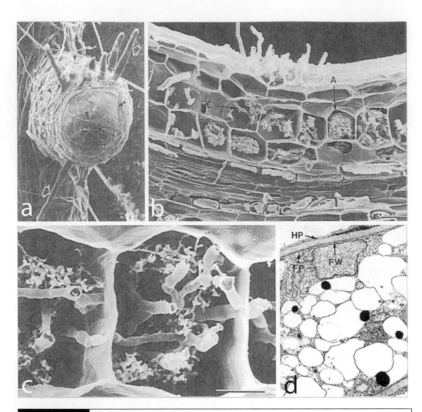

Figure 16.20 (a) Scanning electron micrograph of an endomycorrhiza of *Pinus banksiana*. Note the root tip with unusual apical root hairs (arrows) and a thin mantle of fungal hyphae. A typical endomycorrhiza lacks a hyphal mantle. Magnification × 61. (b) An infected endomycorrhizal root of *Acer saccharum* showing large fungal hyphae (hyphal coils) (C) and arbuscules (much-branched smaller hyphae) (A) in cells of the cortex. Magnification × 198. (c) Close-up of cortical cells showing hyphal coils and arbuscules. Magnification × 1250. (d) Transmission electron micrograph of vacuolated intracellular hyphae. HP, host plasmalemma; FP, fungal plasmalemma; FW, fungal wall. Magnification × 7324. (a) From Scales and Peterson (1991). Used by permission of the National Research Council of Canada. (b–d) From Yawney and Schultz (1990). Used by permission of Blackwell Publishing Ltd.

adaptation to such conditions, mycorrhizae which greatly enhance the root's absorptive capacity have evolved.

A **mycorrhiza** is the structural combination of a fungus and a root (Fig. 16.20a) resulting in a mutually beneficial symbiotic relationship. The fungus may be either an ascomycete or a basidiomycete. According to Bonfante and Perotto (1995) mycorrhizae occur in approximately 90% of terrestrial plant species. They have not been observed in either the Brassicaceae or Cucurbitaceae, and are rare in aquatic land plants and plants that grow in wet substrates such as sedges. Two major types have been recognized, ectomycorrhizae and endomycorrhizae.

In an **ectomycorrhiza** the fungus forms a dense sheath of hyphae around the root tip with numerous branching hyphae extending from this sheath into the surrounding soil and others penetrating the epidermis and cortex. These internal, much-branched, coenocytic hyphae penetrate the parenchyma tissue through intercellular channels forming a network of hyphae, called the **Hartig net**, between, but in intimate contact with, cells of the cortex. The parenchyma cells associated with the Hartig net are thought to function like transfer cells. Photosynthate is transported into the hyphae in this region, providing a source of nutrition for the fungus, and water and minerals are transported from the hyphae into the cortical cells, benefiting the plant by greatly enhancing the supply of water and minerals, especially in mineral-poor, and dry, soils.

A large majority of vascular plants, possibly up to 80% (Bonfante and Perotto, 1995), are characterized by endomycorrhizae. An **endomycorrhiza**, unlike an ectomycorrhiza, lacks a conspicuous, enclosing sheath or mantle of hyphae although in some forms (called **ectendomycorrhizae**) hyphae form a thin enclosing sheath around the root tip (Fig. 16.20a). In both types, hyphae penetrate the root and proliferate within the cortex (Fig. 16.20b) through intercellular channels. Some hyphae penetrate the parenchyma cell walls, but not the plasmalemma. They form complex, much-branched structures called arbuscules (Fig. 16.20b, c) that become almost completely enclosed by the plasmalemma. Within the cell walls and in intimate contact with the cell protoplast through the plasmalemma (Fig. 16.29d) (although not within the protoplast), the **arbuscules** form efficient regions of transfer of photosynthate to the fungus and water and minerals to the plant. Some hyphae form vesicles in the intercellular channels in which glycogen and/or lipids are stored. Thus, the term **vesicular–arbuscular mycorrhiza** is often used in preference to endomycorrhiza. For more detail about this complex and important area, see Yawney and Schultz (1990), Brundrett and Kendrick (1990), Scales and Peterson (1991), Bonfante and Perotto (1995), Smith and Smith (1997), Massicotte et al. (1999), Armstrong and Peterson (2002), Blancaflor et al. (2001), and Yu et al. (2001).

Nitrogen fixation in root nodules

Whereas nitrogen is constantly being taken from the soil it is also constantly being added to the soil by some plants through the process of **nitrogen fixation**. This recycling of nitrogen contributes significantly, with other factors, in preventing the depletion of soil nitrogen. Although nitrogen is an essential element in plant metabolism, atmospheric nitrogen cannot be utilized by most plants. Through time a symbiotic relationship between roots and several soil microorganisms has evolved by which atmospheric nitrogen is converted into a form that plants can use. This process, nitrogen fixation, takes place in root nodules in many taxa of the Fabaceae (Leguminosae) (see Pate et al., 1969; Hirsch, 1992; Subba-Rao et al., 1995) as well as in several other families of dicotyledons. The only gymnosperms known to fix nitrogen are several members of the cycads in which the symbiont is a cyanobacterium (blue-green alga).

In the Fabaceae the endophytic microsymbiont is the bacterium *Rhizobium*. There is a high degree of specificity between *Rhizobium* species and their host plants, thus a species that will infect one legume genus will not infect another. Infection of the root system which occurs through root hairs stimulates the development of root nodules of distinctive structure. The bacteria aggregate within cells in the center of the nodule. This region is surrounded by a parenchymatous tissue containing vascular bundles which, in turn, is enclosed by an endodermis. Exterior to the endodermis is a cortex characterized by a complex system of intercellular air channels. The endodermis retards the entry of oxygen into the center of the nodule preventing denaturation of nitrogenase, essential to the process of nitrogen fixation which occurs in this region. The nitrogenase is provided by the *Rhizobium*, and a protein, hemoglobin, is provided by the host plant. Sugar utilized as a source of energy required in the process is also provided by the host plant. Numerous plasmodesmata connect the *Rhizobium*-containing cells with the surrounding parenchyma cells. These plasmodesmata are thought to be the conduits by which sugar is transported into the cells where nitrogen fixation occurs, as well as the means by which the resulting amino acids are transported out of this region into the surrounding parenchyma.

In other families, the endosymbiont is usually an actinomycete, the most common being *Frankia*. Well-known taxa of woody dicotyledons that fix nitrogen in root nodules include *Alnus* of the Betulacaceae, *Ceanothus* (tea bush) in the Rhamnaceae, *Eleagnus* (oleaster) of the Eleagnaceae, and *Myrica* (sweet gale) of the Myricaceae. Root nodules in which *Frankia* species are the endosymbionts resemble those of the legumes, except that the actinomycete is usually confined to the cortex. The hemoglobin present in cells in which nitrogen fixation occurs may provide oxygen for metabolism while inhibiting oxygen denaturation of nitrogenase.

Several cycads are known to produce root nodules containing the cyanobacteria *Anabaena* or *Nostoc*. Nitrogen fixation in root nodules of the cycad *Macrozamia* is of significance in the ecology of some Australian forests. *Anabaena* is also known to form a symbiotic relationship with the heterosporous fern *Azolla*. In several Asian countries, colonies of *Azolla* infected with *Anabaena* living in rice paddies contribute, upon death, significant quantities of nitrogen which can be utilized subsequently by the rice plants.

Root–stem transition

Primary vascular tissues in the developing roots of many seed plants are radially arranged in separate longitudinal bundles or strands whereas in the stems of the same plant they occur together, most commonly in collateral bundles. Furthermore, the order of maturation (or direction of differentiation laterally) is exarch in the root and endarch in the stem in seed plants. Consequently, in the transition

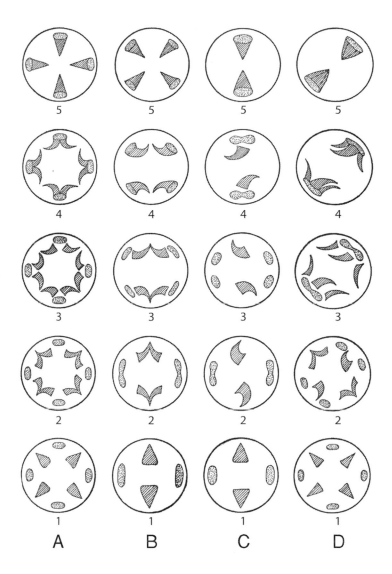

Figure 16.21 Diagrams of four types of root–stem transition (A–D). Diagrams in the lower row, A1–D1, represent roots. Those at level A5–D5 represent stems, with intermediate stages at successive levels between, illustrating separation, rotation, and fusion of primary vascular bundles. Primary xylem is lined; primary phloem is stippled. From Eames and MacDaniels (1925).

region between root and stem there is a reorientation of the primary xylem and phloem. This transition commonly takes place over a short distance in the hypocotyl of the embryo in provascular tissue, but in some species, e.g., *Pisum sativum*, it may extend through several internodes. Esau (1965) indicates that in many species the change in orientation of vascular tissue is completed in the cotyledons. A recent example of a transition of this type has been described in *Arabidopsis thaliana* by Busse and Evert (1999). These changes which would not be conspicuous in the embryo would, however, be clearly expressed upon differentiation of primary vascular tissues in the axis of the developing sporophyte. When the vascular tissue is observed at successive levels in the mature state it becomes apparent that in some taxa the bundles of primary xylem divide and become reoriented in relation to bundles of primary phloem so that collateral bundles with endarch order of maturation are formed. In other taxa, both xylem and phloem bundles divide. One may develop a three-dimensional comprehension of this reorientation of vascular tissues by studying a diagram of vascular tissue orientation at successive levels through the transition region (Fig. 16.21).

REFERENCES

Abeysekera, R. M. and M. E. McCully. 1993a. The epidermal surface of the maize root tip. I. Development in normal roots. *New Phytol.* **125**: 413–429.

1993b. The epidermal surface of the maize root tip. II. Abnormalities in a mutant which grows crookedly through soil. *New Phytol.* **125**: 801–811.

1994. The epidermal surface of the maize root tip. III. Isolation of the surface and characterization of some of its structural and mechanical properties. *New Phytol.* **127**: 321–333.

Armstrong, L. and R. L. Peterson. 2002. The interface between the arbuscular mycorrhizal fungus *Glomus intraradices* and root cells of *Panax quinquefolius*: a Paris-type mycorrhizal association. *Mycologia* **94**: 587–595.

Baluska, F., Kubica, S., and M. Hauskrecht. 1990. Postmitotic "isodiametric" cell growth in the maize root apex. *Planta* **181**: 269–274.

Baluska, F., Parker, J. S., and P. W. Barlow. 1993. A role for gibberellic acid in orienting microtubules and regulating cell growth polarity in the maize root cortex. *Planta* **191**: 149–157.

Baluska, F., Salaj, J., Mathur, J. *et al.* 2000. Root hair formation: F-actin-dependent tip growth is initiated by local assembly of profilin-supported F-actin meshworks accumulated within expansin-enriched bulges. *Devel. Biol.* **227**: 618–632.

Barlow, P. W. 1975. The root cap. In J. G. Torrey and D. T. Clarkson, eds., *The Development and Function of Roots*. London: Academic Press, pp. 21–54.

1992. The meristem and quiescent centre in cultured root apices of the *gib-l* mutant of tomato (*Lycopersicon esculentum* Mill.). *Ann. Bot.* **69**: 533–543.

2002. The root cap: cell dynamics, cell differentiation and cap function. *J. Plant Growth Regul.* **212**: 261–286.

Bell, J. K. and M. McCully. 1970. A histological study of lateral root initiation and development in *Zea mays*. *Protoplasma* **70**: 179–205.

Benkova, E., Michniewicz, M., Sauer, M. *et al.* 2003. Local, efflux-dependent auxin gradients as a common module for plant organ formation. *Cell* **115**: 591–602.

Berger, T., Haseloff, J., Schiefelbein, J., and L. Dolan. 1998. Positional information in root epidermis is defined during embryogenesis and acts in domains with strict boundaries. *Curr. Biol.* **8**: 421–430.

Blancaflor, E. R., Zhao, L., and M. J. Harrison. 2001. Microtubule organization in root cells of *Medicago truncatula* during development of an arbuscular mycorrhizal symbiosis with *Glomus versiforme*. *Protoplasma* **217**: 154–165.

Bonfante, P. and S. Perotto. 1995. Strategies of arbuscular mycorrhizal fungi when infecting host plants. *New Phytol.* **130**: 3–21.

Bowman, J. 1994. Arabidopsis: *An Atlas of Morphology and Development*. Heidelberg: Springer-Verlag.

Brundrett, M. and B. Kendrick. 1990. The roots and mycorrhizas of herbaceous woodland plants. II. Structural aspects of morphology. *New Phytol.* **114**: 469–479.

Busse, J. S. and R. F. Evert. 1999. Vascular differentiation and transition in the seedling of *Arabidopsis thaliana* (Brassicaceae). *Int. J. Plant Sci.* **160**: 241–251.

Casero, P. J., Casimiro, I., and P. G. Lloret. 1996. Pericycle proliferation pattern during the lateral root initiation in adventitious roots of *Allium cepa*. *Protoplasma* **191**: 136–147.

Cho, H. T. and D. J. Cosgrove. 2002. Regulation of root hair initiation and expansin gene expression in *Arabidopsis*. *Plant Cell* **14**: 3237–3253.

Clarkson, D. T. 1991. Root structure and sites of ion uptake. In Y. Waisel and A. Eshel, eds., *Plant Roots*. New York: Marcel Dekker, pp. 417–455.

Clarkson, D. T. and A. W. Robards. 1975. The endodermis, its structural development and physiological role. In J. G. Torrey and D. T. Clarkson, eds., *Development and Function of Roots*. London: Academic Press, pp. 415–437.

Clowes, F. A. L. 1959. Apical meristems of roots. *Biol. Rev. Cambridge Phil. Soc.* **34**: 501–529.

1961. *Apical Meristems*. Oxford, UK: Blackwell.

1976. The root apex. In M. M. Yeoman ed., *Cell Division in Higher Plants*. London: Academic Press, pp. 253–284.

1994. Origin of the epidermis in root meristems. *New Phytol.* **127**: 335–347.

Cosgrove, D. J. 1993. Water uptake by growing cells: an assessment of the controlling roles of wall relaxation, solute uptake, and hydraulic conductance. *Int. J. Plant Sci.* **154**: 10–21.

Coutts, M. P. and B. C. Nicoll. 1991. Orientation of the lateral roots of trees. I. Upward growth of surface roots and deflection near the soil surface. *New Phytol.* **119**: 227–234.

Damus, M., Peterson, R. L., Enstone, D. E., and C. A. Peterson. 1997. Modifications of cortical cell walls in roots of seedless vascular plants. *Bot. Acta* **110**: 190–195.

Dolan, L., Duckett, C., Grierson, C. *et al.* 1994. Clonal relations and patterning in the root epidermis of *Arabidopsis*. *Development* **120**: 2465–2474.

Eames, A. J. and L. H. MacDaniels. 1925. *An Introduction to Plant Anatomy*. New York: McGraw-Hill.

Eleftheriou, E. P. 1990. Monocotyledons. In H. D. Behnke and R. D. Sjölund, eds., *Sieve Elements: Comparative Structure, Induction and Development*. Berlin: Springer-Verlag, pp. 139–159.

Enstone, D. E. and C. A. Peterson. 1997. Suberin deposition and band plasmolysis in the corn (*Zea mays* L.) root exodermis. *Can. J. Bot.* **75**: 1188–1199.

Esau, K. 1965. *Plant Anatomy*, 2nd edn. New York: John Wiley and Sons.

Evans, M. L. 1991. Gravitropism: interaction of sensitivity modulation and effector distribution. *Plant Physiol.* **95**: 1–5.

Evans, M. L. and H. Ishikawa. 1997. Cellular specificity of the gravitropic motor response in plants. *Planta* **203**: S115–S122.

Evert, R. F. 1990. Seedless vascular plants. In H. D. Behnke and R. D. Sjölund, eds., *Sieve Elements: Comparative Structure, Induction and Development*. Berlin: Springer-Verlag, pp. 35–64.

Friml, J. and K. Palme. 2002. Polar auxin transport: old questions and new concepts? *Plant Mol. Biol.* **49**: 273–284.

Friml, J., Benková, E., Blilou, I. *et al.* 2002. AtPIN4 mediates sink-driven auxin gradients and root patterning in *Arabidopsis*. *Cell* **108**: 661–673.

Galway, M. E., Masucci, J. D., Lloyd, A. M. *et al.* 1994. The TTG gene is required to specify epidermal cell fate and cell patterning in the *Arabidopsis* root. *Devel. Biol.* **166**: 740–754.

Halevy, A. H. 1986. The induction of contractile roots in *Gladiolus grandiflorus*. *Planta* **167**: 94–100.

Hirsch, A. M. 1992. Developmental biology of legume nodulation. *New Phytol.* **122**: 211–237.

Hose, E., Clarkson, D. T., Steudle, E., Schreiber, L., and W. Hartung. 2001. The exodermis: a variable apoplastic barrier. *J. Exp. Bot.* **52**: 2245–2264.

Javot, H. and C. Maurel. 2002. The role of aquaporins in root water uptake. *Ann. Bot.* **90**: 301–313.

Jensen, W. A. and L. G. Kalvaljian. 1958. An analysis of cell morphology and the periodicity of division in the root tip of *Allium cepa*. *Am. J. Bot.* **45**: 365–372.

Jernstedt, J. A. 1984a. Seedling growth and root contraction in the soap plant, *Chlorogalum pomeridianum* (Liliaceae). *Am. J. Bot.* **71**: 69–75.

  1984b. Root contraction in hyacinth. I. Effects of IAA on differential cell expansion. *Am. J. Bot.* **71**: 1080–1089.

Kerk, N. and L. Feldman. 1994. The quiescent center in roots of maize: initiation, maintenance and role in organization of the root apical meristem. *Protoplasma* **183**: 100–106.

Knoll, A. H. and K. J. Niklas. 1987. Adaptation, plant evolution, and the fossil record. *Rev. Palaeobot. Palynol.* **50**: 127–149.

Konings, H. 1995. Gravitropism of roots: an evaluation of progress during the last three decades. *Acta Bot. Neerl.* **44**: 195–223.

Lersten, N. R. 1997. Occurrence of endodermis with a Casparian strip in stem and leaf. *Bot. Rev.* **63**: 265–272.

Luxova, M. 1990. Effect of lateral root formation on the vascular pattern of barley roots. *Bot. Acta* **103**: 305–310.

Ma, F. S. and C. A. Peterson. 2001a. Development of cell wall modifications in the endodermis and exodermis of *Allium cepa* roots. *Can. J. Bot.* **79**: 621–634.

  2001b. Frequencies of plasmodesmata in *Allium cepa* L. roots: implications for solute transport pathways. *J. Exp. Bot.* **52**: 1051–1061.

Massicotte, H. B., Melville, L. H., Peterson, R. L., and T. Unestam. 1999. Comparative studies of ectomycorrhiza formation in *Alnus glutinosa* and *Pinus resinosa* with *Paxillus involutus*. *Mycorrhiza* **8**: 229–240.

Masucci, J. D. and J. W. Schiefelbein. 1996. Hormones act downstrean of TTG and GL2 to promote root hair outgrowth during epidermis development in the *Arabidopsis* root. *Plant Cell* **8**: 1505–1517.

McCully, M. E. and J. E. Mallett. 1993. The branch roots of *Zea*. III. Vascular connections and bridges for nutrient recycling. *Ann. Bot.* **71**: 327–341.

Mosiniak, M., Le Rouic, I., and J.-C. Roland. 1995. Croissance pluridirectionnelle des parois hélicoïdales: le raccourcissement cellulaire des raciness tractrices. *Acta Bot. Gallica* **142**: 191–207.

Ottenschläger, I., Wolff, P., Wolverton, C. *et al.* 2003. Gravity-regulated differential auxin transport from columella to lateral root cap cells. *Proc. Natl Acad. Sci. USA* **100**: 2987–2991.

Pate, J. S., Gunning, B. E. S., and L. G. Briarty. 1969. Ultrastructure and functioning of the transport system of the leguminous root nodule. *Planta* 85: 11–34.

Peterson, C. A. 1988. Exodermal Casparian bands: their significance for ion uptake by roots. *Physio. Plant.* **72**: 204–208.

Peterson, C. A. and D. E. Enstone. 1996. Functions of passage cells in the endodermis and exodermis of roots. *Physiol. Plant.* **97**: 592–598.

Peterson, C. A., Murrmann, M., and E. Steudle. 1993. Location of the major barriers to water and ion movement in young roots of *Zea mays* L. *Planta* **190**: 127–136.

Peterson, R. L. 1992. Adaptations of root structure in relation to biotic and abiotic factors. *Can. J. Bot.* **70**: 661–675.

Ponce, G., Lujan, R., Campos, M. E. *et al.* 2000. Three maize root-specific genes are not correctly expressed in regenerated caps in the absence of the quiescent center. *Planta* **211**: 23–33.

Pritchard, J. 1994. The control of cell expansion in roots. *New Phytol.* **127**: 3–26.

Pütz, N. 1991. Measurement of the pulling force of a single contractile root. *Can. J. Bot.* **70**: 1433–1439.

Ridge, R. W. and F. D. Sack. 1992. Cortical and cap sedimentation in gravitropic *Equisetum* roots. *Am. J. Bot.* **79**: 328–334.

Sabatini, S., Beis, D., Wolkenfelt, H. *et al.* 1999. An auxin-dependent distal organizer of pattern and polarity in the *Arabidopsis* root. *Cell* **99**: 463–472.

Scales, P. F. and R. L. Peterson. 1991. Structure and development of *Pinus banksiana–Wilcoxina* ectendomycorrhizae. *Can. J. Bot.* **69**: 2135–2148.

Scheres, B., McKhann, H., van den Berg, C. *et al.* 1996. Experimental and genetic analysis of root development in *Arabidopsis thaliana*. *Plant Soil* **187**: 97–105.

Schiefelbein, J. W. 2000. Constructing a plant cell: the genetic control of root hair development. *Plant Physiol.* **124**: 1525–1531.

Schiefelbein, J. W. and C. Somerville. 1990. Genetic control of root hair development in *Arabidopsis thaliana*. *Plant Cell* **2**: 235–243.

Schiefelbein, J. W., Masucci, J. D., and H. Wang. 1997. Building a root: the control of patterning and morphogenesis during root development. *Plant cell* **9**: 1089–1098.

Schnepf, E. 1993. Golgi apparatus and slime secretion in plants: the early implications and recent models of membrane traffic. *Protoplasma* **172**: 3–11.

Schreiber, L. 1996. Chemical composition of Casparian strips isolated from *Clivia miniata* Reg. roots: evidence for lignin. *Planta* **199**: 596–601.

Schreiber, L., Hartmann, K., Skrabs, M., and J. Zeier. 1999. Apoplastic barriers in roots: chemical composition of endodermal and hypodermal cell walls. *J. Exp. Bot.* **50**: 1267–1280.

Seago, J. L., Peterson, C. A., Instone, D. E., and C. A. Scholey. 1999. Development of the endodermis and hypodermis of *Typha glauca* Godr. and *Typha angustifolia* L. roots. *Can. J. Bot.* **77**: 122–134.

Seago, J. S., Peterson, C. A., and D. E. Enstone. 2000a. Cortical development in roots of the aquatic plant *Pontederia cordata* (Pontederiaceae). *Am. J. Bot.* **87**: 1116–1127.

Seago, J. L., Peterson, C. A., Kinsley, L. J., and J. Broderick. 2000b. Development and structure of the root cortex in *Caltha palustris* L. and *Nymphaea oderato* Ait. *Ann. Bot.* **86**: 631–640.

Shane, M. W., McCully, M. E., and M. J. Canny. 2000. Architecture of branch–root junctions in maize: structure of the connecting xylem and the porosity of pit membranes. *Ann. Bot.* **85**: 613–624.

Skene, K. R. 1998. Cluster roots: some ecological considerations. *J. Ecol.* **86**: 1062–1066.

2000. Pattern formation in cluster roots: some developmental and evolutionary considerations. *Ann. Bot.* **85**: 901–908.

Smith, F. A. and S. E. Smith. 1997. Structural diversity in (vesicular)–arbuscular mycorrhizal symbioses. *New Phytol.* **137**: 373–388.

Steudle, E. and C. A. Peterson. 1998. How does water get through roots? *J. Exp. Bot.* **49**: 775–788.

Stewart, W. N. and G. W. Rothwell. 1993. *Palaeobotany and the Evolution of Plants*, 2nd edn. Cambridge, UK: Cambridge University Press.

Subba-Rao, N. S., Mateos, P. F., Baker, D. *et al.*, 1995. The unique root-nodule symbiosis betweeen *Rhizobium* and the aquatic legume, *Neptunia natans* (L. f.) Druce. *Planta* **196**: 311–320.

Taleisnik, E., Peyrano, G., Cordoba, A., and C. Arias. 1999. Water retention capacity in root segments differing in the degree of exodermis development. *Ann. Bot.* **83**: 19–27.

Torrey, J. G. 1953. The effect of certain metabolic inhibitors on vascular tissue differentiation in isolated pea roots. *Am. J. Bot.* **40**: 525–533.

Troughton, J. and L. A. Donaldson. 1972. *Probing Plant Structure*. Wellington, NZ: New Zealand Ministry of Research, Science and Technology.

van den Berg, C., Willimsen, V., Hendricks, G., Weisbeck, P. and B. Scheres. 1997. Short-range control of cell differentiation in the *Arabidopsis* root meristem. *Nature* **390**: 287–289.

Verdaguer, D. and M. Molinas. 1997. Development and ultrastructure of the endodermis in the primary root of cork oak (*Quercus suber*). *Can. J. Bot.* **75**: 769–780.

Wang, X.-L., Canny, M. J., and M. E. McCully. 1991. The water status of the roots of soil-grown maize in relation to the maturity of their xylem. *Physiol. Plant.* **82**: 157–162.

Wang, X.-L., McCully, M. E. and M. J. Canny. 1995. Branch roots of *Zea*. V. Structural features related to water and nutrient transport. *Bot. Acta* **108**: 209–219.

Watt, M. and J. R. Evans. 1999. Proteoid roots: physiology and development. *Plant Physiol.* **121**: 317–323.

Wilder, G. J. and J. R. Johansen. 1992. Comparative anatomy of absorbing roots and anchoring roots in three species of Cyclanthaceae (Monocotyledoneae). *Can. J. Bot.* **70**: 2384–2404.

Willemsen, V., Wolkenfelt, H., de Vrieze, G., Weisbeek, P. and B. Scheres. 1998. The *HOBBIT* gene is required for formation of the root meristem in the *Arabidopsis* embryo. *Development* **125**: 521–531.

Wilson, K. and J. N. Honey. 1966. Root contraction in *Hyacinthus orientalis*. *Ann. Bot.* **30**: 47–61.

Yawney, W. J. and R. C. Schultz. 1990. Anatomy of a vesicular–arbuscular endomycorrhizal symbiosis between sugar maple (*Acer saccharum* Marsh) and *Glomus etunicatum* Becker & Gerdemann. *New Phytol.* **114**: 47–57.

Yu, T. E. J. C., Egger, K. N. and R. L. Peterson. 2001. Ectendomycorrhizal associations: characteristics and functions. *Mycorrhiza* **11**: 167–177.

Zhu, T., Lucas, W. J., and T. L. Rost. 1998. Directional cell-to-cell communication in the *Arabidopsis* root apical meristem. I. An ultrastructural and functional analysis. *Protoplasma* **203**: 35–47.

FURTHER READING

Barlow, P. W. 1974. Regeneration of the cap of primary roots of *Zea mays. New Phytol.* **73**: 937–954.

1976. Towards an understanding of the behaviour of root meristems. *J. Theor. Biol.* **57**: 433–451.

Barlow, P. W. and F. Baluska. 2000. Cytoskeletal perspectives on root growth and morphogenesis. *Ann. Rev. Plant Physiol. Plant Mol. Biol.* **51**: 289–322.

Barlow, P. W. and J. S. Parker. 1996. Microtubular cytoskeleton and root morphogenesis. *Plant Soil* **187**: 23–36.

Barlow, P. W., Luck, H. B. and J. Luck. 2001. Autoreproductive cells and plant meristem construction: the case of the tomato cap meristem. *Protoplasma* **215**: 50–63.

Baum, S. F. and T. L. Rost. 1996. Root apical organization in *Arabidopsis thaliana*. I. Root cap and protoderm. *Protoplasma* **192**: 178–188.

Bergersen, F. J., Kennedy, G. S., and W. Wittmann. 1965. Nitrogen fixation in the coralloid roots of *Macrozamia communis* L. Johnson.. *Austral J. Biol. Sci.* **18**: 1135–1142.

Bonnett, H. T., Jr. 1968. The root endodermis: fine structure and function. *J. Cell Biol.* **37**: 109–205.

Bonnett, H. T., Jr. and J. G. Torrey. 1966. Comparative anatomy of endogenous bud and lateral root formation in *Convolvulus arvensis* roots cultured in vitro. *Am. J. Bot.* **53**: 496–507.

Byrne, J. M. 1973. The root apex of *Malva sylvestris*. III. Lateral root development and the quiescent center. *Am. J. Bot.* **60**: 657–662.

Carlson, M. C. 1950. Nodal adventitious roots in willow stems of different ages. *Am. J. Bot.* **37**: 555–561.

Chapman, K., Groot, E. P., Nichol, S. A., and T. L. Rost. 2002. Primary root growth and the pattern of root apical meristem organization are coupled. *J. Plant Growth Regul.* **21**: 287–295.

Clowes, F. A. L. 1975. The quiescent centre. In J. G. Torrey and D. T. Clarkson, eds., *The Development and Function of Roots*. London: Academic Press, pp. 3–19.

1981. The difference between open and closed meristems. *Ann. Bot.* **48**: 761–767.

1984. Size and activity of quiescent centres of roots. *New Phytol.* **96**: 13–21.

Erickson, R. O. and K. B. Sax. 1956. Rates of cell division and cell elongation in the growth of the primary root of *Zea mays. Proc. Am. Phil. Soc.* **100**: 499–514.

Esau, K. 1940. Developmental anatomy of the fleshy storage organ of *Daucus carota. Hilgardia* **13**: 175–226.

1943. Vascular differentiation in the pear root. *Hilgardia* **15**: 299–311.

1965. *Vascular Differentiation in Plants*. New York: Holt, Rinehart and Winston.

Fayle, D. C. F. 1975. Distribution of radial growth during the development of red pine root systems. *Can. J. For. Res.* **5**: 608–625.

Feldman, L. J. 1984. The development and dynamics of the root apical meristem. *Am. J. Bot.* **71**: 1308–1314.

Fogel, R. 1983. Root turnover and productivity of coniferous forests. *Plant Soil* **71**: 75–85.

Fontana, A. 1985. Vesicular–arbuscular mycorrhizas of *Ginkgo biloba* L. in natural and controlled conditions. *New Phytol.* **99**: 441–447.

Foster, R. C. and G. C. Marks. 1966. The fine structure of the mycorrhizas of *Pinus radiata* D. Don. Austral. *J. Biol. Sci.* **19**: 1027–1038.

Groot, E. P., Doyle, J. A., Nichol, S. A., and T. L. Rost. 2004. Phylogenetic distribution and evolution of root apical meristem organization in dicotyledonous angiosperms. *Int. J. Plant Sci.* **165**: 97–105.

Haas, D. L. and Z. B. Carothers. 1975. Some ultrastructural observations on endodermal cell development in *Zea mays* roots. *Am. J. Bot.* **62**: 336–348.

Haissig, B. E. 1974. Origins of adventitious roots. *N. Z. J. For. Sci.* **4**: 299–310.

Harley, J. L. and S. E. Smith. 1983. *Mycorrhizal Symbiosis*. London: Academic Press.

Hawes, M. C., Bengough, G., Cassab, G., and G. Ponce. 2002. Root caps and rhizosphere. *J. Plant Growth Regul.* **21**: 352–367.

Hayward, H. E. 1938. *The Structure of Economic Plants*. New York: Macmillan.

Head, G. C. 1973. Shedding of roots. In T. T. Kozlowski, ed., *Shedding of Plant Parts*. New York: Academic Press, pp. 237–293.

Heimsch, C. 1960. A new aspect of cortical development in roots. *Am. J. Bot.* **47**: 195–201.

Iversen, T.-H. and P. Larsen. 1973. Movement of amyloplasts in the statocytes of geotropically stimulated roots: the pre-inversion effect. *Physiol. Plant.* **28**: 172–181.

Ma, F. S. and C. A. Peterson. 2003. Current insights into the development, structure, and chemistry of the endodermis and exodermis of roots. *Can. J. Bot.* **81**: 405–421.

McCully, M. E. 1975. The development of lateral roots. In J. G. Torrey and D. T. Clarkson, eds., *The Development and Function of Roots*. London: Academic Press, pp. 105–124.

Peterson, R. L. 1967. Differentiation and maturation of primary tissues in white mustard root tips. *Can. J. Bot.* **45**: 319–331.

Peterson, R. L. and H. B. Massicotte. 2004. Exploring structural definitions of mycorrhizas, with emphasis on nutrient-exchange interfaces. *Can. J. Bot.* **82**: 1074–1088.

Postgate, J. 1987. *Nitrogen Fixation*, 2nd edn. London: Arnold.

Romberger, J. A., Hejnowicz, Z., and J. F. Hill. 1993. *Plant Structure: Function and Development*. Berlin: Springer-Verlag.

Rost, T. L., Baum, S. F., and S. Nichol. 1996. Root apical organization in *Arabidopsis thaliana* ecotype 'WS' and a comment on root cap structure. *Plant Soil* **187**: 91–95.

Row, H. C. and J. R. Reeder. 1957. Root hair development as evidence of relationships among genera of Gramineae. *Am. J. Bot.* **44**: 596–601.

Samaj, J., Baluska, F., and D. Menzel. 2004. New signaling molecules regulating root hair tip growth. *Trends Plant Sci.* **9**: 217–220.

Shen-Miller, J. and R. R. Hinchman. 1974. Gravity sensing in plants: a critique of the statolith theory. *BioScience* **24**: 643–651.

Timonen, S. and R. L. Peterson. 2002. Cytoskeleton in mycorrhizal symbiosis. *Plant Soil* **244**: 199–210.

Tjepkema, J. D. 1983. Hemoglobins in the nitrogen-fixing root nodules of actinorhizal plants. *Can. J. Bot.* **61**: 2924–2929.

Tjepkema, J. D. and C. S. Yocum. 1974. Measurement of oxygen partial pressure within soybean nodules by oxygen microelectodes. *Planta* **119**: 351–360.

Tomlinson, P. B. 1961. *Anatomy of the Monocotyledons*, vol. 2, *Palmae*. Oxford, UK: Clarendon Press.

Torrey, J. G. 1978. Nitrogen fixation by actinomycete-nodulated angiosperms. *BioScience* **28**: 586–592.

Torrey, J. G. and D. T. Clarkson (eds.) 1975. *The Development and Function of Roots*. London: Academic Press.

Wenzel, C. L. and T. L. Rost. 2001. Cell division patterns of the protoderm and root cap in the "closed" root apical meristem of *Arabidopsis thaliana*. *Protoplasma* **218**: 203–213.

Wilcox, H. 1962. Growth studies of the root of incense cedar, *Libodedrus decurrens*. I. The origin and development of primary tissues. *Am. J. Bot.* **49**: 221–236.

1968. Morphological studies of the roots of red pine, *Pinus resinosa*. II. Fungal colonization of roots and the development of mycorrhizae. *Am. J. Bot.* **55**: 688–700.

Wilson, B. F. 1975. Distribution of secondary thickening in tree root systems. In J. G. Torrey and D. T. Clarkson, eds., *The Development and Function of Roots*. London: Academic Press, pp. 197–219.

Ziegler, H. 1964. Storage, mobilization and distribution of reserve material in trees. In M. H. Zimmermann, ed., *The Formation of Wood in Forest Trees*. New York: Academic Press, pp. 303–320.

Chapter 17

The leaf

Perspective: evolution of the leaf

All vascular plants except their most primitive ancestors are characterized by leaves (see Chapter 1). As the primary photosynthetic organs, leaves are of great significance not only to the plant but also to many other organisms, including humans, that rely on plants as a source of food. Botanists interested in the evolution of plant structures believe that leaves evolved in at least two ways, and in possibly five independent lines in vascular plants (see Niklas, 1997). The leaves of lycophytes are considered **enations** because they are thought to have evolved as simple outgrowths from stems. These leaves, often referred to as **microphylls**, are commonly small although those of some extinct taxa attained great lengths (up to 1 meter in some lepidodendrids). Like all microphylls, however, they were vascularized by only a single midvein. In seed plants and ferns (possibly also in sphenophytes) leaves are thought to represent evolutionarily modified lateral branch systems. This hypothesis (the telome hypothesis) is based on the fact that the earliest seed plant ancestors were leafless, but bore small lateral branch systems. The fossil evidence indicates that over time, three-dimensional branch systems became flattened and subsequently laminate. Seed plant leaves which, on average, are much larger, and much more complex than those of lycophytes in both gross morphology and internal structure, are often referred to as **megaphylls**. For more detailed discussions of the evolution of leaves see Steward and Rothwell (1993) and Taylor and Taylor (1993).

Leaves may be classified in several categories: **foliage leaves** (which function in photosynthesis), **cataphylls** (bud scales and scales on underground stems which function in protection and/or storage; the first cataphylls to develop are often called **prophylls**), **hypsophylls** (floral bracts which are thought to have a protective function), and **cotyledons** (the first leaves produced in the embryo which may be thin, or thick if they function in storage and provide a direct source of nutrition for the developing seedling).

Figure 17.1 Cleared leaf of *Coleus blumei*, a dicotyledon, showing its broad, laminate form and complex system of veins (vascular bundles). M, midrib; S, secondary vein; T, tertiary vein; Q, quaternary vein. From Fisher (1985). Used by permission of the Botanical Society of America.

Basic leaf structure

Although diverse in morphology and anatomy, all leaves share many features in common. The leaves of many seed plants consist of a stalk-like petiole and a relatively thin, broad laminate blade characterized by transectional dorsiventral symmetry (Figs. 17.1, 17.2). Internally a vascular skeleton is enclosed by parenchymatous mesophyll which, in turn, is enclosed by an epidermis (Fig. 17.2a, b). In many ways the anatomy of the leaf resembles that of the stem. In some dicotyledons, for example, the leaf traces that enter the petiole base of large leaves divide and become arranged in a cylinder although other tissues in the petiole have a dorsiventral arrangement, characteristic of most leaves. Furthermore, the petiolar vascular supply as well as that in the leaf proper in some dicotyledons and many conifers is enclosed by a pericycle and endodermis (Fig. 17.3). The mesophyll may contain collenchyma and/or sclerenchyma.

Although similar in anatomy, leaves of most plants differ fundamentally from stems in that they are determinate in growth. That is, their meristems cease to function after a genetically predetermined

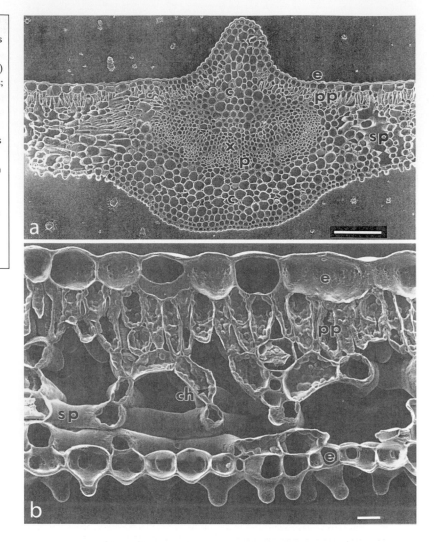

Figure 17.2 Scanning electron micrographs of transverse sections of leaves of *Erythroxylon coca* showing dorsiventral symmetry. (a) Section of the midrib. e, epidermis; pp, palisade mesophyll; sp, spongy mesophyll. x, primary xylem; p, primary phloem. Bar = 100 μm. (b) Section illustrating chloroplasts (ch) in palisade and spongy parenchyma and large air spaces in the spongy mesophyll. Bar = 10 μm. (a, b) From Ferreira *et al.* (1998). Used by permission of the University of Chicago Press. © 1998 The University of Chicago. All rights reserved.

period of growth, and their size is thereby restricted. The leaves of several ferns, however, may continue their development over long periods of time, and are, therefore, essentially indeterminate. Several dicotyledons, e.g., *Guarea* and *Chisocheton* in the Meliaceae, also produce relatively indeterminate leaves characterized by continued apical growth for up to four years (see Fisher and Rutishauser, 1990). Leaves of deciduous plants live for only one growing season. Those of "evergreen" plants are termed persistent, and remain on the plant for two or more years.

Although the leaves of a majority of plants have broad, relatively thin blades, others may be thick and fleshy or, as in some monocotyledons, tubular. In many conifers, there is an absence of specialization into blade and petiole. Whatever the morphology of a leaf, in gymnosperms and angiosperms, **adaxial** and **abaxial** surfaces (the surface oriented toward the stem and the surface oriented away from the stem, respectively) are, with rare exceptions, reflected in its

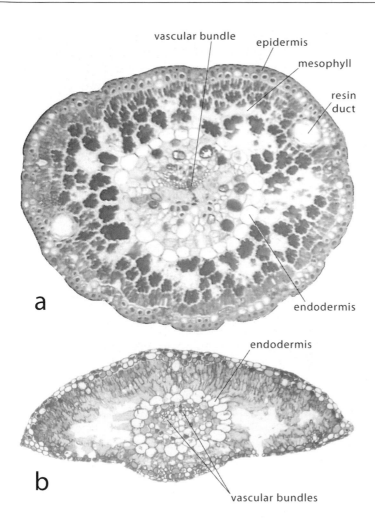

vascular bundle   epidermis

mesophyll

resin
duct

a

endodermis

endodermis

b

vascular bundles

Figure 17.3 Transverse sections of conifer leaves. (a) *Pinus monophylla*, with a single vascular bundle enclosed by an endodermis. Magnification × 99. (b) *Pseudolarix* sp. with two vascular bundles. Magnification × 76.

internal anatomy, especially in the arrangement of primary xylem and phloem in the veins. You will remember that in seed plants the vascular bundles of the stem eustele are commonly **collateral**, that is, xylem comprises the inner part of the bundles and phloem the outer. Consequently, as leaf traces diverge into the leaf, the primary xylem will be oriented toward the adaxial surface of the leaf, the primary phloem toward the abaxial surface (Fig. 17.2a). This information can be especially useful to paleobotanists who study fragments of leaves in the fossil record. Orientation based on position of vascular tissues may be difficult or impossible in some tubular leaves such as that of *Allium* and some other monocotyledons. As we shall see, other aspects of anatomy may also be correlated with position in the leaf.

In seed plants the leaf vascular system originates from one, two, three, five, or more leaf traces that diverge from axial bundles of the stem eustele and enter the leaf base (Beck *et al.*, 1983; see also Chapter 6). In petiolate leaves of dicotyledons, there may be a single bundle in the petiole of small leaves that in the blade will become the midvein. This single bundle may be an extension of the single leaf trace, or may result from the fusion of two leaf traces. In larger leaves the petiole vascular system is usually supplied by three or more leaf traces, a large median trace and two, four, or more smaller lateral traces. The median trace will become the midvein. In leaves supplied by five or more traces, the laterals frequently become arranged in an arc, and in very large leaves, as noted above, following branching of the laterals, a cylinder of vascular bundles may be formed. Even in such cases, however, the median bundle (midvein) with xylem and phloem oriented toward the adaxial (upper) and abaxial (lower) surfaces of the leaf lamina, respectively, is usually easily identified. In some taxa, the midvein may consist of several vascular bundles including the extension of the median trace and branches of the adjacent lateral traces (Beebe and Evert, 1990). In monocotyledons, petioles commonly contain numerous vascular bundles in a parenchymatous ground tissue.

As the result of branching of the leaf traces or petiolar vascular system during development, a complex anastomosing system of veins is produced in the blades of many leaves. In most dicotyledons, a **reticulate venation system** develops (Figs. 17.1a, 17.4a, b) whereas in many monocotyledons a system of interconnected, more or less **parallel veins** is common (Fig. 17.4c, d). In some primitive gymnosperms such as *Ginkgo*, there is an open, **dichotomous system**. In the compound leaves of cycads a single vein services each pinna in *Cycas*, but in some other genera there are many essentially parallel veins that may run the length of the pinnae with some of these dichotomizing. In the cycad *Stangeria* each pinna contains a midvein with laterals. The leaves of conifers are vascularized by a single vein or two veins (Fig. 17.3) that result from the dichotomy of a single leaf trace.

Vascular bundles of the venation system are enclosed by **bundle sheaths**, consisting of parenchyma (sometimes collenchyma or sclerenchyma) one or more cells thick (Fig. 17.5). In some angiosperms cells of the bundle sheaths are characterized by Casparian strips or suberin lamellae similar to those of an endodermis which apparently provide apoplastic barriers to solute transport. In leaves of some taxa, walls or ribs of tissue called **bundle sheath extensions** extend from the veins to the upper and/or lower epidermis (Fig. 17.5). These are supporting structures, often composed of sclerenchyma, that are also thought to be pathways of transport between the veins and the epidermis.

The leaf **mesophyll**, the primary photosynthetic tissue of the leaves of dicotyledons, is predominantly parenchymatous, but in many taxa sclerenchyma and/or collenchyma provide support especially along the leaf margins and around vascular bundles. In the leaves of many angiosperms the mesophyll is divided into upper and lower regions (Fig. 17.2b). The upper region, the **palisade mesophyll**, consists of one or more layers of elongate, tubular cells oriented at right angles to the epidermis. The lower region, the **spongy mesophyll**, is composed of loosely arranged, irregularly shaped cells between which are extensive interconnected **air spaces** and **channels**.

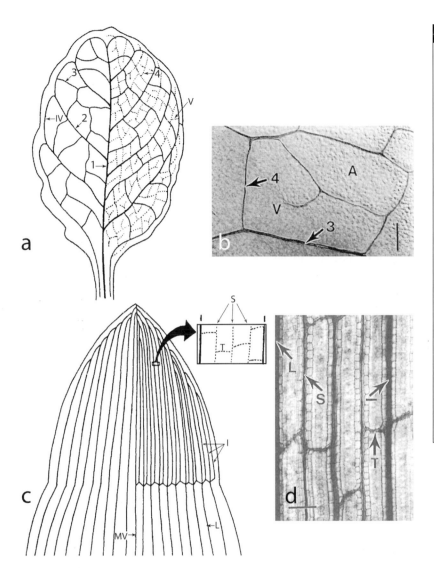

Figure 17.4 Venation patterns in dicotyledon and monocotyledon leaves. (a) Diagram of the reticulate venation pattern in *Arabidopsis thaliana* showing different vein orders. Secondary veins (2) are joined by an intramarginal vein (IV). V, freely ending veinlet. (b) Part of a cleared leaf of *Arabidopsis* showing tertiary (3) and quaternary (4) veins enclosing an areole (A) containing a freely ending veinlet (V). Bar = 100 μm. (c) Diagram of a leaf of *Zea mays* (maize) showing the midvein (MV), large (L), intermediate (I), and small (S) longitudinal veins, and transverse veins (T). (d) Part of a cleared leaf blade of *Zea mays*, showing vein orders as in (c). Bar = 100 μm. Note the parallel pattern of veins in *Zea* as contrasted with the reticulate pattern in *Arabidopsis*. (a–d) From Nelson and Dengler (1997). Used by permission of the American Society of Plant Biologists.

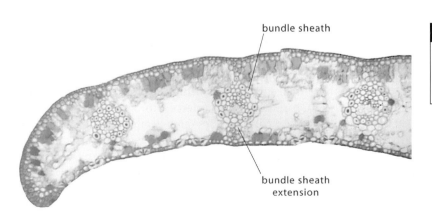

bundle sheath

bundle sheath extension

Figure 17.5 Transverse section of a pinnule of *Zamia* (a cycad) illustrating a vascular bundle sheath and the bundle sheath extension. Magnification × 80.

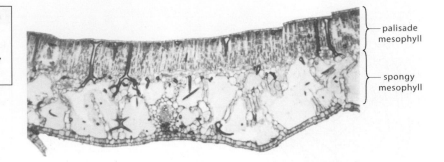

Figure 17.6 Transverse section of the leaf of the aquatic plant *Nymphaea*. Note the very extensive air spaces in the spongy mesophyll. Magnification × 50.

palisade mesophyll

spongy mesophyll

Such tissue is generally referred to as aerenchyma. Air channels also characterize the palisade mesophyll. The number of layers of palisade mesophyll is often correlated with light intensity; the higher the light intensity in which the plant grows, the greater the number of layers. The looseness of the spongy mesophyll, i.e., the volume of air space within it, may be related to the exchange of O_2 and CO_2. For example the floating leaves of aquatic plants (Fig. 17.6) have proportionately larger air spaces in the spongy mesophyll than those in the spongy mesophyll of leaves, the lower surfaces of which are in contact with the atmosphere. In some members of the Leguminosae, e.g. *Glycine* (soybean) and *Calliandra*, and in several other dicotyledon families, the veins are connected by sheets of parenchymatous tissue called **paraveinal mesophyll** (Fig. 17.7a, b). (Kenekordes *et al.*, 1988; Liljebjelke and Franceschi, 1991; Lersten and Curtis, 1993). This tissue, one or two layers thick, in essentially the same plane as the veins, is thought to function in the transport of photosynthate and nitrogenous compounds between the mesophyll and the veins (Franceschi and Giaquinta, 1983a, b).

The leaves of some xeromorphic plants exhibit little or no difference between upper and lower mesophyll regions as, for example, those of many conifers, including *Pinus* (pine). The uniform mesophyll of *Pinus* (Fig. 17.3) and some other conifers is composed of cells characterized by distinctive infoldings of the cell wall. A lack of differentiation of palisade and spongy mesophyll also characterizes some herbaceous perennial dicotyledons and many grasses (Fig. 17.8). Some xerophytes, on the other hand, including the gymnosperms *Araucaria*, *Podocarpus*, and *Callitris*, as well as some angiosperms (e.g., *Atriplex* in the Chenopodiaceae, species of *Acacia* in the Leguminosae, and species of Myrtaceae) have palisade mesophyll in both upper and lower parts of their leaves. Such mesophyll is referred to as being **isobilateral** (Fig. 17.9a). Of 39 xeromorphic species from many families studied by Burrows (2001), the leaves of nearly 50% were isobilateral, and most of these were **amphistomatous** (with stomata in upper and lower epidermises) (see, also, Beerling and Kelly, 1996). The adult leaves of some species of *Hakea* (Proteaceae) (see Groom *et al.*, 1997) are circular in section with a cylindrical, one- to several-layered mesophyll (Fig. 17.9b). The leaf mesophyll may contain various secretory ducts and cavities

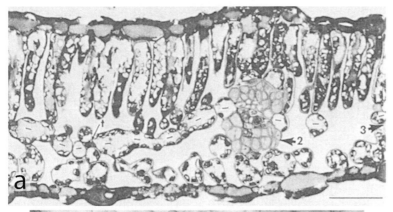

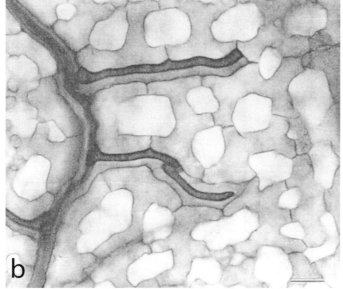

Figure 17.7 Paraveinal mesophyll in leaves of *Calliandra tweedii* (Leguminosae).
(a) Transverse section of a leaf showing a single-layered paraveinal mesophyll (cells containing dashes) in sectional view extending between veins of several orders. Vein order is indicated by numbers. Bar = 50 μm. (b) Paradermal section showing the paraveinal mesophyll in surface view. Note its reticulate nature; also the veins consisting of tracheary elements and parallel parenchyma cells. Bar = 20 μm. From Lersten and Curtis (1993). Used by permission of the Botanical Society of America.

such as resin canals in conifers, oil cavities containing aromatic compounds in *Mentha* and *Eucalyptus*, laticifers in many dicotyledons, etc. (see Chapter 14).

The leaf **epidermis** is a compact parenchyma tissue lacking intercellular spaces (except the openings of stomata between guard cells) (Figs. 17.2b, 17.10a, b). Stomata are interspersed within this tissue, and are especially common in the lower epidermis, a condition termed **hypostomatous**; but in some taxa (e.g., floating aquatic species) stomata are restricted to the upper epidermis (**epistomatous**). Stomata seem to occur randomly in the epidermis of dicotyledons (Fig. 17.10a), but a recent analysis (Larkin *et al.*, 1996) indicates that stomatal spacing may be non-random and controlled by cell lineages associated with stomatal development as well as the cell lineages of neighboring cells. In monocotyledons with parallel venation, stomata typically occur in parallel rows. In mesophytes and hydrophytes stomata occur at the same level as other epidermal cells or, especially

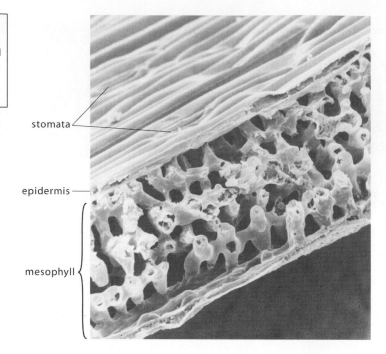

Figure 17.8 Scanning electron micrograph of a grass leaf which lacks differentiation of palisade and spongy parenchyma. Magnification × 87. Photograph by P. Dayanandan.

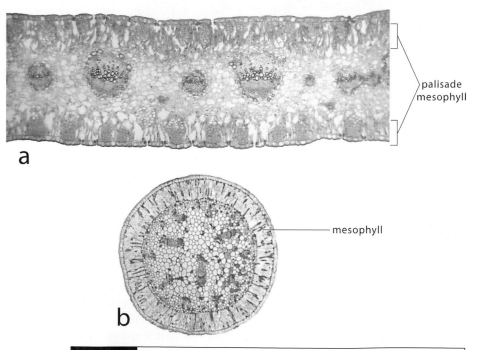

Figure 17.9 (a) Transverse section of a leaf of *Welwitschia mirabilis* with an isobilateral mesophyll (palisade mesophyll in both upper and lower regions of the leaf). Magnification × 52. (b) Transverse section of a leaf of *Hakea erinacea*, circular in section, and containing a cylindrical palisade mesophyll. Magnification × 18. (b) From Groom *et al.* (1997). Reproduced with permission of CSIRO Publishing, Melbourne, Australia. Copyright © CSIRO Australia 1997.

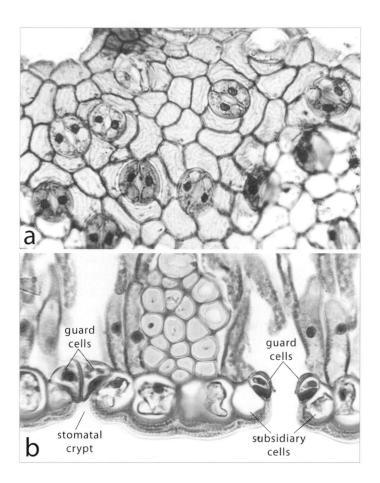

a

b

guard
cells

guard
cells

stomatal
crypt

subsidiary
cells

Figure 17.10 (a) Surface view of a leaf epidermis showing stomata with open apertures. Magnification × 297. (b) Stomata in the lower epidermis of *Welwitschia mirabilis*. In this xerophyte, stomata are sunken in stomatal crypts. Note that the guard cells are overlain by the subsidiary cells; also that there is an air space in the mesophyll beneath each stoma. Magnification × 526.

in aquatic plants, may be raised above this level. The stomata of xerophytes are typically sunken in depressions in the leaf surface called **stomatal crypts** (Fig. 17.10b), and in many species occur in both lower and upper epidermises. In some taxa, the epidermis contains sclereids, as well as (in grasses) cork and silica cells. A variety of unicellular and/or multicellular trichomes (hairs) characterize the leaf epidermis. In leaves that roll or curl under conditions of water stress (e.g., leaves of many grasses), there are rows of large, thin-walled bulliform cells (Fig. 17.11) in the upper epidermis which, through loss of turgor, are thought to facilitate this process. Bulliform cells occur in grooves in the leaf surface that parallel major veins. During leaf development, increase in turgor in bulliform cells may assist in the unrolling of young leaves. In most plants the epidermal cells (with the exception of stomatal guard cells) lack chloroplasts. However, in some highly reduced aquatic species and a few ferns that grow in very low light intensity chloroplasts occur in all epidermal cells. Except in the epidermis of immersed aquatic plants the epidermis is covered by a variably thick cuticle. (For more detail on the epidermis, especially the morphology of stomata and trichomes, see Chapter 8.)

Figure 17.11 Transverse section of part of a leaf of *Saccharum* showing bulliform cells in the upper epidermis. Note also the conspicuous bundle sheaths. Magnification × 141. From Esau (1977). Used by permission of John Wiley and Sons, Inc.

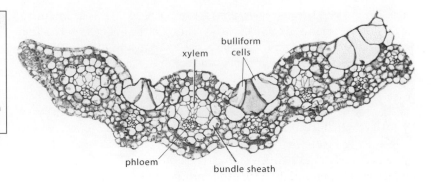

xylem

bulliform cells

phloem

bundle sheath

Leaf development

Leaf primordia, produced by the apical meristem, begin their development by periclinal divisions in a subsurface layer on the flank of the apical dome. Additional periclinal as well as anticlinal divisions in both subsurface and surface layers result in a protuberance commonly called the **leaf primordium buttress**. In pteridophytes, a single apical initial, formed at the tip of the buttress, is the ultimate source of all additional cells produced during the development of the leaf. In many gymnosperms and angiosperms, a small cluster of cells forms the apical meristem of the leaf primordium. Factors that control the initiation of a leaf primordium are, as yet, not clearly understood, but it is widely accepted that growth hormones such as auxin and gibberellin stimulate primordium formation. It has been demonstrated that the application of the protein expansin to the apical meristem can also lead to the formation of primordium-like outgrowths (Fleming *et al.*, 1999). Expansin is thought to cause a loosening of the microfibrils and the extensibility of cell walls at sites of primordium development (Lyndon, 1994; see also Cho and Cosgrove, 2000). This leads to an outward buckling of cells on the surface of the apical meristem, supporting the view of Green (1999) that physical factors may in some degree control the formation of primordia. For a more detailed discussion of the initiation of leaf primordia, please see Chapter 4. Auxin is also known to play a significant role in the continuing development of leaves following primordium formation.

Leaf morphogenesis in dicotyledons can be considered to consist of three phases, **leaf** (or **primordium**) **initiation** (described above), primary morphogenesis, and expansion and secondary morphogenesis (Fig. 17.12) (see Dengler and Tsukaya, 2001). During **primary morphogenesis**, cell division and cell growth in the young leaf primordium result in the formation of a primordial leaf axis, often called a **phyllopodium**, which has a dorsiventral symmetry, and which, ultimately, will become the petiole and midrib of the leaf. Early during this phase, as the phyllopodium increases in thickness, the leaf lamina begins to form as outgrowths on either side resulting from cytokinesis in **marginal meristems**. Some researchers prefer the term **marginal blastozone** over marginal meristems (Hagemann and

Gleissberg, 1996) because it emphasizes the morphogenetic potential of the lateral regions of the phyllopodium. Continued activity in the marginal meristems results in the lateral expansion of the developing leaf blade, each half of which commonly extends upward at an angle on either side of the phyllopodium. In plants with compound leaves, the marginal meristems become subdivided, and each subdivision, from which, ultimately, a leaflet will develop, is characterized by its own phyllopodium with apical and marginal meristems. With continuing cell division in apical and marginal meristems followed by cell expansion, the entire leaf primordium usually curves upward and, in woody perennials, with other immature leaves and bud scales, comprises a vegetative bud. In some plants with petiolate leaves, a basal meristem, proximal to the marginal meristems, develops in the phyllopodium. The activity of this meristem results ultimately in the development of the leaf petiole. In other taxa, the petiole results from the suppression of activity of the marginal meristems. At an early stage, provascular tissue begins to differentiate in the phyllopodium and developing blade in a pattern that will, ultimately, reflect the mature system of veins.

During **expansion and secondary morphogenesis** (Fig. 17.12), the young leaf continues its growth and differentiation, ultimately achieving its mature size and form. During this phase, which covers a much longer time period than primary morphogenesis, there is an increase in surface area and volume of several thousandfold (Dale, 1988; see also Dengler and Tsukaya, 2001) and, according to Dale (1988), about 95% of the cells that comprise the mature leaf are formed. In young leaf primordia and very young leaves of most vascular plants, the marginal meristems are two or more layers thick. The outer layer can be considered a protoderm comparable to that of the stem from which the epidermis develops. The marginal meristems are, however, short-lived and subsequent meristematic activity is intercalary and diffuse (Donnelly *et al.*, 1999). Subsequent growth and differentiation lead to the development of the internal leaf parenchyma. Some parts of the leaf parenchyma will differentiate as mesophyll and other parts as provascular tissue from which the system of veins will ultimately develop. Characteristics of the leaf margins and distinctive features of lobing develop, and the leaves achieve their final form. During this period of expansion and differentiation, the leaf may retain the basic form established during primary morphogenesis (**isometric growth**), or differences in morphology may occur (**allometric growth**) (Fig. 17.12) which is the more common growth pattern (Dengler and Tsukaya, 2001).

The morphology of mature venation systems is highly diverse. It is not surprising, therefore, that the development of venation systems is also highly variable. There are some common patterns of development, however, which we shall consider, using as examples a dicotyledon leaf with reticulate venation and a monocotyledon leaf with parallel venation. In the leaves of many dicotyledons the venation system is initiated by the acropetal differentiation of a central

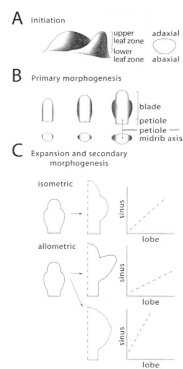

Figure 17.12 Diagrams illustrating three phases of leaf morphogenesis. During leaf initiation (A), the leaf primordium expresses both longitudinal symmetry (upper and lower leaf zones) and dorsiventral symmetry (differences between adaxial and abaxial sides). During primary morphogenesis (B), the marginal meristem (blastozone) (shaded) expresses morphogenetic potential to form the blade, lobes, and leaflets. In the top row is an adaxial view of the phyllopodium (young leaf). The bottom row shows transverse sectional views of the developing blade. During expansion and secondary morphogenesis (C), there is both isometric and allometric expansion of lobes and sinuses produced during primary morphogenesis. From Dengler and Tsukaya (2001). Used by permission of the University of Chicago Press. © 2001 The University of Chicago. All rights reserved.

Figure 17.13 Vascular pattern ontogeny in the leaves of dicotyledons and monocotyledons. (a–d) *Arabidopsis thaliana*, a dicotyledon; (e–h) *Zea mays*, a monocotyledon. See the text for detailed descriptions. From Nelson and Dengler (1997). Used by permission of the American Society of Plant Biologists.

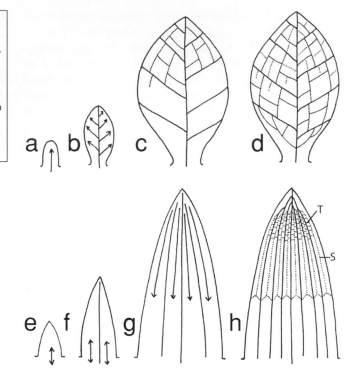

provascular strand in the phyllopodium (Fig. 17.13a). As the laminae expand, the various branch orders of the venation system differentiate in a heirarchical manner with the major (second-order) veins developing first (Fig. 17.13b, c) (Dengler and Kang, 2001), preceded, of course, by second-order provascular strands. These veins, which reflect leaf shape, provide conduits for the transport of nutrients into the developing leaf. Third-, fourth-, and higher-order veins develop subsequently, beginning in the apical regions of the leaf and proceeding basipetally (Fig. 17.13c, d). Last-order veins enclose small regions of parenchyma called areoles into which extend one or several vein endings (Figs. 17.4b, 17.13d). Every mesophyll cell is, thus, in contact with, or very close to, a vascular bundle.

At the stage in which provascular tissue differentiates, the developing lamina consists commonly of five cell layers. The two outer layers are protoderm from which the upper and lower epidermis will differentiate (Fig. 17.14a). The three inner layers are ground meristem, sometime called **promesophyll**. It is in the median layer of this tissue that the successive orders of provascular tissue will differentiate (Fig. 17.14a, b) and from which, ultimately, mature veins containing primary xylem and phloem will develop. Whereas the vascular strands of the venation system are derived solely from the median layer of ground meristem, the adaxial and abaxial layers of this tissue contribute to the development of the bundle sheath and bundle sheath extensions in many dicotyledons.

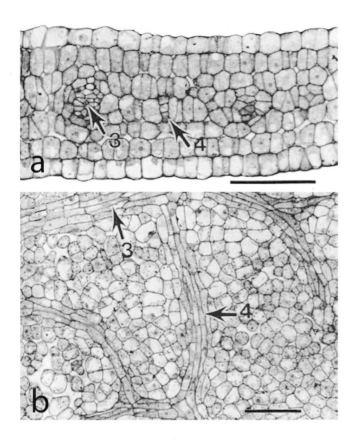

Figure 17.14 Immature leaves of *Arabidopsis thaliana* showing protoderm, ground meristem (promesophyll), and provascular strands. (a) Transverse section. Note provascular strands (arrows) that will develop into third- and fourth-order veins. Bar = 50 μm. (b) Paradermal section showing provascular strands that will develop into third- and fourth-order veins. Bar = 50 μm. From Nelson and Dengler (1997). Used by permission of the American Society of Plant Biologists.

Development of the leaf vascular system in monocots, although generally similar to that of dicotyledons, differs in several important ways. The initial ("midvein") provascular strand differentiates both acropetally and basipetally (Fig. 17.13e) in the base of the developing leaf primordium, followed by provascular strands of large veins which first differentiate acropetally and subsequently basipetally, connecting with vascular bundles of the stem vascular system (Fig. 17.13f, g). Intermediate veins are initiated in the apical region of the young leaf and differentiate basipetally (Fig. 17.13g), some of which will connect with the stem vasculature. Finally, small longitudinal and transverse veins are formed in the apical region of the leaf with transverse vein formation proceeding basipetally (Fig. 17.13h). Strong evidence indicates that polar auxin transport, possibly from leaf margins, influences the differentiation of provascular tissue, and ultimately mature veins (Dengler and Kang, 2001, and references therein). Meristematic activity continues in the more basal region after it has ceased in the more apical region of the developing leaf. Consequently, development of most tissues at this late stage of development is basipetal. DeMason and Chawla (2004) have demonstrated the significant role of auxin in the development of compound leaves of *Pisum sativum*. They conclude that auxin "is the driving force for leaf growth and pinna determination, necessary for pinna initiation, and controls subsequent pinna development."

During leaf development, the frequency (or rate) of cell division within different regions of the developing leaf may be directly related to its mature shape. However, rate of cell division alone may not be a controlling factor since cell growth (i.e., increase in size of individual cells) may also play an important role in leaf morphogenesis (see Fleming, 2002). For example, if the rate of cell division is high, but the new cells remain very small in a part of the leaf (e.g., a leaf lobe), there will be little change in form. Likewise, if the rate is low, but new cells become large, there may also be little change in form. If, however, frequency of cell division is high and the new cells become large the lobe will increase in size and, depending on the relative difference in rate of cell division and degree of growth, the shape of the lobe will also change. An important goal in understanding leaf morphogenesis is determination of the mechanism which controls the integration of frequency of cell division and cell growth (Fleming, 2002). Since in many leaves the greatest frequency of cell division is in their more basal region, such leaves may be broader at the base than at the apex. In many plants the activity of adaxial and abaxial meristems along the developing midrib result in its increase in thickness, especially on the abaxial side. Stipule primordia may develop on either side of the leaf primordia.

Although the patterns of development presented above are generally applicable to the leaves of many vascular plants, significant variations characterize different major taxa. In conifers and some dicotyledons which have leaves that are angular to nearly circular in transverse section, marginal meristems are absent or largely inactive. In some conifers and other taxa that possess linear leaves, a basal meristem, often considered an intercalary meristem, and its derivatives provide most of the tissue of the mature leaf. Differentiation in such leaves is almost entirely basipetal. In many monocotyledons the base of the leaf primordia may encircle the apical meristem resulting, in the mature state, in leaves with leaf sheaths. In the grasses *Zea*, *Avena*, or *Triticum*, for example, a leaf primordium originates as a broad protuberance on one side of the apical meristem (Fig. 17.15). As development continues, the base of the primordium encircles the young stem, and ultimately develops into the leaf sheath. In grasses such as these, characterized by "open" sheaths, the margins of the developing sheaths overlap each other, thus encircling the shoot apex (Fig. 17.15). The regions of overlap of successively formed primordia occur on opposite sides of the shoot apex. Distally, the primordium narrows and, with primarily intercalary growth, the blade elongates and expands, becoming mature prior to the sheath which retains its meristematic potential longer than the blade. In some monocotyledons, e.g., *Allium*, the leaf primordium is "closed," and forms a hood over the apical meristem. Thus, the mature leaf is tubular. The unifacial leaves of some monocotyledons (e.g., *Iris*) result from the extensive activity of adaxial meristems, and little or no activity of marginal meristems. Development of palm leaves is unusually complex involving the formation in the leaf primordium of plications or

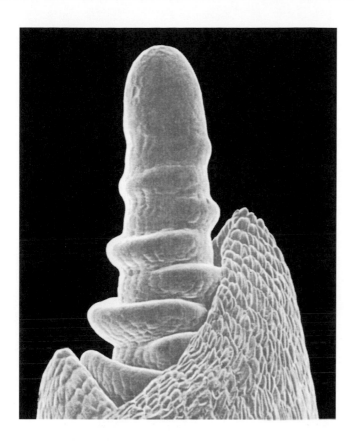

Figure 17.15 The shoot apex of *Triticum* illustrating the form and pattern of development of leaf primordia and the resulting overlapping leaf sheaths which encircle the stem. Magnification × 181. See the text for more detail. From Troughton and Donaldson (1972). Used by permission of the New Zealand Ministry of Research, Science and Technology.

folds which subsequently separate into individual leaflets. The details of this process are controversial. For detailed discussions of development in palm leaves please see Tomlinson (1961) and Periasamy (1962, 1965).

Role of the cytoskeleton in leaf development

In the leaf, as in other regions of the plant, microtubules play an important role in cell growth and the development of cell form. This is especially apparent in the development of cells of the mesophyll and the resulting system of intercellular spaces. Microtubules become oriented just beneath the plasmalemma and cellulose microfibrils are synthesized in an identical pattern in the developing cell wall (Fig. 17.16a). If the microtubules are oriented in rings, the thickened rings of new cell wall resist outward expansion as the cells grow, but the regions of thin wall between the rings bulge out forming tubular cells with regularly spaced constrictions as in *Triticum aestivum* (Jung and Wernicke, 1990). On the other hand, if the system of microtubule bundles and the cell wall thickenings are arranged in a reticulum, as in *Nigella damascena* (Wernicke *et al.*, 1993) and *Adiantum capillis-veneris* (Panteris *et al.*, 1993a) the regions of thin wall between thickenings of the reticulum extend outward (Fig. 17.16b) forming multilobed cells

Figure 17.16 Morphogenesis of mesophyll cells in leaflets of *Adiantum capillis-veneris*. (a) Surface view of a wall thickening in a constricted region (inset) of a mesophyll cell. Orientation of cellulose microfibrils is indicated by the white lines. Microtubules (arrows) just beneath the plasmalemma are oriented parallel to the microfibrils. Magnification × 17025, inset × 934. (b) Outward extension of a lobe of a mesophyll cell between regions of wall thickenings. The arrow indicates the thin wall of the lobe as compared to thicker wall regions on either side of it. Magnification × 2334. (c) A nearly mature mesophyll cell showing the relationship of rings of microtubules (white) to the cell lobes. Magnification × 532. From Panteris *et al.* (1993a). Used by permission of Springer-Verlag Wien.

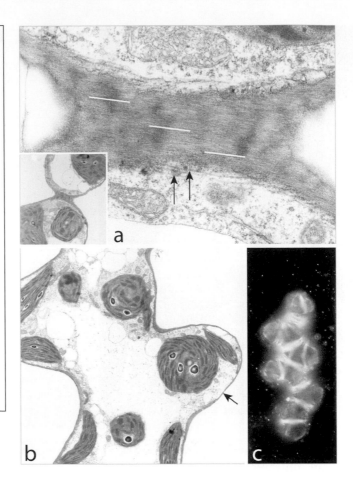

(Fig. 17.16c) and resulting in a tissue with extensive intercellular air spaces.

The mechanism by which the microtubules attain their characteristic pattern in the cells is not fully understood, but it has been suggested that their positioning might be controlled by the actin microfilaments which have been observed in parallel arrangement with them (see Seagull, 1989).

Understanding of the control and integration of factors that lead to leaf development is limited at present. However, recent workers have identified genes that affect dorsiventrality, blade formation, and cell and tissue characteristics of developing leaves (see Waites and Hudson, 1995; Bowman, 2000) as well as pinna form, size, and number in compound leaves (Lu *et al.*, 1996). Genes have also been identified that control the expression of expansin, a protein that controls wall loosening which, with turgor pressure, is required for cell growth (Cho and Cosgrove, 2000). In time, it seems likely that the application of genetics and its integration with the extensive knowledge of aspects of morphology, anatomy, and development will lead to solutions of unsolved problems in leaf morphogenesis.

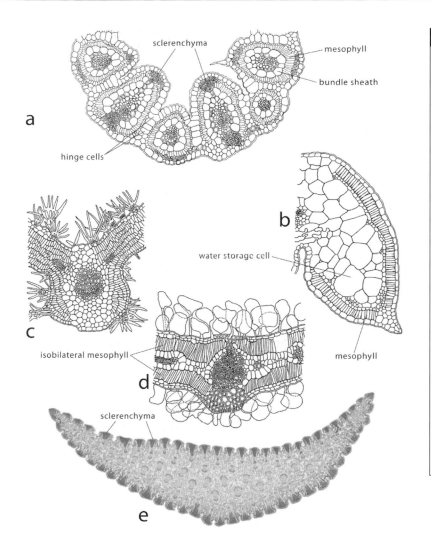

a

sclerenchyma

hinge cells

mesophyll

bundle sheath

b

water storage cell

c

isobilateral mesophyll

d

mesophyll

sclerenchyma

e

Figure 17.17 Drawings of sectional views of leaves of xeromorphic plants. (a) A leaf of the grass *Bouteloua breviseta*, with hinge cells and plates and strands of sclerenchyma. (b) Part of a leaf of the succulent plant *Salsola kali*. Note the large water storage cells. (c) The leaves of *Sphaeralcea incana* are characterized by a dense layer of branched epidermal trichomes and a mesophyll lacking differentiation into palisade and spongy regions. (d) Part of a leaf of *Atriplex canescens* covered by vesicular trichomes, and characterized by an isobilateral mesophyll. (e) Leaf of *Yucca*. The conspicuous strands of sclerenchyma within the ridges are bundle caps of peripheral vascular bundles. Stomata are located in the epidermis that lines the grooves. Magnification × 14. (a) From Shields (1951b). Used by permission of the International Society of Plant Morphologists. (b–d) Redrawn from Shields (1951a) Used by permission of the Botanical Society of America.

Variations in leaf form, structure, and arrangement

Although the basic anatomy and morphology of most leaves are directly related to the process of photosynthesis, some specialized leaves are related to other functions. For example, cotyledons are specialized as food storage organs, bracts and bud scales function in protection and/or storage of photosynthate, and flower parts are related to the process of reproduction. Although the primary function of most leaves is photosynthesis, leaves are, nevertheless, highly diverse in gross morphology, and to a lesser extent in internal anatomy. This diversity is related to the environment in which they have evolved as well as to that in which they develop. For example, **xerophytes**, plants that have evolved in arid (xeric) regions (Fig. 17.17), possess leaves that have structural features such as a heavily cutinized epidermis, sunken stomata, and sclerenchymatous hypodermal layers that contribute to a restriction of water loss. Others have water storage cells, dense coverings of trichomes or isobilateral mesophylls. Xerophytic grasses commonly have bulliform or hinge cells that facilitate the involution (rolling up) of leaves, and many species are characterized by plates or strands of sclerenchyma. Furthermore, because in arid regions light intensity tends to be very high, leaves are often, but not always, small and frequently thick (see, e.g., Groom *et al.*, 1997; Burrows, 2001). On the other hand, **mesophytes**, plants that have evolved in regions of abundant rainfall and where light intensity is lower, have larger and thinner leaves, thinner cuticles, collenchyma more often than sclerenchyma as the supporting tissue in the blade, and stomata at the same level as the rest of the epidermal cells (Fig. 17.2a, b). The leaves of hydrophytes, aquatic plants, have reduced vascular systems, highly aerenchymatous mesophyll (Fig. 17.6), no or relatively small amounts of sclerenchyma, and an epidermis composed of thin-walled cells that often contain chloroplasts.

During development, the morphology of leaves on the same plant may vary depending on factors of the environment such as spectral quality and light intensity. For example leaves that develop in conditions of low light intensity tend to be large and thin, and are called **shade leaves**, whereas those that develop in conditions of high light intensity are smaller and thicker, and called **sun leaves**. Experiments by Buisson and Lee (1993) on the effects of simulated canopy shade demonstrated that leaves of *Carica papaya* (papaya) grown under reduced irradiance were significantly thinner, with lower specific weight, had fewer stomata, produced more chlorophyll per unit area, and were characterized by a larger volume of air spaces in the mesophyll than leaves growing under conditions of high light intensity. In addition, under conditions of low light intensity and low red : far red light, leaf lobing was dramatically reduced. Change in spectral quality also resulted in a reduction in the ratio of chlorophyll a to b.

The leaves of **plagiotropic shoots** (shoots with all leaves oriented in essentially the same plane) are commonly **anisophyllous**, that is, at maturity the leaves on the upper side of the stem are smaller than those on the lower side. Anisophylly is considered an adaptation that facilitates light interception in environments of low light intensity since the smaller leaves and their orientation in relation to the light source minimize shading of the larger leaves on the lower side of the stem (see Dengler, 1991, 1999). Some dorsiventral shoots with distichous phyllotaxy provide a somewhat different variant in leaf arrangement reflected in the anatomy of the buds. In plants characterized by what Charlton (1993) calls the **rotated-lamina syndrome** (some woody species such as *Ulmus, Zelkova, Tilia, Corylus*, etc.), leaf primordia in the bud become oriented in more or less one plane with the upper surface of the laminae facing the axis bearing the bud rather than the axis on which they are borne (Fig. 17.18). Thus, as the leaves mature they occur in one plane on opposite sides of the shoot axis, an adaptation which facilitates light reception. For variations and other examples of the rotated-lamina syndrome, see Charlton (1997). The unusual tropical fern *Teratophyllum rotundifoliatum* provides another interesting

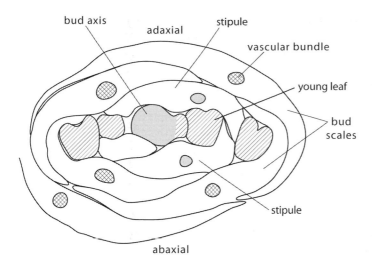

bud axis
adaxial
stipule
vascular bundle
young leaf
bud scales
stipule
abaxial

Figure 17.18 Transverse section of a vegetative bud of *Zelkova serrata* illustrating the "rotated-lamina syndrome." During development, the leaf primordia become oriented in one plane with their upper surfaces facing the axis bearing the bud rather than toward the bud axis. This orientation facilitates light reception of the mature leaves. Drawn from figure 34 in Charlton (1993). Used by permission of the National Research Council of Canada.

example of the effect of low light intensity on leaf anatomy (see Nasrulhaq-Boyce and Duckett, 1991). The major part of the leaf lamina of *Teratophyllum* consists almost entirely of the upper and lower epidermis and the intervening vascular bundles. In its leaves, which essentially lack mesophyll, all epidermal cells contain chloroplasts. Those in the upper, lens-shaped epidermal cells are very large whereas those of the lower epidermis are small and numerous, and identical to those of the stomatal guard cells. The distribution and characteristcs of the chloroplasts and other aspects of anatomy appear to be adaptations that maximize light absorption in conditions of diffuse light of very low intensity (Nasrulhaq-Boyce and Duckett, 1991).

Another type of morphological variation is the acropetal progression of leaves of different morphology during shoot development that comprises what is called a **heteroblastic series**. Such a progression of leaf forms often accompanies development from embryo to juvenile and adult vegetative states and ultimately to the reproductive state. The first leaves to appear in the seedling of a plant that typically has compound leaves when mature may be simple (see Gerrath and Lacroix, 1997). Rarely, the reverse may occur: i.e., the first leaves may be compound and later ones simple. In woody plants, the first leaves that develop on a twig at the beginning of a growth period are bud scales followed later by foliage leaves. Leaves of intermediate morphology often develop between these two extremes. In the transition from a vegetative to a reproductive state, leaf form commonly changes from that of typical foliage leaves with a gradual reduction in size to that of bracts and floral appendages such as sepals and petals (Kerstetter and Poethig, 1998). Leaf form may also change with increasing age of the plant, and in some plants typical foliage leaves may be followed in succession by structures such as tendrils and spines. In addition to transitions in morphology, anatomical changes such as variation in cuticle thickness, changes in epidermal cell shape and size, thickness of the leaf blade, size of bundle sheath cells, variation in the transverse area of veins, distance between veins, number of layers of palisade mesophyll cells, etc., may also characterize leaves in heteroblastic series (see Gould, 1993; Lawson and Poethig, 1995; BongardPierce et al., 1996).

Mechanisms that control heteroblastic development are not clearly understood, but it is apparent that morphogenetic stimuli lead to the formation by the apical meristem of leaf primordia that develop into leaves (and floral appendages) of different morphologies. This stimulus may be a hormone such as auxin or gibberellic acid, and one which is influenced by external factors such as temperature and/or photoperiod. Some workers have suggested that carbohydrate concentration might play a significant role in heteroblastic development (see, e.g., Sussex and Clutter, 1960). Ultimately, understanding of the molecular basis of the control of heteroblasty will be dependent on determining the genes that affect various aspects of leaf development (see Lawson and Poethig, 1995).

Structure in relation to function

Leaves are highly variable in both morphology and anatomy, varying in gross form from simple to compound, in thickness, whether laminate or tubular, in characteristics of lobing and lamina margins, in surface characteristics such as cuticle thickness, type and density of trichomes, position, distribution, and density of stomata, in the presence of dorsiventral or isobilateral mesophyll, presence of toxic compounds, presence of silica, presence or absence and distribution of sclerenchyma, etc. It is likely, therefore, that they have evolved in relation to both biotic and abiotic influences in the environment (see Beerling and Kelly, 1996; Gutschick, 1999; Press, 1999), and that many of these features are directly related to leaf function.

Photosynthesis and assimilate loading

Many of the structural characteristics of leaves are directly related to the process of photosynthesis. All cells of the mesophyll contain chloroplasts, the sites of the process. According to Evans (1999), there are, typically, about 10 million chloroplasts in each square centimeter of leaf! It is well known that the great volume of intercellular space within the mesophyll and the presence in the epidermis of stomata facilitate exchange of O_2 and CO_2. The extensive surface area of exposed cell walls provides for efficient absorption of CO_2 into mesophyll cells where it is utilized in photosynthesis. Cell surfaces abutting on intercellular spaces are lined with a thin layer of "cuticle-like" material which makes them unwettable (Martin and Juniper, 1970). According to Romberger et al. (1993) it is essential that these surfaces be water repellent in order to prevent the intercellular spaces from filling with water during periods of high humidity and, thus, negating their function in aeration.

The laminate form of many leaves and the orientation of the tubular palisade cells at right angles to the leaf surface are adaptations which enhance the penetration of light, the source of energy for the process of photosynthesis. In some plants upper epidermal cells with convex outer cell walls function like lenses, focusing light on the palisade mesophyll as, e.g. in *Medicago sativa* and *Oxalis* (Martin *et al.*, 1989; Poulson and Vogelmann, 1990). The spongy mesophyll scatters the light which enhances its absorption (see Evans, 1999). This scattering effect is especially important because it increases the absorption of the green and yellow wavelengths. Although these wavelengths are the strongest part of the solar spectrum they would be poorly utilized in photosynthesis if it were not for this effect (P. Ray, personal communication). In some plants that grow in conditions of low light intensity the palisade mesophyll cells are cone-shaped with the widest part of the cells located just below the epidermis. In cells of this shape more of the peripherally located chloroplasts are exposed directly to light than in tubular palisade cells typical of plants that grow in conditions of higher light intensity (see Buisson and Lee, 1993), thus increasing the efficiency of light absorption.

It is apparent that the venation system provides pathways for transport of water and minerals (the primary xylem) and of photosynthate (the primary phloem). The presence of companion cells, transfer cells, vascular parenchyma cells, and leaf sheath cells associated with sieve tube members facilitates the transport of photosynthate from the mesophyll into the sieve tube system (phloem loading). Plasmodesmata connecting the mesophyll cells with the bundle sheath cells provide for symplastic transport of photosynthate into the more central regions of the minor veins. In some monocotyledons, e.g. *Zea mays*, there are two types of phloem-loading cells, larger vascular parenchyma cells and smaller, more internal companion cells. The vascular parenchyma cells which are in contact with vessel members are connected to thick-walled sieve tube members by pore–plasmodesmata connections (Evert *et al.*, 1978). They are able to retrieve sucrose from the vessels and transfer it to the thick-walled sieve tubes but are not thought to be involved in long-distance transport of photosynthate (Fritz *et al.*, 1983). The thin-walled sieve tube members which function in long-distance transport are connected to their associated companion cells by numerous pore–plasmodesmata connections, but are essentially isolated from other cells in the leaf by the very low frequency of interconnecting plasmodesmata (Evert *et al.*, 1978).

It is interesting that in many species of seed plants, the minor vein system in which phloem loading and transport take place is characterized by vein endings that contain no sieve tube members. Lersten (1990) found that about a third of the vein endings in *Rudbeckia laciniata* lack a sieve tube, that only 10% have sieve tubes extending to the tip, and that the sieve tubes in about 60% stop at some intermediate point. In an even more extreme case, Fisher (1989) observed that in *Cananga odorata* about 60% of the entire minor vein network lacks

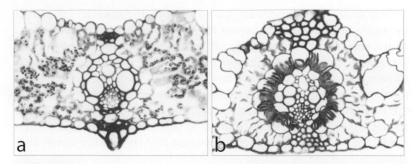

Figure 17.19 Transverse sections of grass leaves showing structural variation in C_3 and C_4 plants. (a) Leaf of *Panicum bisulcatum*, a C_3 plant. The bundle sheath is composed of cells with a few small chloroplasts, and lacks a clearly defined surrounding sheath of mesophyll cells. (b) Leaf of *Eragrostis speciosa*, a C_4 plant. The bundle sheath contains numerous, large chloroplasts, and is enclosed by a well-defined concentric layer of mesophyll cells. Magnification (a, b) × 160. See the text for more detail.

sieve tubes. He concluded that assimilate was transported from the mesophyll through lateral parenchymatous extensions from bundle sheaths, and then along the veins through the bundle sheaths, or vascular parenchyma cells until contact was made with a sieve tube member where phloem-loading could occur. This type of collection of assimilate from the mesophyll has been observed by other workers (see, e.g., Dengler and MacKay, 1975; Franceschi and Giaquinta, 1983b; Russin and Evert, 1984). For a more detailed discussion of phloem-loading in leaves, see Chapter 12 on the phloem.

Because photosynthesis and respiration require passageways for gaseous exchange between the external environment and the interior of the leaf, certain structural features that reduce the loss of water from the leaf have evolved. These are especially apparent in xerophytes. Among these features are the presence of a thick water-impermeable cuticle, a hypodermis, abundant epidermal hairs, sunken stomata, and in some plants, the reduction in number and size of stomata (for detail, see Chapter 11). The latter adaptations are features of many C_4 plants.

Leaf structure of C_3 and C_4 plants

One of the most interesting structural features of leaves is the distinction between the bundle sheaths of C_3 and C_4 plants (Fig. 17.19). The bundle sheaths of C_3 plants (Fig. 17.19a) have few organelles and small chloroplasts, and appear empty at low magnifications. The mesophyll cells surrounding the bundle sheaths show no specific arrangement in relation to the sheath cells. In C_4 plants (Fig. 17.19b), the bundle sheath cells are prominent, of relatively large size and have thick walls. They contain many large chloroplasts (larger than those in cells of the mesophyll) that often (but not always) are located adjacent to the tangential walls in contact with mesophyll cells. The immediately surrounding mesophyll cells are frequently arranged in an orderly array. Because of the prominence of the bundle sheath, its intensely green color, and the concentric layers formed by the sheath and immediately surrounding mesophyll cells, the term "Kranz" (wreath) was applied to this type of anatomy by the German botanist Haberlandt. Of course, such bundle sheaths appear as wreaths only in transverse sections of veins. In recent times, the term, **kranz syndrome**, has been applied to the combination of anatomy and physiology reflected in the processes that incorporate both the C_4 (or Hatch–Slack) pathway and the Calvin cycle during the dark reactions of photosynthesis.

During the **Calvin cycle** in C_3 plants, characteristic of most angiosperms, the first product of CO_2 reduction is the three-carbon compound, 3-phosphoglyceric acid. Following a series of enzymatic reactions, photosynthate (glucose), is formed. By contrast, in C_4 plants utilizing the four-carbon or **Hatch–Slack pathway** in photosynthesis, which takes place in the mesophyll cells, the first product of CO_2 fixation is oxaloacetic acid. Following several intermediate reactions malate or aspartate is formed. The malate or aspartate then moves into the bundle sheaths where, in chloroplasts, it is decarboxylated to yield CO_2. This CO_2 is then utilized in the Calvin cycle with the ultimate synthesis of glucose. C_4 plants thus have two sources of CO_2, the Hatch–Slack pathway and the external atmosphere. Because C_4 plants utilize CO_2 more efficiently than C_3 plants they can maintain a photosynthetic rate comparable to that of C_3 plants with fewer and smaller stomata with a consequent reduction in water loss. This explains why many grasses which are C_4 plants can tolerate very dry conditions. It is not surprising, therefore, to note that many C_4 plants evolved in the tropics in conditions of high temperatures, high light intensity, and low availability of water (but see Press (1999) for a more detailed analysis of the functional significance of the C_4 pathway and some caveats). Recent papers by Nelson and Dengler (1992), Dengler *et al.* (1994, 1997), and Soros and Dengler (2001) present detailed discussions of the development of the vascular system in leaves of C_3 and C_4 plants as well as variations in photosynthetic pathways. Sinha and Kellogg (1996), Kellogg (1999), and Soros and Dengler (2001) discuss the evolution of the C_4 pathway.

Supporting structures in leaves

Certain structural features of leaves such as the hypodermis, sclerenchymatous ribs (Figs. 17.9, 17.17e), bundle sheath extensions (Figs. 17.5, 17.19b), and the system of veins as well as the hydrostatic pressure within the living cells function in providing support. In some mesophytic leaves, however, an important element of support is not immediately apparent. When observed in **paradermal section** (sections cut parallel to the leaf surface), the cells of the spongy mesophyll are characterized by radiating extensions that abut on similar extensions of adjacent cells thus providing a structural net-like system that provides support in the lower part of the leaf (Fig. 17.20).

Figure 17.20 Spongy mesophyll as seen in paradermal section. Note the reticulate system of cells which provides support in the lower part of the leaf. Magnification × 247.

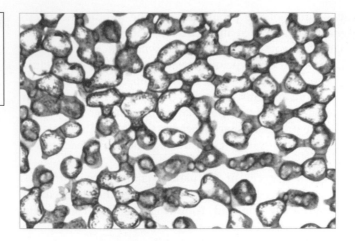

This may be especially important in large, thin laminate leaves, contributing to the several aspects of support that prevent such leaves from collapsing on themselves. Characteristics of the petiole such as length, transverse shape, geometry, and stiffness (provided primarily by the presence of tracheary tissue, collenchyma and sclerenchyma), and the propensity of many laminate leaves "to fold and curl into streamlined objects" reduce the drag forces and allow leaves to resist without damage high wind velocities (Niklas, 1999).

Transfusion tissue in conifers

In conifers and some other gymnosperms the vascular supply in leaves is associated with **transfusion tissue**, a specialized tissue of short, tracheid-like cells, the walls of which contain circular bordered pits intermixed with parenchyma cells (Fig. 17.21a, b). In *Pinus*, the transfusion tissue completely surrounds the vascular bundles (Fig. 17.21a), but varies in quantity and arrangement in other conifers. The vascular bundles and transfusion tissue are enclosed by an endodermis (Fig. 17.3). Since the early studies of Münch and Huber, it has been accepted that water and solutes are transported from the tracheids of the vascular bundles through the tracheid-like cells of the transfusion tissue to the leaf mesophyll, and that photosynthate is transported from the mesophyll by way of the transfusion parenchyma into the phloem of the vascular bundles. It is also believed that the parenchyma cells of the transfusion tissue function as transfer cells. Recently Canny (1993) has suggested that because two-way transport of water and assimilates is characteristic of plants that lack transfusion tissue, this tissue of conifers might have other significant functions. His experiments using fluorescent tracers have led him to propose that "major functions of the transfusion tissue of gymnosperms are (a) the concentration of solutes from the transpiration stream and (b) the retrieval from the stream of selected solutes that are returned to the phloem through the Strasburger cells, or fowarded through

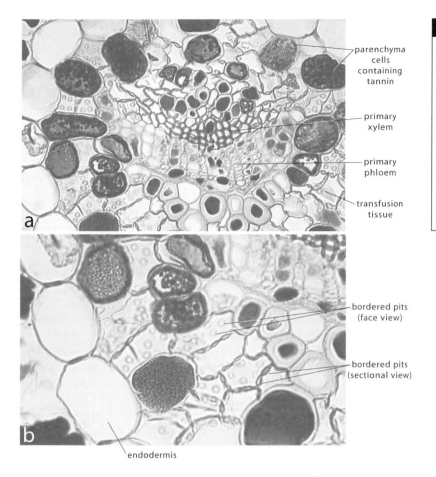

parenchyma
cells
containing
tannin

primary
xylem

primary
phloem

transfusion
tissue

bordered pits
(face view)

bordered pits
(sectional view)

endodermis

Figure 17.21 Transfusion tissue associated with the vascular bundle in a leaf of *Pinus monophylla*. (a) Note the tissue, surrounding the central vascular bundle, composed of large, tannin-filled parenchyma cells interspersed among smaller cells, the walls of which are characterized by circular bordered pits. Magnification × 312. (b) Enlargement of a region of transfusion tissue, showing the bordered pits in face and sectional views. Magnification × 459.

the endodermis to the palisade." Unanswered are the means whereby water and photosynthate cross the endodermis, whether apoplastically or symplastically.

Leaf abscission

Determinate plant structures such as leaves, leaflets (of compound leaves), flowers, flower petals, and fruit are shed from the plant at the end of their functional lifespans. In some plants, e.g., *Populus* (poplar), twigs and branches are also shed. This process of the shedding of plant parts is called **abscission**. It has been known for many years that abscission is controlled by two hormones, auxin and ethylene, and that it is also closely correlated with environmental factors such as photoperiod, ozone, wounding and/or attack by pathogens, water stress, and senescence (see Taylor and Whitelaw, 2001). Recent studies indicate that the protein expansin may also play an important role in abscission through the process of wall loosening and cell expansion (Cho and Cosgrove, 2000). Prior to the release of organs from the plant an **abscission zone** develops, characterized by two anatomically

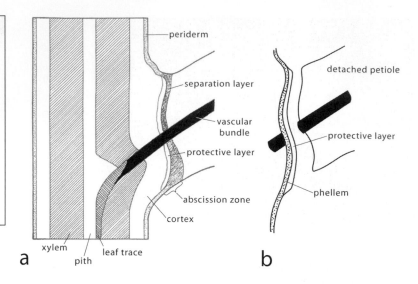

Figure 17.22 (a) Diagram illustrating the abscission zone of a leaf. (a) Median longitudinal section. The abscission zone extends through the vascular bundle(s) in parenchyma cells. (b) Diagram showing the separation from the stem of the leaf of a woody plant. Note that at this stage, the protective layer is underlain by a layer of phellem which provides additional protection against water loss and the entry of pathogens.

distinct layers, a separation layer and a protective layer (Fig. 17.22a). The effects of auxin and ethylene on cells of the developing **separation layer** have been described in detail by Osborne (1976) and Osborne and Sargent (1976a, b). They found that, in *Sambucus nigra*, some cells increase in size in response to ethylene but not auxin, others in response to auxin but not ethylene and in some, growth occurred in response to both hormones. The results of research by numerous workers suggests, however, that in general, ethylene stimulates, and auxin restricts, cell growth in the separation layer. The balance of these two hormones is thought to provide a regulatory mechanism for the control of the size and shape of cells in the separation layer (see Taylor and Whitelaw, 2001). The actual separation of the leaf (Fig. 17.22b), or other organ, from the plant results from the loss of adhesion between cells caused by dissolution of the middle lamella through the action of hydrolytic enzymes such as polygalacturonase and cellulase (see Taylor *et al.*, 1990; Taylor and Whitelaw, 2001). The separation layer consists of parenchyma tissue traversed by vascular bundles, the xylem of which may differ from that on either side of the layer. The metaxylem vessel members in the separation layer of *Acer pseudoplatanus*, for example, have pitted–scalariform, helical–scalariform and scalariform–pitted secondary wall thickenings (Fig. 17.23), but vessel members with pitted walls only on either side of the abscission zone (Andre *et al.*, 1999). In woody plants, the tracheary tissue in the separation layer often consists of very short cells resulting in a weakened zone. Even so, in some plants, e.g., *Quercus*, leaves remain on the tree until the vascular strand is broken by freezing.

The **protective layer**, which prevents water loss and the entrance of pathogens, forms behind the separation layer. The walls of the component cells become impregnated with suberin and wound gum, and the intercellular spaces are often filled with the same substances. In woody plants a layer of phellem (Fig. 17.22b) that enhances or replaces

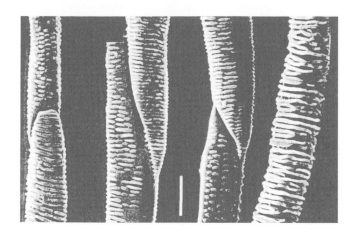

Figure 17.23 Scanning electron micrographs of vessel casts from the abscission zone of *Acer pseudoplatanus*. Bar = 50 μm. From Andre *et al.* (1998). Used by permission of the National Research Council of Canada.

the function of the protective layer develops just before or immediately following separation.

REFERENCES

Andre, J. P., Catesson, A. M., and M. Liberman. 1999. Characters and origin of vessels with heterogeneous structure in leaf and flower abscission zones. *Can. J. Bot.* **77**: 253–261.

Beck, C. B., Schmid, R., and G. W. Rothwell. 1983. Stelar morphology and the primary vascular system of seed plants. *Bot. Rev.* **48**: 691–815.

Beebe, D. U. and R. F. Evert. 1990. The morphology and anatomy of the leaf of *Moricandia arvensis* (L.) DC. (Brassicaceae). *Bot. Gaz.* **151**: 184–203.

Beerling, D. J. and C. K. Kelly. 1996. Evolutionary comparative analyses of the relationship between leaf structure and function. *New Phytol.* **134**: 35–51.

BongardPierce, D. K., Evans, M. M. S., and R. S. Poethig. 1996. Heteroblastic features of leaf anatomy in maize and their genetic regulation. *Int. J. Plant Sci.* **157**: 331–340.

Bowman, J. L. 2000. The *YABBY* gene family and abaxial cell fate. *Curr. Opin. Plant Biol.* **3**: 17–22.

Buisson, D. and D. W. Lee. 1993. The developmental responses of papaya leaves to simulated canopy shade. *Am. J. Bot.* **80**: 947–952.

Burrows, G. E. 2001. Comparative anatomy of the photosynthetic organs of 39 xeromorphic species from subhumid New South Wales, Australia. *Int. J. Plant Sci.* **162**: 411–430.

Canny, M. J. 1993. Transfusion tissue of pine needles as a site of retrieval of solutes from the transpiration stream. *New Phytol.* **123**: 227–232.

Charlton, W. A. 1993. The rotated-lamina syndrome. I. Ulmaceae. *Can. J. Bot.* **71**: 211–221.

1997. The rotated-lamina syndrome. VII. Direct formation of lamina in the rotated position in *Pterospermum* (Sterculiaceae) and the appearance of hyper-rotation. *Can. J. Bot.* **75**: 188–206.

Cho, H. T. and D. J. Cosgrove. 2000. Altered expession of expansin modulates leaf growth and pedicel abscission in *Arabidopsis thaliana*. *Proc. Natl Acad. Sci. USA* **97**: 9783–9788.

Dale, J. E. 1988. The control of leaf expansion. *Annu. Rev. Plant Physiol. Plant Mol. Biol.* **39**: 267–295.

DeMason, D. A. and R. Chawla. 2004. Roles for auxin during morphogenesis of the compound leaves of pea (*Pisum sativum*). *Planta* **218**: 435–448.

Dengler, N. G. 1991. Patterns of leaf development in anisophyllous shoots. *Can. J. Bot.* **70**: 676–691.

1999. Anisophylly and dorsiventral shoot symmetry. *Int. J. Plant Sci.* **160**: S67–S80.

Dengler, N. G. and J. Kang. 2001. Vascular patterning and leaf shape. *Curr. Opin. Plant Biol.* **4**: 50–56.

Dengler, N. G. and L. B. MacKay. 1975. The leaf anatomy of beech, *Fagus grandifolia. Can. J. Bot.* **53**: 2202–2211.

Dengler, N. G. and H. Tsukaya. 2001. Leaf morphogenesis in dicotyledons: current issues. *Int. J. Plant Sci.* **162**: 459–464.

Dengler, N. G., Dengler, R. E., Donnelly, P. M., and P. W. Hattersley. 1994. Quantitative leaf anatomy of C_3 and C_4 grasses (Poaceae): bundle sheath and mesophyll surface area relationships. *Ann. Bot.* **73**: 241–255.

Dengler, N. G., Woodvine, M. A., Donnelly, P. M., and R. E. Dengler. 1997. Formation of vascular pattern in developing leaves of the C_4 grass *Arundinella hirta. Int. J. Plant Sci.* **158**: 1–12.

Donnelly, P. M., Bonetta, D., Tsukaya, H., Dengler, R. E., and N. G. Dengler. 1999. Cell cycling during leaf development in *Arabidopsis. Devel. Biol.* **215**: 407–419.

Esau, K. 1977. *Anatomy of Seed Plants*, 2nd edn. New York: John Wiley and Sons.

Evans, J. R. 1999. Leaf anatomy enables more equal access to light and CO_2 between chloroplasts. *New Phytol.* **143**: 93–104.

Evert, R. F., Eschrich, W., and W. Heyser. 1978. Leaf structure in relation to solute transport and phloem loading in *Zea mays* L. *Planta* **138**: 279–294.

Ferreira, J. F. S., Duke, S. O., and K. C. Vaughn. 1998. Histochemical and immunological localization of tropane alkaloids in *Erythroxylum coca* var. *coca* and *E. novogranatense. Int. J. Plant Sci.* **159**: 492–503.

Fisher, D. G. 1989. Leaf structure of *Cananga odorata* (Annonaceae) in relation to the collection of photosynthate and phloem loading: morphology and anatomy. *Can. J. Bot.* **68**: 354–363.

1985. Morphology and anatomy of the leaf of *Coleus blumei* (Lamiaceae). *Am. J. Bot.* **72**: 392–406.

Fisher, D. G. 1991. Plasmodesmatal frequency and other structural aspects of assimilate collection and phloem loading in leaves of *Sonchus oleraceus* (Asteraceae), a species with minor vein transfer cells. *Am. J. Bot.* **78**: 1549–1559.

Fisher, J. B. and R. Rutishauser. 1990. Leaves and epiphyllous shoots of *Chisocheton* (Meliaceae): a continuum of woody leaf and stem axes. *Can. J. Bot.* **68**: 2316–2328.

Fleming, A. J. 2002. The mechanism of leaf morphogenesis. *Planta* **216**: 17–22.

Fleming, A. J., Caderas, D., Wehrli, E., McQuenn-Mason, S., and C. Kuhlemeier. 1999. Analysis of expansin-induced morphogenesis on the apical meristem of tomato. *Planta* **208**: 166–174.

Franceschi, V. R. and R. T. Giaquinta. 1983a. The paraveinal mesophyll of soybean leaves in relation to assimilate transfer and compartmentation. I. Ultrastructure and histochemistry during vegetative development. *Planta* **157**: 411–421.

1983b. Specialized cellular arrangements in legume leaves in relation to assimilate transport and compartmentation: comparison of the paraveinal mesophyll. *Planta* **159**: 415–422.

Fritz, E., Evert, R. F., and W. Heyser. 1983. Microautoradiographic studies of phloem loading and transport in the leaf of *Zea mays* L. *Planta* **159**: 193–206.

Gerrath, J. M. and C. R. Lacroix. 1997. Heteroblastic sequence and leaf development in *Leea guineensis*. *Int. J. Plant Sci.* **158**: 747–756.

Gould, K. S. 1993. Leaf heteroblasty in *Pseudopanax crassifolius*: functional significance of leaf morphology and anatomy. *Ann. Bot.* **71**: 61–70.

Green, P. B. 1999. Expression of pattern in plants: combining molecular and calculus-based biophysical paradigms. *Am. J. Bot.* **86**: 1059–1076.

Groom, P. K., Lamont, B. B., and A. S. Markey. 1997. Influence of leaf type and plant age on leaf structure and sclerophylly in *Hakea* (Proteaceae). *Austral. J. Bot.* **45**: 827–838.

Gutschick, V. P. 1999. Biotic and abiotic consequences of differences in leaf structure. *New Phytol.* **143**: 3–18.

Hagemann, W. and S. Gleissberg. 1996. Organogenic capacity of leaves: the significance of marginal blastozones in angiosperms. *Plant Syst. Evol.* **199**: 121–152.

Jung, G. and W. Wernicke. 1990. Cell shaping and microtubules in developing mesophyll of wheat (*Triticum aestivum* L.). *Protoplasma* **153**: 141–148.

Kellogg, E. A. 1999. Phylogenetic aspects of the evolution of C_4 photosynthesis. In R. F. Sage and R. K. Monson, eds., *The Biology of C_4 Synthesis*. New York: Academic Press, pp. 411–422.

Kenekordes, K. G., McCully, M. E., and M. J. Canny. 1988. The occurrence of an extended bundle sheath system (paraveinal mesophyll) in the legumes. *Can. J. Bot.* **66**: 94–100.

Kerstetter, R. A. and R. S. Poethig. 1998. The specification of leaf identity during shoot development. *Annu. Rev. Cell Devel. Biol.* **14**: 373–398.

Larkin, J. C., Young, N., Prigge, M., and D. Marks. 1996. The control of trichome spacing and number in *Arabidopsis*. *Development* **122**: 997–1005.

Lawson, J. R. and R. S. Poethig. 1995. Shoot development in plants: time for a change. *Trends Genet.* **11**: 263–268.

Lersten, N. R. 1990. Sieve tubes in foliar vein endings: review and quantitative survey of *Rudbeckia laciniata* (Asteraceae). *Am. J. Bot.* **77**: 1132–1141.

Lersten, N. R. and J. D. Curtis. 1993. Paraveinal mesophyll in *Calliandra tweedii* and *C. emarginata* (Leguminosae; Mimosoideae). *Am. J. Bot.* **80**: 561–568.

Liljebjelke, K. A. and V. R. Franceschi. 1991. Differentiation of mesophyll and paraveinal mesophyll in soybean leaf. *Bot. Gaz.* **152**: 34–41.

Lu, B., Villani, P. J., Watson, J. C., DeMason, D. A., and T. J. Cooke. 1996. The control of pinna morphology in wildtype and mutant leaves of the garden pea (*Pisum sativum* L.). *Int. J. Plant Sci.* **157**: 659–673.

Lyndon, R. F. 1994. Control of organogenesis at the shoot apex. *New Phytol.* **128**: 1–18.

Martin, J. T. and B. E. Juniper. 1970. *The Cuticles of Plants*. New York: St. Martin's Press.

Martin, G., Josserant, S. A., Bornman, J. F., and T. C. Vogelmann. 1989. Epidermal focusing and the light microenvironment within leaves of *Medicago sativa*. *Physiol. Plant.* **76**: 485–492.

Nasrulhaq-Boyce, A. and J. G. Duckett. 1991. Dimorphic epidermal cell chloroplasts in the mesophyll-less leaves of an extreme-shade tropical fern, *Teratophyllum rotundifoliatum* (R. Bonap.) Holtt.: a light and electron microscope study. *New Phytol.* **119**: 433–444.

Nelson, T. and N. G. Dengler. 1992. Photosynthetic tissue differentiation in C$_4$ plants. *Int. J. Plant Sci.* **153**: S93–S105.

1997. Leaf vascular pattern formation. *Plant Cell* **9**: 1121–1135.

Niklas, K. J. 1997. *The Evolutionary Biology of Plants*. Chicago, IL: University of Chicago Press.

1999. A mechanical perspective on foliage leaf form and function. *New Phytol.* **143**: 19–31.

Osborne, D. J. 1976. Auxin and ethylene and the control of cell growth. Identification of three classes of target cells. In P. Pilet, ed., *Plant Growth Regulation*. Berlin: Springer-Verlag, pp. 161–171.

Osborne, D. J. and J. A. Sargent. 1976a. The positional differentiation of ethylene responsive cells in rachis abscission zones in leaves of *Sambucus nigra* and their growth and ultrastructural changes at senescence and separation. *Planta* **130**: 203–210.

1976b. The positional differentiation of abscission zones during development of leaves of *Sambucus nigra* and the response of the cells to auxin and ethylene. *Planta* **132**: 197–204.

Panteris, E., Apostolakos, P., and B. Galatis. 1993a. Microtubule organization, mesophyll cell morphogenesis, and intercellular space formation in *Adiantum capillis-veneris* leaflets. *Protoplasma* **172**: 97–110.

1993b. Microtubules and morphogenesis in ordinary epidermal cells of *Vigna sinensis* leaves. *Protoplasma* **174**: 91–100.

Periasamy, K. 1962. Morphological and ontogenetic studies in palms. I. Development of the plicate condition in the palm-leaf. *Phytomorphology* **12**: 54–64.

1965. Growth pattern of the leaves of *Cocos nucifera* and *Borassus flabellifer* after the initiation of placations. *Austral. J. Bot.* **13**: 225–234.

Poulson, M. E. and T. C. Vogelmann. 1990. Epidermal focusing and effects upon photosynthetic light-harvesting in leaves of *Oxalis*. *Plant Cell Environ.* **13**: 803–811.

Press, M. C. 1999. The functional significance of leaf structure: a search for generalizations. *New Phytol.* **143**: 213–219.

Romberger, J. A., Hejnowicz, Z., and J. F. Hill. 1993. *Plant Structure: Function and Development*. Berlin: Springer-Verlag.

Russin, W. A. and R. F. Evert. 1984. Studies on the leaf of *Populus deltoides* (Salicaceae): morphology and anatomy. *Am. J. Bot.* **71**: 1398–1415.

Seagull, R. W. 1989. The plant cytoskeleton. *CRC Crit. Rev. Plant Sci.* **8**: 131–167.

Shields, L. M. 1951a. Leaf xeromorphy in dicotyledon species from a gypsum sand deposit. *Am. J. Bot.* **38**: 175–190.

1951b. The involution mechanism in leaves of certain xeric grasses. *Phytomorphology* **1**: 225–241.

Sinha, N. R. and E. A. Kellogg. 1996. Parallelism and diversity in multiple origins of C$_4$ photosynthesis in the grass family. *Am. J. Bot.* **83**: 1458–1470.

Soros, C. L. and N. G. Dengler. 2001. Ontogenetic derivation and cell differentiation in photosynthetic tissues of C$_3$ and C$_4$ Cyperaceae. *Am. J. Bot.* **88**: 992–1005.

Steward, W. N. and G. W. Rothwell. 1993. *Palaeobotany and the Evolution of Plants*. Cambridge, UK: Cambridge University Press.

Sussex, I. M. and M. E. Clutter. 1960. A study of the effect of externally supplied sucrose on the morphology of excised fern leaves *in vitro*. *Phytomorphology* **10**: 87–99.

Taylor, J. E. and C. A. Whitelaw. 2001. Signals in abscission. *New Phytol.* **151**: 323–339.

Taylor, J. E., Tucker, G. A., Lasslett, Y., *et al.* 1990. Polygalacturonase expression during leaf abscission of normal and transgenic tomato plants. *Planta* **183**: 133–138.

Taylor, T. M. and E. L. Taylor. 1993. *The Biology and Evolution of Fossil Plants*. Englewood Cliffs, NJ: Prentice-Hall.

Tomlinson, P. B. 1961. *Anatomy of the Monocotyledons*, vol. 2, *Palmae*. Oxford, UK: Clarendon Press.

Troughton, J. and L. A. Donaldson. 1972. *Probing Plant Structure*. Wellington, NZ: New Zealand Ministry of Research, Science and Technology.

Waites, R. and A. Hudson. 1995. *Phantastica*: a gene required for dorsoventrality of leaves in *Antirrhinum majus*. *Development* **121**: 2143–2154.

Wernicke, W., Günther, P., and G. Jung. 1993. Microtubules and cell shaping in the mesophyll of *Nigella damascena* L. *Protoplasma* **173**: 8–12.

FURTHER READING

Appleby, R. F. and W. J. Davies. 1983. The structure and orientation of guard cells in plants showing stomatal responses to changing vapour pressure difference. *Ann. Bot.* **52**: 459–468.

Black, C. C. and H. H. Mollenhauer. 1971. Structure and distribution of chloroplasts and other organelles in leaves with various rates of photosynthesis. *Plant Physiol.* **47**: 15–23.

Brown, W. V. 1958. Leaf anatomy in grass systematics. *Bot. Gaz.* **119**: 170–178.

Campbell, R. 1972. Electron microsopy of the development of needles of *Pinus nigra* var. *maritima*. *Ann. Bot.* **36**: 711–720.

Cronshaw, J., Lucas, W. J., and R. T. Giaquinta (eds.) *Phloem Transport*. New York: Alan R. Liss.

Cross, G. L. 1940. Development of the foliage leaves of *Taxodium distichum*. *Am. J. Bot.* **27**: 471–482.

    1942. Structure of the apical meristem and development of the foliage leaves of *Cunninghamia lanceolata*. *Am. J. Bot.* **29**: 288–301.

DeMason, D. A. and P. J. Villani. 2001. Genetic control of leaf development in Pea (*Pisum sativum*). *Int. J. Plant Sci.* **162**: 493–511.

Edwards, G. and D. Walker. 1983. *C3, C4: Mechanisms, and Cellular Environmental Regulation, of Photosynthesis*. Oxford, UK: Blackwell.

Erickson, R. O. and F. J. Michelini. 1957. The plastochron index. *Am. J. Bot.* **44**: 297–305.

Esau, K. 1965. *Vascular Differentiation in Plants*. New York: Holt, Rinehart, and Winston.

Eschrich, W., Burchardt, R., and S. Essiamah. 1989. The induction of sun and shade leaves of the European beech (*Fagus sylvatica* L.): anatomical studies. *Trees* **3**: 4–10.

Fisher, D. G. 1986. Ultrastructure, plasmodesmatal frequency, and solute concentration in green areas of variegated *Coleus blumei* Benth. leaves. *Planta* **169**: 141–152.

Fleming, A. J. 2003. The molecular regulation of leaf form. *Plant Biol.* **5**: 341–349.

Foster, A. S. 1936. Leaf differentiation in angiosperms. *Bot. Rev.* **2**: 349–372.

1952. Foliar venation in angiosperms from an ontogenetic standpoint. *Am. J. Bot.* **39**: 752–766.

Franck, D. H. 1979. Development of vein pattern in leaves of *Ostrya virginiana* (Betulaceae). *Bot. Gaz.* **140**: 77–83.

Gambles, R. L. and R. E. Dengler. 1982a. The anatomy of the leaf of red pine, *Pinus resinosa*. I. Nonvascular tissues. *Can. J. Bot.* **60**: 2788–2803.

1982b. The anatomy of the leaf of red pine, *Pinus resinosa*. II. Vascular tissues. *Can. J. Bot.* **60**: 2804–2824.

Ghouse, A. K. M. and M. Yunus. 1974. Transfusion tissue in the leaves of *Cunninghamia lanceolata* (Lambert) Hooker (Taxodiaceae). *Bot. J. Linn. Soc.* **69**: 147–151.

Griffith, M. M. 1957. Foliar ontogeny in *Podocarpus macrophyllus*, with special reference to the transfusion tissue. *Am. J. Bot.* **44**: 705–715.

Hall, L. N. and J. A. Langdale. 1996. Molecular genetics of cellular differentiation in leaves. *New Phytol.* **132**: 533–553.

Harris, W. M. 1971. Ultrastructural observations on the mesophyll cells of pine leaves. *Can. J. Bot.* **49**: 1107–1109.

Jarvis, P. G. and T. A. Mansfield (eds.) 1981. *Stomatal Physiology*. Cambridge, UK: Cambridge University Press.

Jones, C. S. and M. A. Watson. 2001. Heteroblasty and preformation in mayapple, *Podophyllum peltatum* (Berberidaceae): developmental flexibility and morphological constraint. *Am. J. Bot.* **88**: 1340–1358.

Kaplan, D. R. 1970. Comparative foliar histogenesis in *Acorus calamus* and its bearing on the phyllode theory of monocotyledonous leaves. *Am. J. Bot.* **57**: 331–361.

1973. The monocotyledons: their evolution and comparative biology. VII. The problem of leaf morphology and evolution in the monocotyledons. *Q. Rev. Biol.* **48**: 437–457.

1984. Alternative modes of organogenesis in higher plants. In R. A. White and W. C. Dickison, eds., *Contemporary Problems in Plant Anatomy*. New York: Academic Press, pp. 261–300.

2001. Fundamental concepts of leaf morphology and morphogenesis: a contribution to the interpretation of molecular genetic mutants. *Int. J. Plant Sci.* **162**: 465–474.

Kaplan, D. R., Dengler, N. G., and R. E. Dengler. 1982. The mechanism of plication inception in palm leaves: histogenic observations on the palmate leaf of *Raphis excelsa*. *Can. J. Bot.* **60**: 2999–3016.

Kausik, S. B. and S. S. Bhattacharya. 1977. Comparative foliar anatomy of selected gymnosperms: leaf structure in relation to leaf form in Coniferales and Taxales. *Phytomorphology* **27**: 146–160.

Kessler, S., Kim, M., Pham, T., Weber, N., and N. Sinha. 2001. Mutations altering leaf morphology in tomato. *Int. J. Plant Sci.* **162**: 475–492.

Laetsch, W. M. 1974. The C_4 syndrome: a structural analysis. *Annu. Rev. Plant Physiol.* **25**: 27–52.

Lee, C. L. 1952. The anatomy and ontogeny of the leaf of *Dacrydium taxoides*. *Am. J. Bot.* **39**: 393–398.

Lersten, N. R. and J. D. Curtis. 2001. Idioblasts and other unusual internal foliar secretory structures in Scrophulariaceae. *Plant Syst. Evol.* **227**: 63–73.

Lersten, N. R. and H. T. Horner. 2000. Calcium oxalate crystal types and trends in their distribution patterns in leaves of *Prunus* (Rosaceae: Prunoideae). *Plant Syst. Evol.* **224**: 83–96.

Maksymowych, R. 1973. *Analysis of Leaf Development*. Cambridge, UK: Cambridge University Press.

Mansfield, T. A., Hetherington, A. M., and C. J. Atkinson. 1990. Some current aspects of stomatal physiology. *Annu. Rev. Plant Physiol. Plant Mol. Biol.* **41**: 55–75.

Marcotrigiano, M. 2001. Genetic mosaics and the analysis of leaf development. *Int. J. Plant Sci.* **162**: 513–525.

Merrill, E. K. 1979. Comparison of ontogeny of three types of leaf architecture in *Sorbus* L. (Rosaceae). *Bot. Gaz.* **140**: 328–337.

Metcalfe, C. R. and L. Chalk. 1950. *Anatomy of the Dicotyledons*, 2 vols. Oxford UK: Clarendon Press.

Millington, W. F. and J. E. Gunckel. 1950. Structure and development of the vegetative shoot tip of *Liriodendron tulipifera* L. *Am. J. Bot.* **37**: 326–335.

Nelson, T. and J. A. Langdale. 1989. Patterns of leaf development in C_4 plants. *Plant Cell* **1**: 3–13.

Njoku, E. 1971. The effect of sugars and applied chemicals on heteroblastic development in *Ipomoea purpurea* grown in aseptic culture. *Am. J. Bot.* **58**: 61–64.

Pate, J. S. and B. E. S. Gunning. 1969. Vascular transfer cells in angiosperm leaves: a taxonomic and morphological survey. *Protoplasma* **68**: 135–156.

Philpott, J. 1953. A blade tissue study of leaves of forty-seven species of *Ficus*. *Bot. Gaz.* **115**: 15–35.

Pray, T. R. 1955a. Foliar venation of angiosperms. II. Histogenesis of the venation of *Liriodendron*. *Am. J. Bot.* **42**: 18–27.

  1955b. Foliar venation of angiosperms. IV. Histogenesis of the venation of *Hosta*. *Am. J. Bot.* **42**: 698–706.

  1963. Origin of vein endings in angiosperm leaves. *Phytomorphology* **13**: 60–81.

Romberger, J. A. 1963. Meristems, growth, and development in woody plants. *US Dept. Agric. Tech. Bull.* **1293**: 1–214.

Russin, W. A. and R. F. Evert. 1985. Studies on the leaf of *Populus deltoides* (Salicaceae): ultrastructure, plasmodesmatal frequency, and solute concentrations. *Am. J. Bot.* **72**: 1232–1247.

Steeves, T. A. and I. M. Sussex. 1989. *Patterns in Plant Development*, 2nd edn. Cambridge, UK: Cambridge University Press.

Stevens, R. A. and E. S. Martin. 1978. A new ontogenetic classification of stomatal types. *Bot. J. Linn. Soc.* **77**: 53–64.

Williams, R. F. 1975. *The Shoot Apex and Leaf Growth: A Study in Quantitative Biology*. London: Cambridge University Press.

Wylie, R. B. 1939. Relations between tissue organization and vein distribution in dicotyledon leaves. *Am. J. Bot.* **26**: 219–225.

  1951. Principles of foliar organization shown by sun–shade leaves from ten species of deciduous dicotyledonous trees. *Am. J. Bot.* **38**: 355–361.

  1952. The bundle sheath extension in leaves of dicotyledons. *Am. J. Bot.* **39**: 645–651.

Ziegler, H. 1987. The evolution of stomata. In E. Zeiger, G. D. Farquhar, and I. R. Cowan, eds., *Stomatal Function*. Stanford, CA: Stanford University Press, pp. 29–57.

Chapter 18

Reproduction and the origin of the sporophyte

Perspective: the plant life cycle

Reproduction in higher plants is relatively complex, involving a life cycle consisting of two phases, a diploid sporophyte phase and a haploid gametophyte phase, comprising what is called an **alternation of generations**. The prominent bodies of angiosperm trees, shrubs, perennials, and annuals as well as those of gymnosperms, ferns, sphenophytes, and lycophytes are **sporophytes**, having developed from fertilized egg cells (zygotes). The gametes which fused to form the zygotes, however, were produced by **gametophytes**, very small plant bodies, parasitic on the sporophytes in seed plants, but somewhat larger and free-living in pteridophytes (except in heterosporous species in which gametophytes when mature remain, at least in part, within the the walls of the spores from which they develop).

The sporophyte in pteridophytes is dominant, and although dependent initially for its nutrition on the gametophyte, soon becomes independent. The gametophyte is much reduced in size but is free-living and either autotrophic or saprophytic. In seed plants, the sporophyte is also dominant and initially dependent on the gametophyte, but soon becomes independent. The gametophyte is greatly reduced, however, and parasitic on the sporophyte. In angiosperms it is exceptionally small, consisting in many taxa of only seven cells and eight nuclei, and can be observed only with a microscope.

The life cycle of a vascular plant can be summarized as follows. The sporophyte produces specialized cells called **sporocytes** that undergo meiosis producing **haploid spores**. The spores germinate to form the gametophytes in which **gametes** are produced. The gametes fuse to form a **diploid zygote** from which the **embryo** (young sporophyte) develops.

In most pteridophytes, spores of only one size are produced. Plants of this type are described as being **homosporous**. Each spore has the potential to develop into a gametophyte that produces both egg cells and sperms. In contrast, the sporophytes of a few pteridophytes and all seed plants produce spores of two sizes, with different potentials, called microspores and megaspores. These plants are, thus,

Figure 18.1 Female and male cones of *Pinus* (pine), the sites of development of ovules and microsporangia, respectively. (a) Megasporangiate cones of *Pinus banksiana* (jack pine) in which ovules are produced. (b) Microsporangiate cones of *Pinus resinosa* (red pine) in which microsporangia are produced.

heterosporous. **Microspores** develop into gametophytes that produce sperms, and **megaspores** develop into gametophytes that produce egg cells.

Reproduction in gymnosperms

Although reproduction in gymnosperms and angiosperms is basically similar, there are major differences in the location on the sporophyte of the ovules (containing the **megasporangia** in which megaspores, and ultimately, egg cells are produced) and the **microsporangia** (in which microspores and ultimately, pollen grains and sperms develop), the complexity of the gametophytes, mechanisms of pollination and fertilization, and nutrition of the developing embryo, among others. In general, the structures involved in angiosperm reproduction, contained in the flower, are much reduced in comparison with those in gymnosperms. The ovules and microsporangia of extant gymnosperms are produced in female (**megasporangiate**) and male (**microsporangiate**) cones (or strobili) respectively (Fig. 18.1a, b). Exceptions are the ovules of *Ginkgo biloba* which are borne terminally, usually in pairs, on long stalks called peduncles, and the ovules of *Gnetum* which occur laterally on branched axes. Whereas we shall emphasize reproduction in conifers in this book, interesting research on *Ginkgo, Ephedra*, and *Gnetum* has been published recently (Friedman, 1990, 1994, 1998; Friedman and Gifford, 1997; Friedman and Carmichael, 1998). This research is especially important because of the demonstration of double fertilization in both *Gnetum* and *Ephedra*, and the light it throws on the phylogenetic relationship of these taxa to the angiosperms.

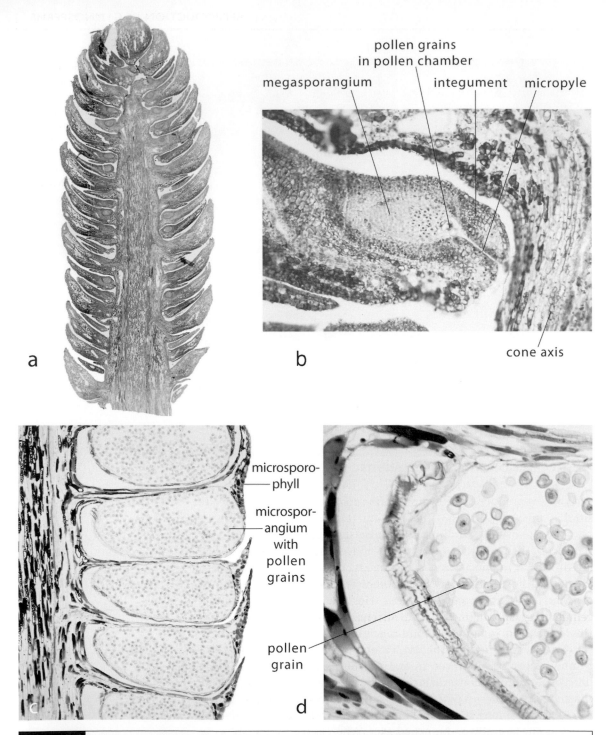

Figure 18.2 (a) Median longitudinal section of a female cone of *Pinus wallichiana* (Himalayan pine) showing bracts, ovuliferous scales and adaxial ovules. Magnification × 5. (b) Enlargement of an ovule from (a) showing the integument and megasporangium. Note that the micropylar end of the ovule faces the cone axis; also the presence of pollen grains in the pollen chamber. Magnification × 74. Compare with Fig. 18.3a. (c) Median longitudinal section of part of a microsporangiate (male) cone of *Pinus* sp. showing microsporophylls bearing, abaxially, two microsporangia. Only one of the pair can be seen in this longitudinal section. Magnification × 28. (d) Enlargement of a part of a microsporangium from (c) illustrating the pollen grains which have developed from microspores. Magnification × 15.

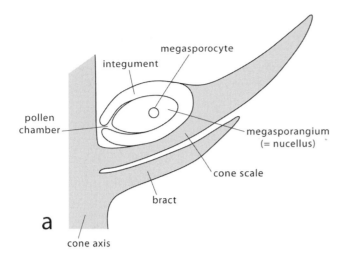

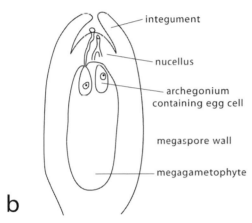

Figure 18.3 (a) Diagram illustrating the megasporocyte in the megasporangium of a conifer ovule (b) Mature ovule containing a megagametophyte. Following meiosis three of the megaspores abort, and the remaining spore develops into the megagametophyte. Please see the text for a more detailed description.

In conifers, ovules are borne on the adaxial surface of **ovuliferous scales** (Fig. 18.2a) with the **micropylar end** of the ovules near to, and facing, the cone axis (Fig. 18.2b). Microsporangia are borne on the abaxial surface of **microsporophylls** (Fig. 18.2c). In conifers a multi-cellular megasporangium (called nucellus in older literature) and a single enclosing integument comprise the ovule. Meiosis occurs in the **megasporocyte** located within the megasporangium (Fig. 18.3a) result-ing, typically, in a **linear tetrad of megaspores**, three of which abort, the remaining one developing into the **megagametophyte** (Fig. 18.3b). As the functional megaspore enlarges, the nucleus divides repeatedly, in some species, over a period of several months, accompanied by increase in the volume of cytoplasm and in the size of the develop-ing gametophyte. During the following spring in, e.g., *Pinus virginiana*, walls develop between the many nuclei, and one or more **archegonia** develop near the micropylar end of the gametophyte, each contain-ing a large egg cell (Fig. 18.3b). Archegonia vary in number among gymnosperms from one to as many as 50. During its development and growth the megagametophyte derives nutrition symplastically via plasmodesmata from the enclosing megasporangial tissue. In the

Figure 18.4 (a) A pollen grain of *Pinus* at the time of pollination. (b) Germinating pollen grain (male gametophyte). The stalk cell and body cell were derived from the generative cell. The body cell will divide to form two sperms. Note the branching of the pollen tube which in many conifer species becomes haustorial. From Coulter and Chamberlain (1917).

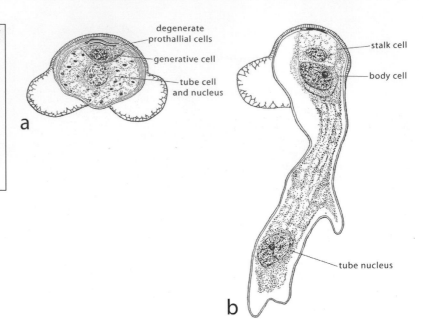

mature ovule (Fig. 18.3b) at the time of fertilization, the megagametophyte is enclosed by the enlarged **megaspore wall** and a layer of disintegrated and compressed sporangial tissue called the **tapetal wall**.

Following **microsporogenesis** (meiosis and spore development) microspores develop into pollen grains (Fig. 18.2d) which are immature male gametophytes. At the time of pollination each pollen grain contains two degenerate prothallial cells, a generative cell and a tube cell (Fig. 18.4a). The pollen grain, wind-dispersed, lands in a pollination droplet which retracts and draws the pollen grain through the micropyle and in to contact with the surface of the megasporangium (Fig. 18.2b). Here it germinates producing a **pollen tube**, often branched (Fig. 18.4b), which enters megasporangial tissue where it functions as a **haustorium**. After a variable period of time, in different species, during which the megagametophyte develops to maturity (e.g., several months in *Picea abies* and over a year in several species of *Pinus*), the **generative cell** divides, forming a stalk cell and a body cell (Fig. 18.4b). At this stage, the pollen tube containing the body cell begins to grow, and the **body cell** divides to form two sperm nuclei. By secreting proteases (see Pettitt, 1985) the tube digests its way through gametophyte tissue and into contact with the egg cell where the tip bursts, releasing the sperm nuclei, one of which fuses with the egg nucleus, forming the zygote, the other disintegrating. In some gymnosperms, several egg cells may be fertilized with the result that several embryos will develop (Fig. 18.5a), a process called **polyembryony**. In many conifers, the **proembryo**, which represents an early stage in sporophyte development, becomes subdivided into several (usually four) embryos, a process called **cleavage polyembryony**. In either

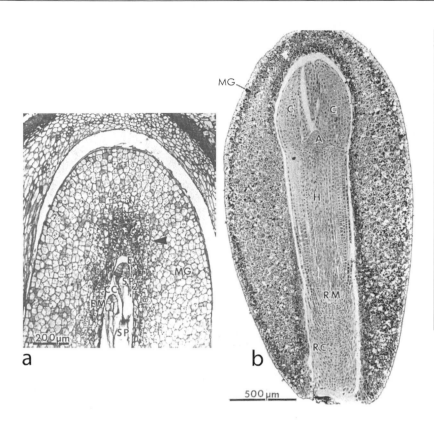

Figure 18.5 Conifer embryos. (a) Section through the ovule of *Picea glauca* (white spruce) showing the megagametophyte containing two proembryos, an example of polyembryony. MG, megagametophyte; EM, proembryo; SP, suspensor; CC, corrosion cavity; arrowhead indicates cells containing starch grains. (b) A mature embryo in the seed of *Picea laricio* (Corsican pine). A, apical meristem; C, cotyledon; H. hypocotyl; MG, megagametophyte; RM, root meristem; RC, root cap. From Krasowski and Owens (1993). Used by permission of the National Research Council of Canada.

case, only one of the embryos continues its development (Fig. 18.5b), the others abort. During the growth and development of the embryo within the seed, the young sporophyte obtains nutrients from storage products, predominantly lipids and proteins (Krasowski and Owens, 1993), in the tissue of the megagametophyte in which it is enclosed. After release from the female cones the seeds may germinate, followed, ultimately, by development of independent sporophytes. For a detailed presentation of many aspects of reproduction and embryogeny in gymnosperms, see Doyle (1963), Singh (1978), Owens and Morris (1991), Romberger *et al.* (1993), Owens *et al.* (1998), Runions and Owens (1999a, b), Bruns and Owens (2000), and Owens and Bruns (2000).

Reproduction in angiosperms

The **flower** (Fig. 18.6) is the site of sexual reproduction in angiosperms. Whereas it consists of fewer component parts than the cones of most gymnosperms, the parts, usually grouped in whorls, or shallow spirals, are more highly specialized, and each has evolved with a specific function. Evidence from the fossil record indicates that the various floral parts evolved from the leaves of their ancestors. The major evidence for this conclusion is the form and venation pattern

Figure 18.6 (a) The flower of *Helleborus*. This specimen has been dissected in order to show the separate floral components. The large leaf-like structures, forming the basal whorl are sepals (SE), followed distally by petals (P) (in this taxon, highly modified), stamens (S) and carpels (simple pistils) (C). Part of the ovary wall of one carpel has been removed to show the ovules (O). (b) Stamens of *Senna artemisioides*. Note the broad filaments and large anthers which contrast with the slender filaments and smaller anthers of *Helleborus*. (a) From Jaeger (1961). Used by permission of Chambers Harrap Publishers Ltd. © Chambers Harrap Publishers 1961. (b) From Tucker (1997). Used by permission of the University of Chicago Press. © 1997 The University of Chicago. All rights reserved.

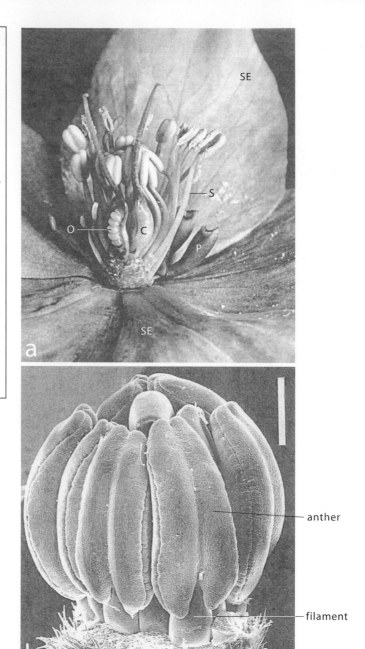

of the sepals and petals and, to a lesser extent, carpels and stamens (see Stewart and Rothwell, 1993). The **sepals**, called collectively the **calyx**, and the lowermost whorl of floral parts, are the most leaf-like components of the flower. In many flowers they are green and photosynthetic. Following, distally, are the **petals**, collectively called the **corolla** which with the calyx comprises the **perianth**. The petals typically are not photosynthetic, but in form and vasculature are usually, but not always, very leaf-like. They often contain brightly colored

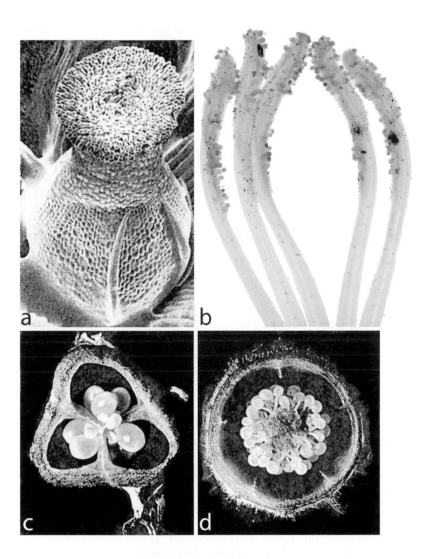

Figure 18.7 (a) The compound pistil of *Alyssum*. Note the short, broad style and prominent, papillate stigma. Magnification × 98. (b) The branched stigma of *Lilium grandiflorum* to which adhere numerous pollen grains, many of which have germinated. Magnification × 32. (c, d) Transverse sections through the ovaries of compound pistils. (c) A tricarpellate ovary of *Endymion* sp. with axile placentation. (d) A five-carpellate ovary of *Lychnis* sp. with free central placentation. Note loss of septa between the carpels. (a, c, d) From Jaeger (1961). Used by permission of Chambers Harrap Publishers Ltd. © Chambers Harrap Publishers 1961.

pigments and sometimes have specialized shapes that attract pollinators. The **stamens** (Fig. 18.6a, b) which comprise the next more distal whorl of floral parts may also be of foliar origin, but evidence from the fossil record in support of this conclusion is meager. Stamens consist of a **filament**, usually (but not always) long and slender, which terminates in an **anther** which contains, typically, four microsporangia. Its major function is the production of pollen grains, which upon germination produce pollen tubes, each containing two sperms. The terminal whorl of floral parts is made up of the **carpels** (Fig. 18.6a). They are also considered to have a foliar origin, but the details of their evolution as well as the several hypotheses regarding their origin are highly controversial. Carpels consist of the **ovary**, containing the ovules (Figs. 18.6a, 18.7c, d) which ultimately develop into **seeds**, and the **style** (lacking in some taxa), a usually slender apical extension of the ovary (Fig. 18.6a), terminated by the **stigma** (Fig. 18.7a, b). The stigma is the receptor of pollen and the structure upon which the

Figure 18.8 A flower of *Acer platanoides* (Norway maple) containing a prominent nectary surrounding the ovary and enclosing the bases of the stamen filaments. (b) Flowers of *Cornus sanguinea*, each with a nectary situated upon the inferior ovary and enclosing the base of the style. (a, b) From Jaeger (1961). Used by permission of Chambers Harrap Publishers Ltd. © Chambers Harrap Publishers 1961.

pollen germinates. The parts of each floral whorl may remain separate or they may fuse during development. A single carpel is called a **simple pistil** (Fig. 18.6a) whereas fused carpels comprise a **compound pistil** (Fig. 18.7c, d). As the structure in which reproduction occurs in angiosperms, the flower has been modified during evolution in several significant ways that attract pollinators such as insects and birds. One such innovation is the presence in most flowers of one or more nectaries (Fig. 18.8a, b), specialized secretory glands of diverse form that produce **nectar**, a liquid often containing a high concentration of sugar. Nectar is utilized as a food source by some insects and birds (especially hummingbirds). Nectaries commonly occur on the receptacle of the flower where they may surround the base of the pistil and even enclose the base of the stamen filaments as in *Acer* (Fig. 18.8a). In flowers with inferior ovaries, however, nectaries are usually located upon the ovary, and may surround the base of the style, as in *Cornus* (dogwood) (Fig. 18.8b). Bird-pollinated plants often produce copious quantities of nectar. It has been reported that the flowers of *Eucalyptus* produce so much nectar that they may overflow. A tree producing several hundred thousands of flowers would, therefore, provide a food supply for innumerable bird pollinators (Jaeger, 1961).

Meiosis occurs in two different parts of the flower, anthers and ovules. In anthers meiosis occurs in **microsporocytes** (sometimes called meiocytes) contained in microsporangia, producing microspores (Fig. 18.9a–c). Microspores develop into pollen grains (Fig. 18.9d, e) which upon germination develop into male gametophytes, commonly called **microgametophytes**. Each pollen grain consists of two cells, a generative cell and a tube nucleus (Fig. 18.9e). Upon germination, the generative cell nucleus divides to form two sperms (Fig. 18.10a, b) each comprising a nucleus and some surrounding cytoplasm. It was long assumed that the two sperms were essentially identical and that it was a matter of chance which one fused with the egg cell. Recent research has shown, however, that the sperms in some taxa differ in size and content of cytoplasmic organelles such as plastids and mitochondria, and that preferential fertilization may

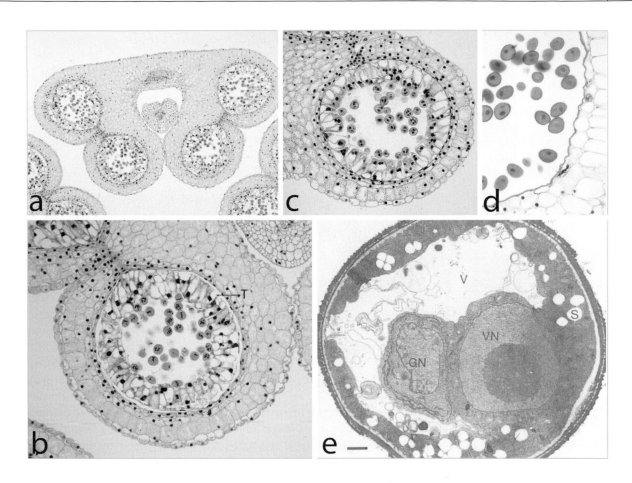

occur in species in which the sperms are different (Russell, 1984, 1985; Knox *et al.*, 1993). Sperms of some other species, for example, *Nicotiana tabacum* (tobacco), are approximately the same size and have similar distributions of cytoplasmic organelles. Consequently, they are characterized as being isomorphic (Yu *et al.*, 1992). Prior to fertilization, the two sperms are in contact with each other, and the leading sperm is intimately associated with the tube (vegetative) nucleus (Fig. 18.10a, b).

Meiosis also occurs in **megasporocytes**, one of which is contained in each of the developing ovules (Fig. 18.11). The megasporocyte is enclosed in a vegetative tissue, the **nucellus**, bounded by one or two ovular integuments. Most commonly, in angiosperms, as in conifers, meiosis results in the formation of a linear tetrad of megaspores, oriented in a plane parallel to the long axis of the ovule (Fig. 18.12a, b). Three of these spores degenerate and the remaining megaspore (Fig. 18.12c) develops into the female or **megagametophyte**, called in angiosperms the **embryo sac**. Three mitotic divisions within this cell result in eight nuclei (Fig. 18.12d–f). As these nuclear divisions are occurring, the original cell expands and elongates, and four of the nuclei migrate to each end of the developing embryo sac. At the **micropylar end** (the end of the embryo sac

Figure 18.9 Microsporogenesis in *Lilium* sp. (a) A section through an anther with four microsporangia. Magnification × 29. (b) Microsporangium showing the tapetum (T) and the first meiotic division in microsporocytes. Magnification × 87. (c) Tetrads of microspores resulting from meiosis. Magnification × 76. (d) Mature pollen grains. Magnification × 94. (e) Transmission electron micrograph of a pollen grain illustrating the generative cell (GN) and the tube (vegetative) nucleus (VN). V, vacuole; S, starch. Bar = 1 μm. (e) From Polowick and Sawhney (1993). Used by permission of the National Research Council of Canada.

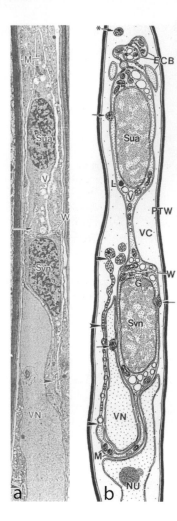

a   b

Figure 18.10 (a) Transmission electron micrograph of sperm cells and the vegetative nucleus (VN) in a pollen tube of *Nicotiana tabacum*. The two sperm cells travel in tandem down the pollen tube behind the vegetative nucleus. V, vacuole; M, mitochondrion; Sua, trailing sperm cell; Svn, leading sperm cell. Magnification × 3375. (b) A diagrammatic reconstruction of the male germ unit (sperm cells plus vegetative nucleus) in a pollen tube of *Nicotiana tabacum*. ECB, enucleated cytoplasmic bodies; L, lipid body; arrows and arrowheads, vesicle-containing bodies; W, wall between sperm cells; G, Golgi bodies; NU nucleolus; PTW, pollen tube wall; VN, tube nucleus; W, region of contact between sperm cells. From Yu *et al.* (1992). Used by permission of Springer-Verlag Wien.

adjacent to the micropyle in the ovule) (Fig. 18.12g) one nucleus becomes at least partially enclosed by a cell wall and functions as the egg cell, while two others differentiate into **synergids**. These cells, in contact with the egg cell, are distinctive in possessing a **filiform apparatus** (Fig. 18.13), a much-branched system of haustoria that extends from the synergid walls into the surrounding cytoplasm (Jensen, 1965; Jensen and Fisher, 1968). The function of the filiform apparatus is unclear, but it may be a transfer structure, similar to the highly branched wall ingrowths of transfer cells. The egg cell and the two synergids may be homologous with the archegonia in the megagametophytes of gymnosperms. Three of the nuclei at the other end of the developing embryo sac, the **chalazal end**, differentiate as **antipodal cells**. The remaining (fourth) nucleus at each end migrates to the center of the embryo sac. These two nuclei, **polar nuclei**, and the cytoplasm remaining after wall formation around the antipodal cells, the synergids and the egg cell, comprise the **central cell**. The three antipodal cells, the central cell with its polar nuclei, the two synergids and the egg cell comprise the mature female gametophyte (Fig. 18.12g). In **megagametogenesis** of the type just described the embryo sac develops from a single spore and, thus, is referred to as **monosporic**. This most common type of embryo sac development in angiosperms was first described in *Polygonum* and, consequently, was called the *Polygonum* type by Maheshwari (1950). There are several variations in embryo sac development, the next best known probably being the **tetrasporic type**, often designated as the *Fritillaria* type in which, following meiosis, there is no degeneration of megaspores, all four becoming incorporated into the embryo sac.

Figure 18.11 (a) Transverse section of a tricarpellate ovary of *Lilium* showing young ovules. Magnification × 26. (b) Enlargement of an immature ovule containing a megasporocyte. Note the immature integuments. Magnification × 116.

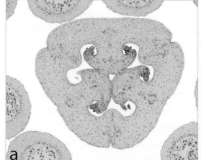

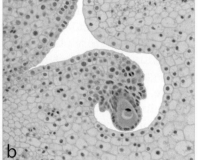

a   b

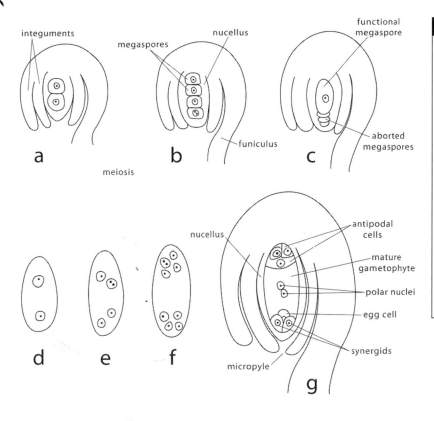

integuments

megaspores

nucellus

functional megaspore

meiosis

funiculus

aborted megaspores

a   b   c

nucellus

antipodal cells

mature gametophyte

polar nuclei

egg cell

synergids

micropyle

d   e   f   g

Figure 18.12 Diagrams representing stages in the development of the megagametophyte in *Lilium*. (a, b) Meiosis results in the formation of a linear tetrad of megaspores, three of which abort leaving the single functional megaspore (c). (d–f) A series of three mitotic divisions results in eight nuclei. (g) Three nuclei migrate to each end and two migrate to the center of the developing gametophyte. The three at the micropylar end develop into the egg cell and two synergids. Those at the opposite end (the chalazal end) differentiate as antipodal cells, and the central pair function as polar nuclei. For more detail, please see the text.

Following pollination, and upon germination of the pollen grain on the stigma (Fig. 18.14a–c), the pollen tube will grow down through the style into the locule of the carpel and through the ovular micropyle. As the pollen tube approaches the embryo sac, one of the synergids begins to deteriorate, in preparation for entrance of the tip of the pollen tube. Upon entry, the two sperms are released into the synergid, the plasmalemma of which has disintegrated (Jensen and Fisher, 1968).

The mechanism whereby non-motile sperms are transported down the pollen tube has been of great interest for many years. Recent studies have demonstrated that myosin, adsorbed to the surfaces of the sperms, interacts with microfibrils of F-actin promoting their transport down the pollen tube (Zhang and Russell, 1999). The possibility that microtubules associated with the sperms may also contribute to their mobility has been proposed recently (Heslop-Harrison and Heslop-Harrison, 1997), but the mechanism is unknown. Upon entry into the embryo sac (female gametophyte), the leading sperm of the pair (Fig. 18.10), associated with the tube nucleus, will fuse preferentially with the polar nuclei forming the triploid **endosperm nucleus**. The trailing sperm fuses preferentially with the egg cell (Zhang and Russell, 1999), forming the diploid **zygote**, completing the process of **double fertilization**. At this stage the ovule consists of the embryo sac, enclosed by the nucellus, and one or two integuments. It is attached to the wall of the carpel by a stalk, the **funiculus**. At

Figure 18.13 Transmission electron micrograph of the egg cell (E) and the two associated synergids (Sy) in the female gametophyte of *Nicotiana tabacum*. Extending from the filiform apparatus (FA) of one of the synergids into the surrounding cytoplasm of the gametophyte are much-branched haustoria, seen here as circular to irregularly shaped light areas. Int, cells of the integument; P, plastids; V, vacuole; N, nucleus; ER, endoplasmic reticulum. Bar = 5 μm. From Huang and Russell (1994). Used by permission of Springer-Verlag GmbH & Co. KG. © Springer-Verlag Berlin Heidelberg.

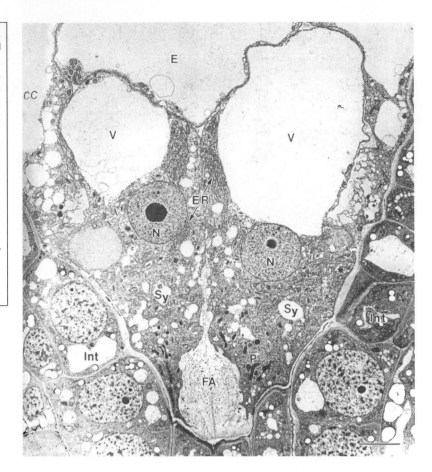

least one primary vascular bundle, which differentiated from provascular tissue prior to formation of endosperm, serves the ovule. This strand, which enters the funiculus from vasculature in the carpel wall, or sometimes several strands and their branches, extends to the chalazal end of the ovule through the outer integument.

Development of the seed in angiosperms

In most taxa of dicotyledons, following double fertilization the endosperm begins a relatively rapid development with numerous free-nuclear divisions. As development continues, **nuclear domains** are defined by systems of microtubules which mark the sites of initial cell wall formation (Nguyen *et al.*, 2001). Upon completion of wall formation the endosperm becomes a cellular tissue, enclosing the developing embryo (Fig. 18.15b–d), and becoming its direct source of nutrition. Transport of nutrients into the embryo sac and the endosperm from sporophyte tissues (e.g., from ovular integuments) is typically apoplastic, by way of transfer cells. Nutrient transfer between

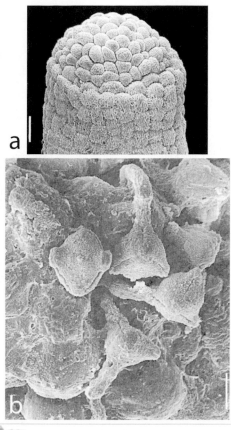

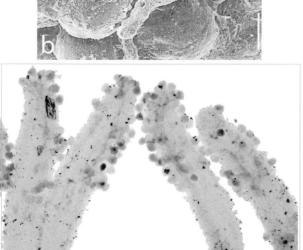

Figure 18.14 (a) The stigma of *Smyrnium perfoliatum*. Bar = 20 μm. (b) Germinated pollen grains on the stigma of *S. perfoliatum*. Individual cells of the stigma are obscured by a covering of exudate which facilitates adhesion of the pollen grains. Bar = 10 μm. (c) The branched stigma of *Lilium grandiflorum* to which many pollen grains adhere. Note the pollen tubes within the stigmatic arms. Magnification × 56. (a, b) From Weber (1994). Used by permission of the University of Chicago Press. © 1994 The University of Chicago. All rights reserved.

endosperm and embryo, however, can be either apoplastic or symplastic through plasmodesmata (see Johansson and Walles, 1993a, b).

The ovule and embryo sac increase in size concomitantly during this period. The first division of the zygote, usually transverse, but sometimes oblique, results in an apical and a basal cell. The **basal cell** is usually relatively large and highly vacuolate. Its progeny give rise to the suspensor which is attached to the micropylar end of the

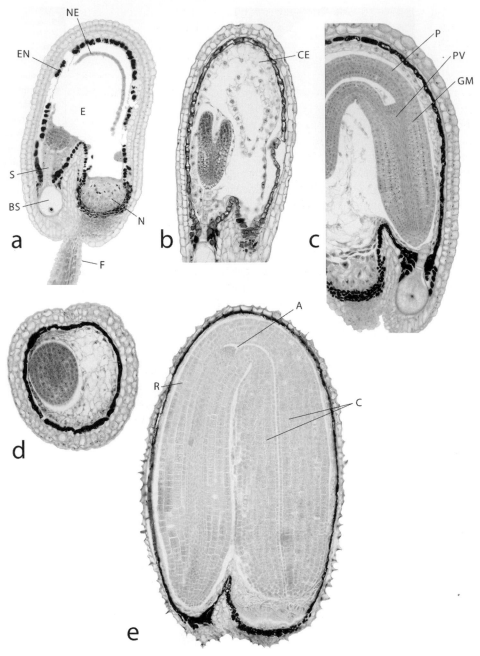

Figure 18.15 Sections of ovules of *Capsella bursa-pastoris* illustrating development of the embryo and endosperm. (a) Longitudinal section of a young, globular proembryo (E). Note the large, vacuolated, basal cell (BS) of the filamentous suspensor (S). EN, endothelium; N, nucellus; NE, nuclear endosperm; F, funiculus. (b) A young seed containing an embryo with developing cotyledons. Note the cellular endosperm (CE). (c) A young seed containing a nearly mature embryo with bent cotyledons. Protoderm (P), ground meristem (GM), and provascular tissue (PV) are conspicuous in the embryo at this stage. (d) Transverse section of an immature seed showing the radicle (young root) of the embryo enclosed by cellular endosperm. (e) A mature seed in which the embryo occupies most of the space within the seed coat. A, apical meristem of the shoot; C, cotyledons; R, radicle. Magnification (a–e) × 127.

embryo sac (Fig. 18.15a). The **suspensor** is thought to function primarily as a conduit for the translocation of nutrients from surrounding tissues to the developing embryo. Cells derived from divisions of the apical cell and its progeny form a globular mass of very small, densely cytoplasmic and meristematic cells called the **proembryo** (Fig. 18.15a). As development of the proembryo continues, it enlarges and elongates (Fig. 18.15b, c). At the same time protoderm and ground meristem differentiate, and a peripheral region that will become the cortex is delimited. A central region differentiates as provascular tissue, and ultimately extends into the **cotyledons**. Cotyledon primordia develop on either side of the distal end of the dome-shaped proembryo which, with continued cytokinesis, becomes expanded and in some taxa heart-shaped (Fig. 18.15b). An undifferentiated apical region between the cotyledon primordia becomes the apical meristem of the incipient shoot, and a similar region just above the suspensor differentiates as the apical meristem of the incipient root (Fig. 18.15c). As the embryo increases in size the endosperm decreases in volume, in many dicotyledons ultimately being entirely utilized by the growing embryo which fills the embryo sac (Fig. 18.15e).

Prior to seed dormancy, the embryo of dicotyledons consists of an **epicotyl**, bearing an apical meristem, and two cotyledons (rarely three or four as, e.g., in *Degeneria*), the **hypocotyl**, and the **radicle**, or immature root (Fig. 18.15e). At this stage the suspensor has disintegrated. In many seeds photosynthate in the endosperm has been transferred symplastically to the cotyledons which have become much expanded and function as food storage organs (Fig. 18.15e). The early stages of embryogenesis in monocotyledons is essentially like that of dicotyledons. As development continues, however, the young embryo becomes elongate and columnar, lacking the heart-shaped stage of many dicotyledons. This columnar form is related to the presence of a single cotyledon which, with continued growth, becomes a dominant part of the embryo. Within the seed, the mature embryo may be curved or more or less straight. The embryo of members of the Gramineae (grasses) differs conspicuously from that of other monocotyledons (Fig. 18.16). The large cotyledon, called the **scutellum**, appears to be laterally attached to the axis of the embryo. In addition, the embryo consists of an **epicotyl**, consisting of the apical meristem and several leaf primordia enclosed in a sheath called the **coleoptile**, and a radicle (young root) enclosed in a sheath called the **coleorhiza**.

During development of the embryo other profound changes take place within the ovule as it develops into a seed (see Boesewinkel and Bouman, 1984; Bouman, 1984). In dicotyledons, the seed coat, or **testa**, develops from the integuments of the ovule, and at maturity is usually hard and dry. The number of integuments usually parallels that of the ovule although in some taxa the inner integument may disintegrate (as, for example, in *Cucurbita*). Thickness of the mature seed coat depends to some extent on the original thickness of the ovular integuments, but largely on developmental changes that occur

Figure 18.16 The embryo of *Zea mays*.

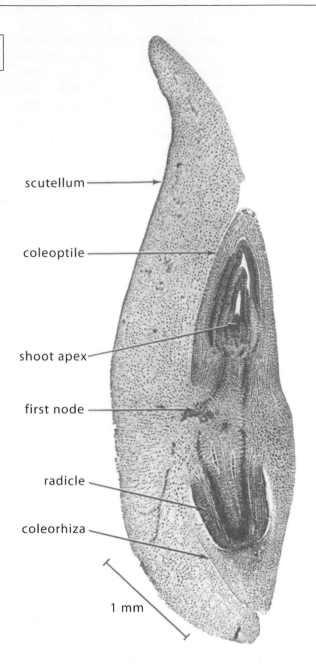

scutellum

coleoptile

shoot apex

first node

radicle

coleorhiza

1 mm

as the seed matures including cell division and cell growth within the integuments. In general, cells of the integuments become more thick-walled and some differentiate into sclereids of several types. However, interspersed layers of aerenchyma and chlorenchyma also characterize the seed coats of some taxa. Typically, the seed coat is covered by a thick water-impermeable cuticle.

For more detail on the embryogeny of angiosperms, and references to the extensive literature on the subject, please see Maheshwari (1950), Bouman (1984), Romberger *et al.* (1993), and Jensen (1998) and references therein. For a recent discussion of research on the genetic control of fertilization see Faure and Dumas (2001) and for research on signal transduction and its role in plant reproduction, see Brownlee (1994).

Fruit development and the role of fruits in seed dispersal

Since seeds are contained in fruits, a primary means of their distribution is the distribution of the fruits. A **fruit** is a matured ovary and any attached floral parts such as the receptacle, calyx, or bracts. A **simple fruit** is one that consists of a single carpel, or several fused carpels, without any attached floral parts. Examples are bean, tomato, and peach. Such fruits develop from **hypogynous** flowers in which the ovaries are superior. An **aggregate fruit** is one that consists of several to many separate carpels of a single flower such as strawberry or raspberry. A **multiple fruit** consists of the fused ovaries of several to many flowers such as mulberry or pineapple. If, in addition, these fruits are composed of floral parts other than the carpels they are termed **accessory fruits**. For example, an apple is a **simple accessory fruit** because it develops from an inferior ovary enclosed by the receptacle which becomes a part of the mature fruit. A strawberry is an **aggregate accessory fruit** consisting of an enlarged, fleshy receptacle in the surface of which many small ovaries are embedded. A mulberry is a **multiple accessory fruit** because, in addition to the ovaries of many flowers, it is also composed of the perianths and receptacles of these flowers.

At maturity, fruits may be fleshy or dry, and if the latter, dehiscent or indehiscent. The fruit wall, whether or not it develops from only the ovary wall or the ovary wall plus accessory parts, is termed **pericarp**. The pericarp may contain several histologically distinct layers: the outer layer, the **exocarp**, the middle layer, the **mesocarp**, and the inner layer, the **endocarp**. In fleshy fruits, all three layers, if they are distinct at any stage of development, may become fleshy, that is, composed of thin-walled parenchyma as in tomato. In this as well as in other berries, the septa between locules also become fleshy. In other taxa, the exocarp and mesocarp usually become the fleshy part of the pericarp, the endocarp maturing into a stony layer consisting entirely of sclereids as in drupes such as cherry and peach. Citrus fruits and members of the Cucurbitaceae, among many others, are characterized by an outer rind, usually consisting of the exocarp and at least part of the mesocarp. The inner fleshy tissue is usually derived from the endocarp, although mesocarp may also be involved in some taxa. In many cases, subdivisions of the pericarp are not clear. The rind, if soft as in citrus fruits, is usually composed of parenchyma and/or collenchyma and aerenchyma; if hard, as in cucurbits, of parenchyma, collenchyma, and sclerenchyma, often in layers. Vascularization of these fruits reflects that in the carpels of which the ovary is composed, although during fruit development branching of vascular bundles may occur. Prior to the completion of fruit development, the tracheary cells of the primary xylem in the vascular bundles are modified by enzymatic action, resulting in a thinning and softening of the cell walls. In the mature fruit, the vascular tissue is probably no longer functional.

Dry fruits are far more abundant than fleshy fruits. Among dehiscent dry fruits are **legumes** characteristic of members of the Leguminosae (bean family), **follicles** as in *Paeonia* (peony) and *Aquilegia* (columbine), and **capsules** as in *Papaver* (poppy) and *Koelreuteria* (golden rain tree). Legumes and follicles develop from single carpels whereas capsules usually develop from three or five carpels, but occasionally two as in the siliques of some members of the Cruciferae. When mature, the dry pericarp of these fruits consists usually of one or more outer layers of thick-walled parenchyma cells, sometimes highly lignified, and at least one inner layer of sclerenchyma, but sclerenchyma may be absent. Dehiscence results from differential shrinkage in layers of the pericarp as the fruit becomes dry. It has been suggested that differential shrinkage, and ultimate dehiscence in some legumes is directly related to the difference in the angle of microfibrils in S2 wall layers of sclerenchyma cells in different regions of the pericarp.

Dehiscence occurs variably in different fruit types. For example, in legumes, derived from single carpels, dehiscence occurs along both sutures whereas in follicles, also derived from single carpels, dehiscence occurs along the dorsal suture only (i.e., the suture formed by fusion of carpel margins). Dehiscence in capsules varies, depending in large part on the type of placentation of the ovaries. If placentation is axile, the split may occur between the carpels along the septa (**septicidal dehiscence**) or in the outer walls of the carpels, opening the locules (**loculicidal dehiscence**). When dehiscence is septicidal the carpels usually pull away from the central column of the ovary. When placentation is parietal, dehiscence may occur between contiguous carpels, or midway between carpel margins. The capsules of some plants that produce very small seeds, such as *Papaver*, release their seeds through apical pores.

Among indehiscent dry fruits are the **achene** and **caryopsis**. The pericarp of each is fused to the seed coat, and each contains a single seed. The seed coat usually remains thin and parenchymatous and, in some taxa, essentially disintegrates prior to maturity of the fruit. Achenes are common in the Compositae and Ranunculaceae. Large, winged achenes (**samaras**) characterize *Acer*, *Fraxinus*, and *Ulmus* among others. The fruits of the Compositae as well as of *Fraxinus* and *Ulmus* are composed of two carpels, but only one seed develops. Those of the Ranunculaceae consist of a single one-seeded carpel. The caryopsis is characteristic of members of the Gramineae (grasses). Caryopses differ from achenes in that their pericarp is fused to the seed coat whereas in achenes the seed is free from the pericarp except at the site of attachment of the funiculus. The cellular composition of the pericarp of dry, indehiscent fruits is variable, but usually contains largely sclerenchyma (achenes) and/or thick-walled, pitted parenchyma (caryopses).

As noted above, dispersal of seeds is frequently the result of their being contained in fruits that are distributed by various means. Fleshy fruits are commonly eaten by animals, and their seeds pass unharmed through the digestive tract of the disperser. Some fruits dehisce suddenly, scattering seeds some distance from the parent plant. Others such as the achenes of composites which bear a pappus, and the winged samaras of maples, ashes, and elms, are wind-dispersed as are some very small fruits. Fruits may also be dispersed by water as, for example, the coconut.

Once released from dry, dehiscent fruits, many seeds also possess modifications such as wings, pappus-like attachments, and extensive hairs (e.g., cotton seeds) that facilitate wind dispersal. Seeds that can pass intact through the digestive system of animals are dispersed by animal migration. Some seeds are characterized by the release of sticky substances such as mucilage, contained in the epidermis, upon being exposed to water. The adherence of such seeds to the coats of animals results in their distribution. The seeds of some plants, for example, the orchids, are so light and small that they are wind-dispersed. For more detail on the functional morphology of fruits, please see Weberling (1989).

Seed germination and development of the seedling

Following seed distribution and a period of dormancy, variably long in different taxa, seed germination occurs when conditions are favorable. The protective seed coat contains inhibitors that prevent germination during unfavorable conditions for seedling growth. It is also typically impermeable to water by virtue of its being covered by a thick cuticle. When the impermeability is removed by exposure to the elements, by bacterial or fungal activity, or being passed through the digestive tract of animals, water is absorbed, the inhibitors are neutralized, and oxygen enters the seed, the young sporophyte resumes its growth, and germination ensues. During germination, the young sporophyte begins a transition from dependence on its parent to existence as an independent entity.

At the beginning of the germination process the young sporophyte is composed of primarily meristematic regions, protoderm, ground meristem, and provascular tissue as well as apical meristems of the epicotyl and hypocotyl. If the cotyledons have become storage organs, they usually no longer contain meristematic tissues. If they remain thin and become photosynthetic upon germination, they usually maintain their meristematic potential.

During germination of dicotyledon seeds the radicle is often the first part of the young sporophyte to begin active growth. At the same time, provascular tissue begins to differentiate into functional xylem and phloem. The radicle may elongate initially by general cytokinesis and cell elongation throughout. Subsequently, in some dicotyledonous taxa characterized by **epigeous germination**, cytokinesis of

Figure 18.17 Diagrams of flowers showing diversity of form. See the text for descriptions. From Lawrence (1951). Reprinted by permission of Pearson Education, Inc., Upper Saddle River, NJ.

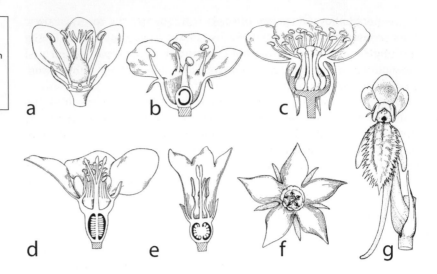

great frequency in the basal region of the hypocotyl results in the development of a hook that extends upward, pulling the seed out of the soil. The epicotyl then expands, becomes free from the seed coat, and cytokinesis in the apical meristem followed by cell and tissue differentiation leads to the formation of the young shoot. As the young sporophyte continues its growth, the transitional tissue regions, ground meristem, provascular tissue, and protoderm, in both root and shoot systems differentiate into pith (in some but not most roots), cortex, vascular tissues, and epidermis.

In many monocotyledons, whether germination is epigeous or **hypogeous** (cotyledons remaining below the soil), the young sporophyte is extracted from the seed (or fruit, in the case of grasses) by extension growth of the single cotyledon. In others, e.g., *Zea mays* and other members of the Gramineae, growth of the coleorhiza and coleoptile push through the enclosing pericarp followed by elongation of the hypocotyl and epicotyl which push through the enclosing sheaths. The cotyledon remains in the soil within the kernel, the radicle extends deeper within the soil, and the epicotyl extends upward above the soil.

Floral morphogenesis

Development of the flower has attracted the interest and energy of many botanists. Although flower form is highly diverse, certain morphological features are common to many taxa, e.g., variation in the number of floral parts (Fig. 18.17a–c), whether the flower is **hypogynous** (Fig. 18.17a–c) or **epigynous** (Fig. 18.17d, e) (i.e., whether floral parts such as sepals and petals are attached below the ovary or upon it), whether organ "whorls" are free or fused to others (Fig. 18.17a–e), and whether flower symmetry is **radial** (Fig. 18.17f) or

zygomorphic (i.e., bilaterally symmetrical) (Fig. 18.17g) (Tucker, 1997). Tucker notes, further, that whereas these characters may be stable in individual families there may be "intriguing divergences" such as "shifts in number of organ whorls, loss of some organs, and tendencies toward unisexuality."

Flowers are determinate structures and flowering in annuals terminates the vegetative growth of the plant. In perennials, however, new floral apices are formed repeatedly and flowering occurs throughout the life of the plant. Prior to **flower induction**, internodes of the vegetative shoot apex typically elongate, and numerous lateral buds may be initiated below the apical meristem. The floral meristem usually becomes much broader than in its vegetative state, and the rate of cell division increases. Genetic analyses have identified about 80 genes that function in multiple genetic pathways that control the transition from the vegetative to the floral state (Araki, 2001), a transition that he describes as "the most dramatic phase change in plant development." The induction of the various floral components is initiated by one or several forms of signal transduction, factors in the external and internal environments such as photoperiod, temperature, and the production, often in the leaves, of various hormones and other chemical compounds which are transported, possibly through the phloem, to the apical meristems (see O'Neill, 1992; Lejeune *et al.*, 1993; O'Neill *et al.*, 1994; Bradley *et al.*, 1996; Chang *et al.*, 2001; Ruiz-Medrano *et al.*, 2001; Hamano *et al.*, 2002).

Tucker (1997) describes floral ontogeny as "a continuous succession of events, a cascade in which later events build on earlier ones." Among the many events that occur during floral morphogenesis are determination of the number, sites, and timing of initiation of floral parts of different types, the differentiation of form during which organ primordia become recognizable, increase in size and, in some taxa, fuse with adjacent primordia, and the development of specialized features often related to pollination such as nectaries, stigmatic papillae, specially shaped petals, etc. Research on the control by regulatory genes of floral patterning, the number and recognition of floral organs, and the development of these and other floral structures is destined to provide a much clearer understanding of the underlying mechanisms of floral morphogenesis (see Weigel, 1995; Frolich and Meyerowitz, 1997; Running and Hake, 2001).

The several floral parts are typically arranged in sequence beginning, at the base of the flower, with the sepals and continuing through petals, stamens, and carpels. Similarly, during ontogeny, initiation of the various floral components is typically acropetal, often with sepals appearing first, followed in sequence by petals, stamens, and carpels (Fig. 18.18a–e). However, this developmental sequence is highly variable and different sequences may characterize different species. Whereas the components of particular "whorls" may initially appear as separate entities, as they grow during development and contact those adjacent, they may fuse, as noted above, forming

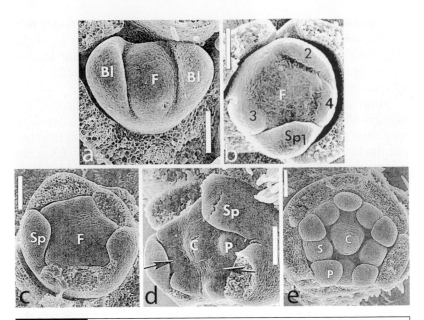

Figure 18.18 Early stages in the ontogeny of various floral organs of *Amherstia nobilis*, demonstrating the acropetal sequence of development from bracteoles to stamens. (a) A floral apex (F) with two opposite bracteole (B) primordia. (b) Bracteoles have been removed. Four of the five sepal primordia (Sp) have been initiated in a helical sequence. (c) Floral apex showing the last of the sepal primordia to be initiated. (d) The five petal primordia (P) have now appeared around the central carpel primordium (C). (e) The five petal primordia are prominent as is the carpel primordium. Stamen primordia (S) alternate with the petal primordia. Bars = 100 μm. From Tucker (1997). Used by permission of the University of Chicago Press. © 1997 The University of Chicago. All rights reserved.

compound structures. In the flowers of more primitive angiosperms the floral parts commonly remain separate at maturity whereas in flowers of more evolutionarily advanced taxa the floral parts often fuse. Fusion of carpels (forming a pistil) and petals (forming a corolla) is especially common. Although fusion of plant parts has been observed for many years, the actual process whereby epidermal cells of the initially separate parts dedifferentiate following contact has only begun to be understood in the last few years. Siegel and Verbeke (1989), studying *Catharanthus roseus* (Madagascar periwinkle, often referred to by the synonym *Vinca rosea*), demonstrated that movement of unidentified "diffusible factors (morphogens)" between carpel primordia leads to a dedifferentiation of the epidermal cells and their redifferentiation into parenchyma cells in the contact zone. Although it has been widely accepted that fusion of floral parts in angiosperms occurs without transcellular cytoplasmic continuity, the presence of secondary plasmodesmata between fusing carpels has been demonstrated recently in *Vinca* (Van der Schoot *et al.*, 1995). They suggested that these plasmodesmata might facilitate the transcellular

 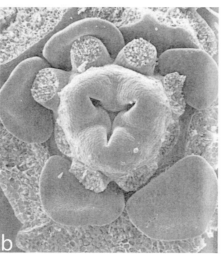

Figure 18.19 Open, ovulate carpels. (a) *Amherstia nobilis*. Outer floral parts have been removed. Magnification × 74. (b) View from above of the flower of *Koelreuteria elegans* showing a tricarpellate gynoecium before carpel closure. Sepals and stamens have been removed. Magnification × 96. From Tucker and Kantz (2001). Used by permission of the University of Chicago Press. ⓒ 2001 The University of Chicago. All rights reserved.

movement of hormones or proteins essential for continued gynoecial development. Much is yet to be learned about the process of fusion of floral parts, the identity and specific activity of the diffusible factors mentioned above, the significance of the plasmodesmata that develop between the contacting cells of carpel primordia during the development of the gynoecium, and whether or not what is becoming known in *Vinca* also characterizes other angiosperms.

One of the defining distinctions between gymnosperms and angiosperms is the presence of ovules on the abaxial surface of cone scales in the former (Figs. 18.2, 18.3) and in enclosed carpels in the latter (Figs. 18.6a, 18.7b, c). As has been shown by Tucker and Kantz (2001), however, young carpels containing ovules in some members of the legume family, Fabaceae, are open (Fig. 18.19) and become closed only during later stages of ontogeny. In their survey of the literature they found 44 species in 20 families with open, ovulate carpel primordia. They note that whereas open carpel primordia containing ovules are fairly uncommon in angiosperms, open carpel primordia without visible ovules have been reported in 180 taxa in 140 angiosperm families. During the final stages of development and prior to maturity, carpel closure occurs in a variety of ways, including simple appression, interdigitation of epidermal cells, elimination of cuticles and fusion of the contacting cells, and cell divisions and reorientation of cells in the appressed tissues (Tucker and Kantz, 2001). Recently Endress and Igersheim (2000) have reported that, in most basal (i.e., very primitive) angiosperms (e.g., Amborellaceae, Cabombaceae, Nymphaeaceae, Trimeniaceae, Illiciaceae, Schisandraceae, and Chloranthaceae), carpel closure is accomplished "from the outside by secretion." Apparently, secretory cells or canals in the tissue on either side of the gap between carpel margins secrete a substance that fills the gap, thus closing the carpel.

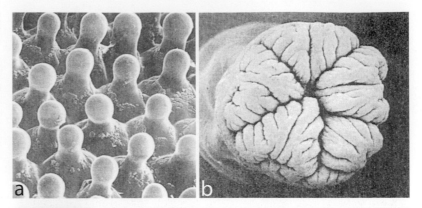

Figure 18.20 (a) Papillate epidermal cells of the stigma of *Primula* sp. (primrose). The epidermis is a glandular tissue that secretes a sugar-containing fluid that facilitates the adhesion of pollen grains to the stigma. It may also play a role in pollen hydration and germination. Magnification × 500. (b) The stigma of *Rhododendron intranervatum*, composed of five major lobes separated by grooves through which pollen tubes enter the stylar canal. Magnification × 20. (a) From Troughton and Donaldson (1972). Used by permission of the New Zealand Ministry of Research, Science and Technology. (b) From Palser *et al.* (1992). Used by permission of the National Research Council of Canada.

Much of the material presented heretofore in this chapter is widely accepted, common knowledge among botanists. During the past decade, however, with the application of techniques of cellular and molecular biology, several important and exciting areas of research have produced information that is expanding our understanding of reproductive biology in plants. Among these areas are pollen–pistil interactions, including pollen adhesion, hydration and pollen tube growth, self-incompatibility, and the role of the cytoskeleton in various aspects of reproduction and embryogeny. We shall consider each of these in some detail.

Pollen–pistil interactions

The nature of the **stigmatic surface**, whether smooth or papillate, wet or dry, is important in the process of pollination (Figs. 18.14a, 18.20a, b). A **wet stigmatic surface** is covered by exudates, secreted granulocrinously (i.e., by fusion of Golgi and/or ER vesicles with the plasma membrane of stigmatic cells; see Weber, 1994). In families such as the Solanaceae and Leguminosae, a wet surface facilitates the adhesion of pollen grains, and is known to be receptive to pollen in many species (Wheeler *et al.*, 2001). In families such as the Gramineae and Brassicaceae, however, characterized by **dry stigmatic surfaces**, adhesion of pollen is controlled by an interaction between the pollen coat and the stigmatic surface (Heslop-Harrison, 1979; Elleman and Dickinson, 1990, 1996; Elleman *et al.*, 1992; Wheeler *et al.*, 2001).

At the time of contact with a stigmatic surface, most pollen is highly dehydrated and must become hydrated in order to germinate.

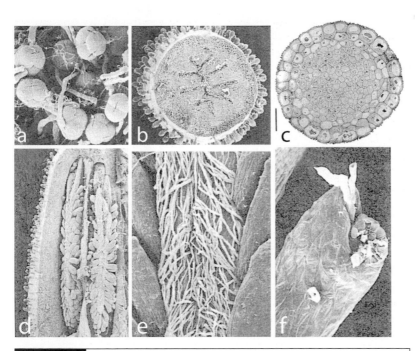

Figure 18.21 (a) Germinated pollen tetrads of *Rhododendron fortunei* on the stigmatic surface. Magnification × 158. (b) A transverse section of the style of *R. fortunei*, showing the multi-armed stylar canal containing pollen tubes. Magnification × 18. (c) A transverse section of the style of *Smyrnium perfoliatum* (Apicaceae) containing a solid core of transmitting tissue. Bar = 20 μm. (d) Two locules of an ovary of *R. fortunei* showing ovules attached to placentae. Numerous pollen tubes are visible on the placental surfaces. Magnification × 8. (e) Englargement showing pollen tubes on the placental surface of *R. fortunei*. Magnification × 83. (f) Ovule of *R. intranervatum*. Two pollen tubes have entered the ovule through the micropylar end. Magnification × 464. (a, b, d–f) From Palser *et al.* (1992). Used by permission of the National Research Council of Canada. (c) From Weber (1994). Used by permission of the University of Chicago Press. © 1994 The University of Chicago. All rights reserved.

The exudate on the surface of wet stigmas facilitates pollen hydration as well as adhesion and germination (Figs. 18.14b, 18.21a). On dry stigmas such as those of *Brassica oleracea*, Elleman and Dickinson (1990, 1996) have demonstrated that the coat of a pollen grain on a receptive stigma interacts with the subtending stigmatic surface cells eliciting an expansion of their outer wall layers and a "loosening of the wall matrix." Associated with these modifications of wall structure is uptake of water, a granulocrinous secretion (Elleman and Dickinson, 1990, 1996), and the extension of the pollen coat to form an adhesive "foot." Several proteins have been implicated in the adhesion of pollen grains in *Brassica* (see, e.g., Stephenson *et al.*, 1997). The presence of long-chain lipids (see Wheeler *et al.*, 2001) as well as proteins such as aquaporin (Ikeda *et al.*, 1997) in the stigmatic exudate play significant roles in regulating hydration. Proteins such as aquaporin are thought to form molecular "channels" in the plasma membrane, thus facilitating the transport of water from cells of the stigma onto the stigmatic

Figure 18.22 (a) A pollen tube penetrating the stylar transmitting tissue of *Smyrnium perfoliatum*. Bar = 10 μm. (b) Pollen tube in an exudate-filled intercellular canal of the stylar transmitting tissue. Bar = 5 μm. (a, b) From Weber (1994). Used by permission of the University of Chicago Press.

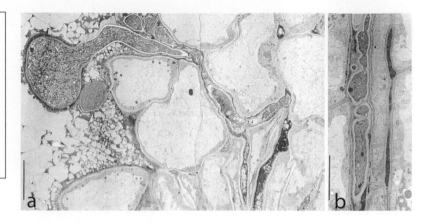

surface and hydrating the pollen grains, or in self-incompatible taxa, transporting water away from the stigmatic surface, thus preventing hydration and germination of pollen (for more detail, see Ikeda *et al.*, 1997; Wheeler *et al.*, 2001). In contrast to that of many other families, grass pollen is transmitted to the stigmatic surfaces in a hydrated condition, and can germinate and penetrate the stigma within less than 1 minute (Heslop-Harrison, 1979, 1987, 2000). For more detail on pollen–pistil interactions and other interesting examples see Endress (1994).

Styles are characterized either by a hollow **stylar canal** (Fig. 18.21b) filled with exudate, composed of various combinations of polysaccharides, lipids, phenolics, pectins, and proteins, or a **central core** of specialized **transmitting tissue** (Fig. 18.21c). In the latter case, the transmitting tissue contains longitudinal, intercellular channels filled with secretions (Weber, 1994). In both types of styles, pollen tubes grow through the secretions toward the ovule. Upon germination the pollen tube penetrates the cuticle, grows through the stigma and into the stylar canal or transmitting tissue (Figs. 18.14b, 18.21a, 18.22a, b). Some workers (e.g., Lush *et al.*, 1998, 2000; Wolters-Arts *et al.*, 1998) believe that the presence of triglycerides in the stigmatic exudate establishes a gradient of water that functions as a guidance cue for the growing pollen tube. Other workers have provided evidence of the presence of other chemical guidance cues that direct the pollen tubes toward the ovules. Cheung (1995) and Wang *et al.* (1993) have proposed that proteins in the intercellular secretion in the transmitting tissue of solid styles provide directional cues for growing pollen tubes. There is also evidence that, in some species, the embryo sac (female gametophyte) secretes substances that attract the pollen tube to the ovules (Ray *et al.*, 1997; Hererro, 2000). In contrast to these proposals, Heslop-Harrison and Reger (1988) suggested that the direction of pollen tube growth resulted simply from the presence of longitudinal, intercellular channels between cells of the transmitting tissue; in other words, that the control of direction of pollen tube growth is physical, related directly to the structure of the tissue in the style. It seems that control of the direction of pollen tube growth is either multifaceted, or that it varies in different taxonomic groups. Further research will, no doubt, lead to clarification. Many pollen tubes may enter the locules of the carpels (Fig. 18.21d, e), and one or more will enter an ovule through the micropyle (Fig. 18.21f).

Self-incompatibility

Two types of pollen incompatibility have evolved in angiosperms. Pollen from a different species may be rejected because it is too dissimilar to that of the recipient species whereas pollen from the same plant or from the same species may be rejected because it is genetically too similar to that of the recipient, a mechanism referred to as **self-incompatibility**. This rejection by some plants of their own pollen was noticed by Darwin (1877) and described by him "as one of the most surprising facts I have ever observed". Self-incompatibility which forces cross-pollination and fertilization is important because it results in the maintenance of a high degree of heterozygosity in a species, and prevents the potentially deleterious effect on progeny that often results from selfing. On the other hand, some taxa are regularly self-pollinating and self-fertilizing (see Lloyd and Schoen (1992) for a general discussion and more detail).

Between 30% and 50% of angiosperm species are self-incompatible (Wheeler et al., 2001). Rejection of pollen may occur at any of several stages in the reproductive process including hydration, germination, during growth of the pollen tube through the style, in the ovule, or even post-fertilization in some species (Wheeler et al., 2001). Genetic studies during the 1980s and 1990s demonstrated that the incompatibility response in the Brassicaceae, Solanaceae, Papaveraceae, Rosaceae, and Scrophulariaceae is controlled by a single, multigene locus (see de Nettancourt, 2001; Dixit and Nasrallah, 2001) but in some other taxa, e.g., the Poaceae, by more than one "recognition locus" (Dixit and Nasrallah, 2001). When this so-called S (self-incompatibility) locus occurs in both pollen and pistil, the incompatibility response is initiated, preventing self-fertilization, often by inhibition of pollen germination or pollen tube growth. In members of the Solanaceae, the degradation of rRNA by cytotoxic proteins restricts the growth of the pollen tube through the style (Wheeler et al., 2001). In Brassica which has a dry stigma, it has been shown that self-incompatibility of pollen is related to the regulation of the transfer of water from the stigma to the pollen grains (see pp. 374–375). These brief summaries of several of the self-incompatibility mechanisms in flowering plants are taken largely from Wheeler et al. (2001) and Dixit and Nasrallah (2001). The interested student should consult these references as well as Linskens (1988), Stephenson et al. (1997), and de Nettancourt (2001) for more detailed discussions of the genetic control of self-incompatibility and other aspects of pollen/pistil interactions as well as for comprehensive bibliographies.

The role of the cytoskeleton in pollen tube growth

The mechanism of growth of pollen tubes has long fascinated plant anatomists and other plant scientists. It is widely accepted that actin microfilaments of the cytoskeleton play a significant role in this process (see Miller *et al.*, 1996; Vidali and Hepler, 2001). In order for the pollen tube to grow, there must be a continuous supply of cellulose and wall matrix precursors provided at the tip (i.e., the distal region) where extension growth occurs. Golgi vesicles containing these materials are moved to the tip by cytoplasmic streaming under the influence, possibly the control, of actin microfilaments. As turgor pressure results in the extension of the thin-walled distal region of the pollen tube, the Golgi vesicles fuse with the plasmalemma, expelling their contents into the apoplast where wall synthesis takes place.

It has been reported by some workers that the bundles of microfilaments extend longitudinally throughout the pollen tube (Heslop-Harrison and Heslop-Harrison, 1991), and others have observed dense accumulations in the distal end (Perdue and Parthasarathy, 1985; Pierson, 1988). Recent studies of pollen tube growth in *Lilium longiflorum*, however, have provided strong evidence that, at least in this taxon, the very tip of the pollen tube is free of microfilaments (Miller *et al.*, 1996; Vidali and Hepler, 2001). In addition to the role of microfilaments in transporting Golgi vesicles to the growing region of the pollen tube, Vidali and Hepler (2001) conclude that growth is dependent on the continual polymerization of new actin microfilaments by which their presence is maintained close to the tip, thus assuring a constant supply of Golgi vesicles in the region of tip growth. They also suggest that a gradient of calcium ions and the presence of actin-binding proteins have important roles in determining the polarization and location of actin microfilaments thus assuring the presence of vesicles in the distal end of the pollen tube where cell wall synthesis is occurring. They conclude that "although many issues and details remain to be deciphered, it is nevertheless evident that understanding the regulation of actin polymerization will be a major factor in providing a complete description of the growth mechanism of the pollen tube [in angiosperms]." In addition to actin microfilaments, microtubules, arranged longitudinally or helically, are also found throughout the pollen tubes of angiosperms but, as in the case of microfilaments, they are rare or non-existent in the very tips (see Anderhag *et al.*, 2000). Furthermore, they are not radially (horizontally) arranged even in the region of tip growth and their role in pollen tube growth has not been determined.

Recent studies indicate some differences in the cytoplasmic organization of pollen tubes of conifers and angiosperms, especially in the distal, growing regions. In *Picea abies* Anderhag *et al.* (2000) observed that microtubules and actin microfilaments are abundant and oriented longitudinally throughout the length of the major part of the pollen tube. Microfilaments are not present in the growing distal end, but microtubules there are arranged radially (i.e., horizontally) to the direction of growth. Anderhag *et al.* (2000) believe, therefore, that, as has been well established for the growth of cell walls in vegetative tissue (see Chapter 4), the peripheral microtubules probably control the synthesis of cellulose microfibrils in the wall of the elongating region of the pollen tube. Because chemical disruption of both microfilaments and microtubules inhibits tip growth and causes pollen tube branching, they conclude on the basis of this and other evidence that a coordination of microtubules and microfilaments drives tip extension in conifer pollen tubes. They are at a loss to explain the absence of radially arranged microtubules in the growing region of angiosperm pollen tubes.

REFERENCES

Anderhag, P., Hepler, P. K., and M. D. Lazzaro. 2000. Microtubules and microfilaments are both responsible for pollen tube elongation in the conifer *Picea abies* (Norway spruce). *Protoplasma* **214**: 141–157.

Araki, T. 2001. Transition from vegetative to reproductive phase. *Curr. Opin. Plant Biol.* **4**: 63–68.

Boesewinkel, F. D. and F. Bouman. 1984. The seed: structure. In B. M. Johri, ed., *Embryology of Angiosperms*. Berlin: Springer-Verlag, pp. 567–610.

Bouman, F. 1984. The ovule. In B. M. Johri, ed., *Embryology of Angiosperms*. Berlin: Springer-Verlag, pp. 123–157.

Bradley, D., Vincent, C., Carpenter, R., and E. Coen. 1996. Pathways for inflorescence and floral induction in *Antirrhinum*. Development **122**: 1535–1544.

Brownlee, C. 1994. Signal transduction during fertilization in algae and vascular plants. *New Phytol.* **127**: 399–423.

Bruns, D. and J. N. Owens. 2000. Western white pine (*Pinus monticola* Dougl.) reproduction. II. Fertilization and cytoplasmic inheritance. *Sex. Plant Reprod.* **13**: 75–84.

Chang, S. T., Chen, W. S., Koshioka, M., *et al.* 2001. Gibberellins in relation to flowering in *Polianthes tuberosa*. *Physiol. Plant.* **112**: 429–432.

Cheung, A. Y. 1995. Pollen–pistil interactions in compatible pollen. *Proc. Natl Acad. Sci. USA* **92**: 3077–3080.

Coulter, J. M. and C. J. Chamberlain. 1917. *Morphology of Gymnosperms*. Chicago, IL: University of Chicago Press.

Darwin, C. 1877. *The Different Forms of Flowers on Plants of the Same Species*. London: John Murray.

de Nettancourt, D. 2001. *Incompatibility and Incongruity in Wild and Cultivated Plants*, 2nd edn. Berlin: Springer-Verlag.

Dixit, R. and J. B. Nasrallah. 2001. Recognizing self in the self-incompatibility response. *Plant Physiol.* **125**: 105–108.

Doyle, J. 1963. Proembryogeny in *Pinus* in relation to that in other conifers: a survey. *Proc. Roy. Irish Acad.* **62B**: 181–216.

Elleman, C. J., and H. G. Dickinson. 1990. The role of the exine coating in pollen-stigma interactions in *Brassica oleracea* L. *New Phytol.* **114**: 511–518.

1996. Identification of pollen components regulating pollination-specific responses in the stigmatic papillae of *Brassica oleracea*. *New Phytol.* **133**: 197–205.

Elleman, C. J., Frankin-Tong, V., and H. G. Dickinson. 1992. Pollination in species with dry stigmas: the nature of the early stigmatic response and the pathway taken by pollen tubes. *New Phytol.* **121**: 413–424.

Endress, P. K. 1994. *Diversity and Evolutionary Biology of Tropical Flowers*. Cambridge, UK: Cambridge University Press.

Endress, P. K. and A. Igersheim. 2000. Gynoecium structure and evolution in basal angiosperms. *Int. J. Plant Sci.* **161**: S211–S223.

Faure, J.-E. and C. Dumas. 2001. Fertilization in flowering plants: new approaches for an old story. *Plant Physiol.* **125**: 102–104.

Friedman, W. E. 1990. Double fertilization in *Ephedra*, a nonflowering seed plant: its bearing on the origin of angiosperms. *Science* **247**: 951–954.

1994. The evolution of embryogeny in seed plants and the developmental origin and early history of endosperm. *Am. J. Bot.* **81**: 1468–1486.

1998. The evolution of double fertilization and endosperm: an "historical" perspective. *Sex. Plant Reprod.* **11**: 6–16.

Friedman, W. E. and J. S. Carmichael. 1998. Heterochrony and developmental innovation: evolution of female gametophyte ontogeny in *Gnetum*, a highly apomorphic seed plant. *Evolution* **52**: 1016–1030.

Friedman, W. E. and E. M. Gifford. 1997. Development of the male gametophyte of *Ginkgo biloba*: a window into the reproductive biology of early seed plants. In T. Hori, R. W. Ridge, W. Tuleke, *et al.*, eds., Ginkgo biloba: *A Global Treasure*. Berlin: Springer-Verlag, pp. 29–49.

Frolich, M. W. and E. M. Meyerowitz. 1997. The search for flower homeotic gene homologs in basal angiosperms and Gnetales: a potential new source of data on the evolutionary origin of flowers. *Int. J. Plant Sci.* **158**: S131–S142.

Hamano, M., Yamato, Y., Yamazaki, H., and H. Miura. 2002. Endogenous gibberellins and their effects on flowering and stem elongation in cabbage (*Brassica oleracea* var. *capitata*). *J. Hort. Sci. Biotech.* **77**: 220–225.

Hererro, M. 2000. Changes in the ovary related to pollen tube guidance. *Ann. Bot.* **85**: 79–87.

Heslop-Harrison, J. 1979. An interpretation of the hydrodynamics of pollen. *Am. J. Bot.* **66**: 737–741.

1987. Pollen germination and pollen-tube growth. *Int. Rev. Cytol.* **107**: 1–78.

Heslop-Harrison, J. and Y. Heslop-Harrison. 1991. The actin cytoskeleton in unfixed pollen tubes following microwave-accelerated DMSO-permeabilization and TRITC-phalloidin staining. *Sex. Plant Reprod.* **4**: 6–11.

1997. Intracellular motility and the evolution of the actin cytoskeleton during development of the male gametophyte of wheat (*Triticum aestivum* L.). *Phil. Trans. Roy. Soc. London* B**352**: 1985–1993.

Heslop-Harrison, Y. 2000. Control gates and micro-ecology: the pollen-stigma interaction in perspective. *Ann. Bot. (Suppl. A)* **85**: 5–13.

Heslop-Harrison, Y. and B. J. Reger. 1988. Tissue organization, pollen receptivity and pollen tube guidance in normal and mutant stigmas of the grass *Pennisetum pyphoides* (Burm) Stap. et Hubb. *Sex. Plant Reprod.* **1**: 182–183.

Huang, B.-Q. and S. D. Russell. 1994. Fertilization in *Nicotiana tabacum*: cytoskeletal modifications in the embryo sac during synergid degeneration. *Planta* **194**: 200–214.

Ikeda, S., Nasrallah, J. B., Dixit, R., Preiss, S., and M. E. Nasrallah. 1997. An aquaporin-like gene required for the *Brassica* self-incompatibility response. *Science* **276**: 1564–1566.

Jaeger, P. 1961. *The Wonderful Life of Flowers*. New York: E. P. Dutton.

Jensen, W. A. 1965. The ultrastructure and histochemistry of the synergids of cotton. *Amer. J. Bot.* **52**: 238–256.

_____ 1998. Double fertilization: a personal view. *Sex. Plant Reprod.* **11**: 1–5.

Jensen, W. A. and D. Fisher. 1968. Cotton embryogenesis: the entrance and discharge of the pollen tube in the embryo sac. *Planta* **78**: 158–183.

Johansson, M. and B. Walles. 1993a. Functional anatomy of the ovule in broad bean (*Vicia faba* L.). I. Histogenesis prior to and after pollination. *Int. J. Plant Sci.* **154**: 80–89.

_____ 1993b. Functional anatomy of the ovule in broad bean, *Vicia faba* L. II. Ultrastructural development up to early embryogenesis. *Int. J. Plant Sci.* **154**: 535–549.

Knox, R. B., Zee, S. Y., Blomstedt, C., and M. B. Singh. 1993. Male gametes and fertilization in angiosperms. *New Phytol.* **125**: 679–694.

Krasowski, M. J. and J. N. Owens. 1993. Ultrastructural and histochemical postfertilization megagametophyte and zygotic embryo development of white spruce (*Picea glauca*) emphasizing the deposition of seed storage products. *Can. J. Bot.* **71**: 98–112.

Lawrence, G. H. M. 1951. *Taxonomy of Vascular Plants*. New York: Macmillan.

Lejeune, P., Bernier, G., Requier, M. C., and J. M. Kinet. 1993. Sucrose increase during floral induction in the phloem sap collected at the apical part of the shoot of the long-day plant *Sinapis alba* L. Planta **190**: 71–74.

Linskens, H. F. 1988. Present status and future prospects of sexual reproduction research in higher plants. In M. Cresti, P. Gori, and E. Pacini, eds., *Sexual Reproduction in Higher Plants*. Berlin: Springer-Verlag, pp. 451–458.

Lloyd, D. G. and D. J. Schoen. 1992. Self- and cross-fertilization in plants. I. Functional dimensions. *Int. J. Plant Sci.* **153**:358–369.

Lush, W. M., Grieser, F., and M. Wolters-Arts. 1998. Directional guidance of *Nicotiana alata* pollen tubes in vitro and on the stigma. *Plant Physiol.* **118**: 733–741.

Lush, W. M. Spurck, T., and R. Joosten. 2000. Pollen tube guidance by the pistil of a solanaceous plant. *Ann. Bot. (Suppl. A)* **85**: 39–47.

Maheshwari, P. 1950. *An Introduction to the Embryology of Angiosperms*. New York: McGraw-Hill.

Miller, D. D., Lancelle, S. A., and P. K. Hepler. 1996. Actin microfilaments do not form a dense meshwork in *Lilium longiflorum* pollen tube tips. *Protoplasma* **195**: 123–132.

Nguyen, H., Brown, R. C., and B. E. Lemmon. 2001. Patterns of cytoskeletal organization reflect distinct developmental domains in endosperm of *Coronopus didymus* (Brassicaceae). *Int. J. Plant Sci.* **162**: 1–14.

O'Neill, S. D. 1992. The photoperiodic control of flowering: progress toward understanding the mechanism of induction. *Photochem. Photobiol.* **56**: 789–801.

O'Neill, S. D., Zhang, X. S., and C. C. Zheng. 1994. Dark and circadian regulation of messenger-RNA accumulation in the short-day plant *Pharbitis-nil*. *Plant Physiol.* **104**: 569–580.

Owens, J. N. and D. Bruns. 2000. Western white pine (*Pinus monticola* Dongl.) reproduction. I. Gametophyte development. *Sex. Plant Reprod.* **13**: 61–74.

Owens, J. N. and S. J. Morris. 1991. Cytological basis for cytoplasmic inheritance in *Pseudotsuga menziesii*. II. Fertilization and proembryo development. *Am. J. Bot.* **78**: 1515–1427.

Owens, J. N., Takaso, T., and C. J. Runions. 1998. Pollination in conifers. *Trends Plant Sci.* **3**: 479–485.

Palser, B. F., Rouse, J. L., and E. G. William. 1992. A scanning electron microscope study of the pollen tube pathway in pistils of *Rhododendron*. *Can. J. Bot.* **70**: 1039–1060.

Perdue, T. D. and M. V. Parthasarathy. 1985. *In situ* localization of F-actin in pollen tubes. *Eur. J. Cell Biol.* **39**: 13–20.

Pettitt, J. M. 1985. Pollen tube development and characteristics of the protein emission in conifers. *Ann. Bot.* **56**: 379–397.

Pierson, E. S. 1988. Rhodamine-phalloidin staining of F-actin in pollen after dimethyl sulphoxide permeabilization: a comparison with the conventional formaldehyde preparation. *Sex. Plant Reprod.* **1**: 83–87.

Polowick, P. L. and V. K. Sawhney. 1993. An ultrastructural study of pollen development in tomato (*Lycopersicon esculentum*). II. Pollen maturation. *Can. J. Bot.* **71**: 1048–1055.

Ray, S., Park, S.-S., and A. Ray. 1997. Pollen tube guidance by the female gametophyte. *Development* **124**: 2489–2498.

Romberger, J. A., Hejnowicz, Z., and J. F. Hill. 1993. *Plant Structure: Function and Development*. Berlin: Springer-Verlag.

Ruiz-Medrano, R., Xoconostle-Cazares, B., and W. J. Lucas. 2001. The phloem as a conduit for inter-organ communication. *Curr. Opin. Plant Biol.* **4**: 202–209.

Runions, C. J. and J. N. Owens. 1999a. Sexual reproduction of interior spruce (Pinaceae). I. Pollen germination to archegonial maturation. *Int. J. Plant Sci.* **160**: 631–640.

1999b. Sexual reproduction of interior spruce (Pinaceae). II. Fertilization to early embryo formation. *Int. J. Plant Sci.* **160**: 641–652.

Running, M. P. and S. Hake. 2001. The role of floral meristems in patterning. *Curr. Opin. Plant Biol.* **4**: 69–74.

Russell, S. D. 1984. Ultrastructure of the sperm of *Plumbago zeylanica*. II. Quantitative cytology and three-dimensional organization. *Planta* **162**: 385–391.

1985. Preferential fertilization in *Plumbago*: ultrastructural evidence for gamete-level recognition in an angiosperm. *Proc. Natl Acad. Sci. USA* **82**: 6129–6132.

Siegel, B. A. and J. A. Verbeke. 1989. Diffusible factors essential for epidermal cell redifferentiation in *Catharanthus roseus*. *Science* **244**: 580–582.

Singh, H. 1978. *Encyclopedia of Plant Anatomy*, vol. 10, part 2, *Embryology of Gymnosperms*. Berlin: Gebruder Borntraeger.

Stephenson, A. G., Doughty, J., Dixon, S., *et al.* 1997. The male determinant of self-incompatibility in *Brassica oleracea* is located in the pollen coating. *Plant J.* **12**: 1351–1359.

Steward, W. N. and G. W. Rothwell. 1993. *Paleobotany and the Evolution of Plants*. Cambridge, UK: Cambridge University Press.

Troughton, J. and L. A. Donaldson. 1972. *Probing Plant Structure*. Wellington, NZ: New Zealand Ministry of Research, Science and Technology.

Tucker, S. C. 1997. Floral evolution, development, and convergence: the hierarchical-significance hypothesis. *Int. J. Plant Sci.* **158**: S143–S161.

Tucker, S. C. and K. E. Kantz. 2001. Open carpels with ovules in Fabaceae. *Int. J. Plant Sci.* **162**: 1065–1073.

Van der Schoot, C., Dietrich, M. A., Storms, M., Verbeke, J. A., and W. J. Lucas. 1995. Establishement of a cell-to-cell communication pathway between separate carpels during gynoecium development. *Planta* **195**: 450–455.

Vidali, L. and P. K. Hepler. 2001. Actin and pollen tube growth. *Protoplasma* **215**: 64–76.

Wang, H., Wu, H.-M., and A. Y. Cheung. 1993. Development and pollination regulated accumulation and glycosylation of a stylar transmitting tissue-specific proline-rich protein. *Plant Cell* **5**: 1639–1650.

Weber, M. 1994. Stigma, style, and pollen tube pathway in *Smyrnium perfoliatum* (Apiaceae). *Int. J. Plant Sci.* **155**: 437–444.

Weberling, F. 1989. *Morphology of Flowers and Inflorescences*. Cambridge, UK: Cambridge University Press.

Weigel, D. 1995. The genetics of flower development: from floral induction to ovule morphogenesis. *Annu. Rev. Genet.* **29**: 19–39.

Wheeler, M. J., Franklin-Tong, V. E., and F. C. H. Franklin. 2001. The molecular and genetic basis of pollen-pistil interactions. *New Phytol.* **151**: 565–584.

Wolters-Arts, M., Lush, W. M., and C. Mariani. 1998. Lipids are required for directional pollen tube growth. *Nature* **392**: 818–821.

Yu, H.-S, Hu, S.-Y, and S. D. Russell. 1992. Sperm cells in pollen tubes of *Nicotiana tabacum* L.: three-dimensional reconstruction, cytoplasmic diminution, and quantitative cytology. *Protoplasma* **168**: 172–183.

Zhang, Z. and S. D. Russell. 1999. Sperm cell surface characteristics of *Plumbago zeylanica* L. in relation to transport in the embryo sac. *Planta* **208**: 539–544.

FURTHER READING

Barnard, C. 1957a. Floral histogenesis in the monocotyledons. I. The Gramineae. *Austral. J. Bot.* **5**: 1–20.

    1957b. Floral histogenesis in the monocotyledons. II. The Cyperaceae. *Austral. J. Bot.* **5**: 115–128.

Barton, L. V. 1965. Dormancy in seeds imposed by the seed coat. *Handb. Pflanzenphysiol.* **15**: 727–745.

Bernier, G. 1971. Structural and metabolic changes in the shoot apex in transition to flowering. *Can. J. Bot.* **49**: 803–819.

Brown, R. C. and H. L. Mogensen. 1972. Late ovule and early embryo development in *Quercus gambelii*. *Am. J. Bot.* **59**: 311–316.

Brown, W. V. 1960. The morphology of the grass embryo. *Phytomorphology* **10**: 215–223.

    1965. The grass embryo: a rebuttal. *Phytomorphology* **15**: 274–284.

Camefort, H. 1969. Fécondation et proembryogenese chez les Abietacées (notion de néocytoplasme). *Rev. Cytol. Biol. Vég.* **32**: 253–271.

Carlquist, S. 1969. Toward acceptable evolutionary interpretations of floral anatomy. *Phytomorphology* **19**: 332–352.

Chesnoy, L. and M. J. Thomas. 1971. Electron microscopy studies on gametogenesis and fertilization in gymnosperms. *Phytomorphology* **21**: 50–63.

Davis, G. L. 1966. *Systematic Embryology of the Angiosperms*. New York: John Wiley and Sons.

Dickinson, H. G., Elleman, C. J., and J. Doughty. 2000. Pollen coatings: chimaeric genetics and new functions. *Sex. Plant Reprod.* **12**: 302–309.

Dunbar, A. 1973. Pollen ontogeny in some species of Campanulaceae: a study by electron microscopy. *Bot. Notiser* **126**: 277–315.

Edwards, M. M. 1968. Dormancy in seeds of charlock. III. Occurrence and mode of action of an inhibitor associated with dormancy. *Ann. Bot.* **19**: 601–610.

Elleman, C. J. and H. G. Dickinson. 1999. Commonalities between pollen/stigma and host/pathogen interactions: calcium accumulation during stigmatic penetration by *Brassica oleracea* pollen tubes. *Sex. Plant Reprod.* **12**: 194–202.

Esau, K. 1977. *Anatomy of Seed Plants*, 2nd. edn. New York: John Wiley and Sons.

Foster, A. S. and E. M. Gifford, Jr. 1974. *Comparative Morphology of Vascular Plants*, 2nd edn. San Francisco, CA: W. H. Freeman.

Greyson, R. I. 1994. *The Development of Flowers*. New York: Oxford University Press.

Hayward, H. E. 1938. *The Structure of Economic Plants*. New York: Macmillan.

Hepler, P. K., Vivaldi, L., and A. Y. Cheung. 2001. Polarized cell growth in higher plants. *Annu. Rev. Cell Devel. Biol.* **17**: 159–187.

Heslop-Harrison, J. (ed.) 1971. *Pollen Development and Physiology*. London: Butterworth.

Hoefert, L. L. 1969. Fine structure of sperm cells in pollen grains in *Beta*. *Protoplasma* **68**: 237–240.

Holdaway-Clarke, T. L. and P. K. Hepler. 2003. Control of pollen tube growth: role of ion gradients and fluxes. *New Phytol.* **159**: 539–563.

Hori, T., Ridge, R. W., Tulecke, W., *et al.* (eds.) 1997. Ginkgo biloba: *A Global Treasure*. Tokyo: Springer-Verlag.

Hulme, A. C. (ed.) 1970. *The Biochemistry of Fruits and their Products*, vol. 1. London: Academic Press.

Hyde, B. B. 1970. Mucilage-producing cells in the seed coat of *Plantago ovata*: developmental fine structure. *Am. J. Bot.* **57**: 1197–1206.

Jensen, W. A. 1965. The ultrastructure and composition of the egg and central cell of cotton. *Am. J. Bot.* **52**: 781–797.

1968. Cotton embryogenesis: the zygote. *Planta* **79**: 346–366.

1969. Cotton embryogenesis: pollen tube development in the nucellus. *Can. J. Bot.* **47**: 383–385.

1973. Fertilization in flowering plants. *BioScience* **23**: 21–27.

Johri, B. M. (ed.) 1984. *Embryology of Angiosperms*. Berlin: Springer-Verlag.

Justus, C. D., Anderhag, P., Goins, J. L., and M. C. Lazzaro. 2004. Microtubules and microfilaments coordinate to direct a fountain streaming pattern in elongating conifer pollen tube tips. *Planta* **219**: 103–109.

Kaplan, D. R. 1969. Seed development in *Downingia*. *Phytomorphology* **19**: 253–278.

Kozlowski, T. T. (ed.) 1972. *Seed Biology*, 2 vols. New York: Academic Press.

Lloyd, D. G. and S. C. H. Barrett (eds.) 1996. *Floral Biology*. New York: Chapman and Hall.

Luckwill, L. C. 1959. Factors controlling the growth and form of fruits. *J. Linn. Soc. London, Bot.* **56**: 294–302.

Maheshwari, P. and H. Singh. 1967. The female gametophyte of gymnosperms. *Biol. Rev.* **42**: 88–130.

Mahlberg, P. G. 1960. Embryogeny and histogenesis in *Nerium oleander* L. I. Organization of primary meristematic tissues. *Phytomorphology* **10**: 118–131.

Netolitzky, F. 1926. *Handbuch der Pflanzenanatomie, Anatomie der Angiospermen-Samen*, vol. 10, *Lief.* Berlin: Bornträger.

Norstog, K. 1972. Early development of the barley embryo: fine structure. *Am. J. Bot.* **59**: 123–132.

O'Brien, T. P. and K. V. Thimann. 1967. Observations on the fine structure of the oat coleoptile. III. Correlated light and electron microscopy of the vascular tissues. *Protoplasma* **63**: 443–478.

Periasamy, K. 1977. A new approach to the classification of angiosperm embryos. *Proc. Indian Acad. Sci.* **86B**: 1–13.

Pijl, L. van der 1972. *Principles of Dispersal in Higher Plants*, 2nd edn. Berlin: Springer-Verlag.

Puri, V. 1952. Placentation in angiosperms. *Bot. Rev.* **18**: 603–651.

Raghavan, V. 1976. *Experimental Embryogenesis in Vascular Plants*. London: Academic Press.

   1986. *Embryogenesis in Angiosperms: A Developmental and Experimental Study*. Cambridge, UK: Cambridge University Press.

Rost, T. L. 1975. The morphology of germination in *Setaria lutescens* (Gramineae): the effects of covering structures and chemical inhibitors on dormant and non-dormant florets. *Ann. Bot.* **39**: 21–30.

Russell, S. D. 1991. Isolation and characterization of sperm cells in flowering plants. *Annu. Rev. Plant Physiol. Plant. Mol. Biol.* **42**: 189–204.

   1994. Fertilization in higher plants. In A. B. Stephenson and T. H. Kao, eds., *Pollen–Pistil Interactions and Pollen Tube Growth*. Rockville, MD: American Society of Plant Physiology, pp. 140–152.

   1996. Attraction and transport of male gametes for fertilization. *Sex. Plant Reprod.* **9**: 337–342.

Schulz, P. and W. A. Jensen. 1977. Cotton embryogenesis: the early development of the free nuclear endosperm. *Am. J. Bot.* **64**: 384–394.

   1968. *Capsella* embryogenesis: the egg, zygote, and young embryo. *Am. J. Bot.* **55**: 807–819.

   1969. *Capsella* embryogenesis: the suspensor and the basal cell. *Protoplasma* **67**: 138–163.

Singh, D. and A. S. R. Dathan. 1972. Structure and development of seed coat in Cucurbitaceae. VI. Seeds of *Cucurbita*. *Phytomorphology* **22**: 29–45.

Sporne, K. R. 1958. Some aspects of floral vascular systems. *Proc. Linn. Soc. London B* **169**: 75–84.

Srivastava, L. M. and R. E. Paulson. 1968. The fine structure of the embryo of *Lactuca sativa*. II. Changes during germination. *Can. J. Bot.* **46**: 1447–1453.

Steeves, T. A. and I. M. Sussex. 1989. *Patterns in Plant Development*, 2nd edn. Cambridge, UK: Cambridge University Press.

Tepfer, S. S. 1953. Floral anatomy and ontogeny in *Aquilegia formosa* var. *truncata* and *Ranunculus repens*. *Univ. Calif. Publ. Bot.* **25**: 513–648.

Thimann, K. V. and T. P. O'Brien. 1965. Histological studies of the coleoptile. II. Comparative vascular anatomy of coleoptiles of *Avena* and *Triticum*. *Am. J. Bot.* **52**: 918–923.

Tian, H. Q., Zhang, Z. J., and S. D. Russell. 2001. Sperm dimorphism in *Nicotiana tabacum* L. *Sex. Plant Reprod.* **14**: 123–125.

Tucker, S. C. 1959. Ontogeny of the inflorescence and the flower of *Drimys winteri* var. *chilensis*. *Univ. Calif. Publ. Bot.* **30**: 257–336.

    1999. Evolutionary lability of symmetry in early floral development. *Int. J. Plant Sci.* **160**: S25–S39.

    2003. Floral development in legumes. *Plant Physiol.* **131**: 911–926.

Tucker, S. C. and J. Grimes. 1999. The inflorescence: introduction. *Bot. Rev.* **65**: 303–316.

Wardlaw, C. W. 1955. *Embryogenesis in Plants*. New York: John Wiley and Sons.

Webb, M. C. and B. E. S. Gunning. 1991. The microtubular cytoskeleton during development of the zygote, proembryo and free-nuclear endosperm in *Arabidopsis thaliana* (L.) Heynh. *Planta* **184**:187–195.

Weterings, K. and S. D. Russell. 2004. Experimental analysis of the fertilization process. *Plant Cell* **16**: S107–S118.

Yeung, E. C. 1980. Embryogeny of *Phaseolus*: the role of the suspensor. *Z. Pflanzenphysiol.* **96**: 17–28.

Zhang, Z., Tian, H. Q., and S. D. Russell. 1999. Localization of myosin on sperm-cell-associated membranes of tobacco (*Nicotiana tabacum* L.). *Protoplasma* **208**: 123–128.

Glossary

This glossary is based on that of Esau, K. (1977) *Anatomy of Seed Plants*, 2nd edn. Some terms have been added, some deleted, and the definitions of some have been modified to reflect recent research and/or the usage in this book. Used by permission of John Wiley and Sons, New York.

abaxial Directed away from the axis. Opposite of *adaxial*.

abscission The shedding of leaves, flowers, fruit, or other plant parts, usually after formation of an *abscission zone*.

abscission layer In abscission zone, layer of cells the disjunction or breakdown of which causes the shedding of a plant part. Other term: *separation layer*.

abscission zone Zone at base of leaf, flower, fruit, or other plant part that contains an *abscission* (or *separation*) *layer* and a *protective layer*, both involved in the abscission of the plant part.

accessory bud A bud located above or on either side of the main *axillary bud*.

accessory cell See *subsidiary cell*.

accessory parts in fruit Parts not derived from the ovary but associated with it in fruit.

accessory transfusion tissue *Transfusion tissue* located within the mesophyll rather than associated with vascular bundle. In leaves of certain gymnosperms.

acicular crystal Needle-shaped crystal.

acropetal development (or differentiation) Produced or becoming differentiated in a succession toward the apex of an organ. The opposite of *basipetal development* but means the same as *basifugal development*.

actinomorphic Having a flower that can be divided in two equal parts in more than one longitudinal plane, i.e., a radially symmetrical or *regular flower*. Opposite of *zygomorphic*.

actinostele *Protostele* with star-shaped outline in transverse section.

adaxial Directed toward the axis. Opposite of *abaxial*.

adaxial meristem Meristematic tissue on the adaxial side of a young leaf that contributes to the increase in thickness of the petiole and midrib.

adenosine triphosphate The major source of usable chemical energy in metabolism; commonly abbreviated as ATP.

adnation In a flower, union of members of different whorls, as stamens and petals.

adventitious Of structures, arising not at their usual sites, e.g., roots originating on stems or leaves instead of on other roots, or buds developing on leaves or roots instead of in leaf axils on shoots.

aerenchyma Parenchyma tissue containing particularly large intercellular spaces of *schizogenous, lysigenous,* or *rhexigenous* origin.

aggregate fruit A fruit developing from a single *gynoecium* (single flower) composed of separate carpels, as the strawberry or raspberry fruits.

aggregate ray In secondary vascular tissues, a group of small rays arranged so as to appear to be one large ray.

albedo White tissue of the rind in citrus fruit.

albuminous cells In gymnosperm phloem, certain ray and phloem–parenchyma cells spatially and functionally associated with the sieve elements, thus resembling the companion cells of angiosperms but usually not originating from the same precursory cells as the sieve elements. Also called *Strasburger cells*.

albuminous seed A seed that contains endosperm in mature state.

aleurone grains Granules of protein present in seeds, usually restricted to the outermost layer, the *aleurone layer* of the endosperm. (*Protein bodies* is the preferred term for aleurone grains.)

aleurone layer Outermost layer of endosperm in cereals and many other taxa which contains protein bodies and enzymes concerned with endosperm digestion.

aliform paratracheal parenchyma In secondary xylem, vasicentric groups of axial parenchyma cells having tangential wing-like extensions as seen in transverse sections. See also *paratracheal parenchyma* and *vasicentric paratracheal parenchyma*.

alternation of generations A reproductive cycle in which the haploid ($1n$) phase, the gametophyte, produces gametes which fuse to form a zygote (diploid; $2n$) which develops into a sporophyte. Meiosis in the sporophyte results in the production of haploid spores which germinate, forming new gametophytes.

alternate pitting In tracheary elements, pits in diagonal rows.

amoeboid tapetum In anther locules, tapetum assuming amoeboid form when it disintegrates during pollen wall development.

amphicribral vascular bundle Concentric vascular bundle in which the phloem surrounds the xylem.

amphiphloic siphonostele A stele in which the vascular system appears as a tube and has phloem both external and internal to the xylem.

amphivasal vascular bundle Concentric vascular bundle in which the xylem surrounds the phloem.

amyloplast A colorless *plastid* (*leucoplast*) that forms starch grains.

analogous Having the same function as, but a different phylogenetic origin from, another entity.

anastomosis Type of structure in which cells or strands of cells are interconnected with one another as, for example, the veins in a leaf.

anatomy The study of structure.

androecium Collective term for the stamens in a flower of an angiosperm; part of the flower in which male gametogenesis is initiated or also carried to completion.

angiosperm A member of a group of plants the seed (or seeds) of which are borne within a matured ovary (fruit).

angstrom A unit of length equal to one-tenth of a millimicrometer (mμ), or one-tenth of a nanometer (nm). Symbol A or Å.

angular collenchyma A form of collenchyma in which the primary wall thickening is most prominent in the angles where several cells are joined.

anisocytic stoma A stomatal complex in which three subsidiary cells, one distinctly smaller than the other two, surround the stoma.

anisotropic Having different properties along different axes; optical anisotropy causes polarization and double refraction of light.

annual ring In secondary xylem, growth ring formed during one season. The term is deprecated because more than one growth ring may be formed during a single year.

annular cell wall thickening In tracheary elements of the xylem, secondary wall deposited in the form of rings.

anomalous secondary growth A term of convenience referring to types of secondary growth that differ from the more familiar ones.

anomocytic stoma A stoma without subsidiary cells.

anther The pollen-bearing part of the stamen.

anthesis The time of full expansion of the flower, from development of a receptive stigma to fertilization.

anthocyanin A water-soluble blue, purple, or red flavonoid pigment occurring in the cell sap.

anticlinal Having the orientation of the cell wall or plane of cell division perpendicular to the nearest surface. Opposite of *periclinal*.

antipodals Three or more cells located at the chalazal end of the mature embryo sac in angiosperms.

aperture in pollen grain A depressed region in the wall in which thick intine is covered by thin exine; the pollen tube emerges through the aperture.

apex (pl. **apices**) Tip, topmost part, pointed end of anything. In shoot or root the tip containing the apical meristem.

apical cell The single cell that occupies the distal position in the shoot or root apex of many pteridophytes (and is usually interpreted as the initial cell from which other cells and tissues are derived).

apical meristem A group of meristematic cells at the apex of the root or shoot which by cell division produce the precursors of the primary tissues of the root and shoot; may be *vegetative*, initiating vegetative tissues and organs, or *reproductive*, initiating reproductive tissues and organs.

apocarpy Condition in the flower characterized by lack of union of carpels (free carpels).

apomixis Vegetative reproduction without meiosis or fusion of gametes.

apoplast The interconnected system of plant cell walls. Compare with *symplast*.

apotracheal parenchyma In secondary xylem, axial parenchyma typically lacking contact with vessel members. Includes *boundary* (or *terminal*), *banded*, and *diffuse* apotracheal parenchyma.

apposition Growth of the cell wall by successive depositions of wall material, layer upon layer. Opposite of *intussusception*.

areole A small area of mesophyll in a leaf delimited by intersecting veins.

aril A fleshy outgrowth enveloping the seed and usually arising at the base of the ovule.

articulated laticifer A system of uniseriate cells produced by the apical meristem, with common walls intact or partly or entirely removed and containing latex; anastomosing or non-anastomosing.

aspirated pit In gymnosperm wood, bordered pit in which the pit membrane is laterally displaced with the torus covering the aperture.

astrosclereid A branched sclereid.

atactostele A stele in which the vascular bundles appear, in transverse section, to be scattered within the ground tissue.

ATP See *adenosine triphosphate*.

axial parenchyma Parenchyma cells in the axial system of secondary vascular tissues, as contrasted with ray parenchyma cells in the radial system.

axial system All vascular cells derived from the fusiform cambial initials and oriented with their longest dimension parallel to the main axis of the stem or root.

axial tracheid Tracheid in the axial system of secondary xylem, as contrasted with ray tracheid.

axil The upper angle between a stem and a twig or a leaf.

axillary bud Bud in the axil of a leaf.

axillary meristem Meristem located in the axil of a leaf and giving rise to an axillary bud.

banded apotracheal parenchyma In secondary xylem, axial parenchyma in concentric bands as seen in transverse section, typically lacking contact with vessel members. See also *apotracheal parenchyma*.

bark A non-technical term applied collectively to all tissues outside the vascular cambium or the xylem; in older trees it may be divided into dead outer bark and living inner bark which consists of secondary phloem. See also *rhytidome*.

bars of Sanio See *crassulae*.

basifugal development See *acropetal development*.

basipetal development Produced or becoming differentiated in a succession toward the base of an organ. The opposite of *acropetal* and *basifugal development*.

bast fiber Originally phloem fiber; now any extraxylary fiber.

bicollateral vascular bundle A vascular bundle with primary phloem along the inner and outer surfaces of the primary xylem.

bifacial leaf A leaf with palisade parenchyma on one side of the blade and spongy parenchyma on the other. A *dorsiventral leaf*. Conceived ontogenetically, a leaf that develops continuously from the original leaf primordium apex and includes tissues from both adaxial and abaxial sides of the primordium. Compare with *unifacial* leaf.

bilateral symmetry Of a flower, having two corresponding or complementary sides and which, thus, can be divided by a single longitudinal plane through the floral axis into two halves that are mirror images of one another. Contrasted with *radial symmetry*.

biseriate ray A vascular ray two cells wide.

blind pit A pit without a complementary pit in an adjacent wall.

bordered pit A pit in which the secondary wall arches over the pit membrane except in the region of the *pit aperture*.

bordered pit-pair Two bordered pits opposite each other in adjacent cell walls.

boundary apotracheal parenchyma In secondary xylem, axial parenchyma cells occurring either singly or in a layer at the end or the beginning of a season's growth. Also called *terminal apotracheal parenchyma*.

brachysclereid A short, roughly isodiametric sclereid, resembling a parenchyma cell in shape. Also called a *stone cell*.

branch gap In the nodal region of a non-seed plant stem, a region of parenchyma in the primary vascular cylinder through which branch traces extend toward a lateral branch. It may be confluent with a subtending *leaf gap*.

branch traces Vascular bundles connecting the primary vascular system of the branch and that of the main stem.

bulliform cell An enlarged epidermal cell present with other similar cells in longitudinal rows in leaves of grasses; thought to function in the rolling and unrolling of leaves.

bundle cap Sclerenchyma or collenchyma appearing in transverse section like a cap on the outer surface of a vascular bundle.

bundle sheath Layer or layers of cells enclosing a vascular bundle in a leaf; may consist of parenchyma or sclerenchyma.

bundle sheath extension A plate of tissue extending from a bundle sheath to the epidermis in a leaf. May be present on one or on both sides of the bundle and may consist of parenchyma or sclerenchyma.

callose A polysaccharide, β-1,3-glucan, yielding glucose on hydrolysis. Common wall constituent in the sieve areas of sieve elements; also develops rapidly in reaction to injury in sieve elements and parenchyma cells.

callus A tissue composed of large thin-walled cells developing as a result of injury, as in wound-healing or grafting, and in tissue culture.

calyptrogen In the root apex, meristem giving rise to the root cap.

calyx The sepals collectively, which with the corolla comprise the perianth.

cambial initials Cells in the *vascular cambium* or *phellogen* which through periclinal divisions contribute cells toward the inside or toward the outside of the axis; in the vascular cambium classified as either *fusiform initials* (source of axial cells of secondary phloem or secondary xylem) and *ray initials* (source of ray cells).

cambium A meristem with products of periclinal divisions commonly contributed in two directions and arranged in radial files. The term is preferably applied only to the two lateral meristems, the *vascular cambium* and the *cork cambium*, or *phellogen*.

carpel Leaf-like organ in angiosperms enclosing one or more ovules; a constituent of the *gynoecium*, the female part of the flower.

caruncle A fleshy protuberance near the hilum of a seed.

Casparian band (Casparian strip in older literature) A band-like wall formation within primary walls that contains suberin and lignin; typical of endodermal and exodermal cells in roots in which it occurs in the radial and transverse walls.

cataphylls Leaves inserted at low levels of the plant or shoot, as bud scales, rhizome scales, and others. Contrasted with *hypsophylls*.

cauline Belonging to the stem or arising from it.

caulis Stem.

cell A structural and physiological unit of a living organism. The plant cell consists of protoplast and cell wall; in non-living state, of cell wall only.

cell plate A partition appearing at telophase between the two nuclei formed during mitosis (and some meioses) and indicating the early stage of the division of a cell (*cytokinesis*) by means of a new cell wall; is formed in the *phragmoplast*.

cell wall A more or less rigid membrane enclosing the protoplast of a cell and, in higher plants, composed of cellulose and other organic and inorganic substances.

cellulose A polysaccharide, β-1,4-glucan, the main component of cell walls in most plants; consists of long chain-like molecules the basic units of which are anhydrous glucose residues of the formula $C_6H_{10}O_5$.

central cylinder A term of convenience applied to the vascular tissues and associated ground tissue in the stem and root. Refers to the same part of the stem and root that is designated as *stele*.

central mother cells Large vacuolated cells in the subsurface position in the apical meristem of the shoot in gymnosperms.

centric mesophyll A modification of isobilateral mesophyll in which the adaxial and abaxial palisade layers form a continuous layer; found in narrow or cylindrical leaves.

centrifugal development Produced or developing successively farther away from the center.

centripetal development Produced or developing successively closer to the center.

chalaza Region in the ovule where the integuments and the nucellus merge with the *funiculus*.

chimera A plant consisting of a combination of tissues of different genetic composition. In a periclinal chimera, cells of different composition are arranged in periclinal layers.

chlorenchyma Parenchyma tissue containing chloroplasts; leaf mesophyll and other green parenchyma.

chlorophyll The green pigment of plant cells required for photosynthesis.

chloroplast A chlorophyll-containing *plastid* with thylakoids organized into grana, and embedded in a stroma.

chromoplast A *plastid* containing pigments other than chlorophyll, usually yellow and orange carotenoid pigments.

cicatrice The scar left by a wound or by the separation of one plant part from another (as a leaf from a stem) and characterized by substances protecting the exposed surface.

circular bordered pit A circular pit with an overarching border; usually forms a pit-pair with a pit in a contiguous cell wall.

cisterna (pl. **cisternae**) A flattened, sac-like membranous compartment as in endoplasmic reticulum, a Golgi body, or a thylakoid.

cladophyll A branch resembling a foliage leaf.

closed venation Leaf venation characterized by anastomosing veins.

closing layer One of the compact layers of cells formed in alternation with the loose filling tissue (*complementary cells*) in a lenticel.

coenocyte A multinucleate organism or a multinucleate component of an organism; sometimes applied to multinucleate cells in seed plants.

cohesion In a flower, union of members of the same whorl, as sepals with sepals and petals with petals.

coleoptile The sheath enclosing the *epicotyl* of the grass embryo; sometimes interpreted as the first leaf of the epicotyl.

coleorhiza The sheath enclosing the *radicle* of the grass embryo.

collateral vascular bundle A bundle with phloem on only one side of the xylem, usually the abaxial side.

collenchyma A living, supporting tissue composed of generally elongate cells with unevenly thickened non-lignified primary walls. Common in the peripheral regions of stems and leaves.

colleter A multicellular appendage or a multicellular trichome producing a sticky secretion. Found on buds of many woody species.

columella The central part of a root cap in which the cells are arranged in longitudinal files.

commissural vascular bundle A small bundle interconnecting larger parallel bundles as in leaves of grasses.

companion cell A parenchyma cell in the phloem of an angiosperm associated with a sieve tube member and originating jointly with the latter from the same mother cell; some have the structure of a *transfer cell*.

compitum A region in the style of a syncarpous gynoecium where stylar canals are joined into one cavity.

complementary cells Cells of the loose tissue formed by the lenticel phellogen toward the outside; may or may not be suberized.

complementary tissue Loose tissue between closing layers in a lenticel. See *filling tissue*.

complete flower A flower having all types of floral parts: sepals, petals, stamens, and carpels, or tepals, stamens, and carpels.

compound laticifer Term sometimes applied to articulated laticifer.

compound middle lamella A collective term applied to two primary walls and the middle lamella; usually used when the true middle lamella is not distinguishable from the primary walls.

compound sieve plate A sieve plate composed of several sieve areas.

compression wood The reaction wood in conifers which is formed on the lower sides of branches and leaning or crooked stems and characterized by dense structure, strong lignification, and certain other features. See also *reaction wood* and *tension wood*.

concentric vascular bundle A vascular bundle with either the phloem surrounding the xylem (*amphicribral*) or the xylem surrounding the phloem (*amphivasal*).

conducting tissue See *vascular tissue*.

confluent paratracheal parenchyma In secondary xylem, coalesced aliform groups of axial parenchyma cells forming irregular tangential or diagonal bands, as seen in transverse section. See also *paratracheal parenchyma* and *aliform paratracheal parenchyma*.

conjunctive tissue Secondary parenchyma tissue interspersed with vascular tissue where the latter does not form a solid cylinder, as in monocotyledons and in dicotyledons with anomalous secondary growth.

connate Condition in which parts of the same whorl in a flower are united, as petals united into a corolla tube. See also *cohesion*.

connective The tissue between the two lobes of an anther.

contact cell An axial parenchyma or a ray cell physically as well as physiologically associated with a tracheary element. Analogous to a companion cell in the phloem. Also a cell next to a stoma.

contractile root A root that undergoes contraction at some time during its development and thereby effects a change in position of the shoot with regard to the ground level.

convergent evolution The independent evolution of similar structures in species that are unrelated or very distantly related; frequently characteristic of organisms living in similar environments.

coordinated growth Growth of cells in a manner that involves no separation of walls, as opposed to *intrusive growth*; also called *symplastic growth*.

copal A resinous substance that exudes from various tropical trees and hardens in air into roundish or irregular pieces. May be colorless, yellow, red, or brown.

cork See *phellem.*

cork cambium See *phellogen.*

cork cell A phellem cell derived from the phellogen, non-living at maturity, and having suberized walls; protective in function because the walls are highly impervious to water.

corolla A collective term for the petals of a flower.

corolla tube The tube-like part of a corolla resulting from congenital or ontogenetic union of petals.

corpus The core in an apical meristem covered by the *tunica.*

cortex The primary ground tissue region between the vascular system and the epidermis in stem and root. The term is also used with reference to the peripheral region of a cell protoplast.

cotyledon The leaf or leaves of an embryo within a seed.

crassulae (sing. **crassula**) Thickenings of intercellular material and primary wall along the upper and lower margins of a bordered pit-pair in the tracheids of gymnosperms; also called *bars of Sanio.*

cristae (sing. **crista**) Crest-like infoldings of the inner membrane in a mitochondrion.

cross-field A term of convenience for the rectangle formed by the walls of a ray cell against an axial tracheid, as seen in radial section of the secondary xylem of conifers.

crystalloid A protein crystal that is less angular than a mineral crystal and swells in water.

cuticle A layer consisting of cutin and waxes located on the outer walls of epidermal cells.

cutin A complex fatty substance that impregnates cell walls in some plant tissues including the epidermis and which comprises a layer called the cuticle on the outer surface of cell walls.

cutinization The process of impregnation with cutin.

cyclosis The streaming of cytoplasm in a cell.

cystolith A concretion of calcium carbonate on an outgrowth of a cell wall. Occurs in a cell called a *lithocyst.*

cytokinesis The process of division of a cell as distinguished from the division of the nucleus, or *karyokinesis.*

cytological zonation Presence of regions in the apical meristem, or other parts of shoot and root apices, having distinctive cytological characteristics that form the basis for a subdivision into distinguishable tissue regions.

cytology The study of cell structure and function.

cytoplasm In a strict sense, the visibly least differentiated part of the protoplasm of a cell that constitutes the groundmass enclosing all other components of the protoplast. Also called *hyaloplasm.*

decussate Arrangement of leaves in pairs that alternate with each other at right angles.

dedifferentiation A reversal in differentiation of a cell or tissue which is assumed to occur when a more or less completely differentiated cell resumes meristematic activity.

dehiscence The opening of a structure such as an anther or a fruit, allowing the release of reproductive structures contained therein.

dermal tissue system The outer covering tissue of a plant, epidermis, or periderm.

dermatogen The meristem forming the epidermis and arising from independent apical initials. One of the three histogens, *plerome, periblem,* and *dermatogen,* according to Hanstein.

desmotubule Tubule connecting the two endoplasmic reticulum cisternae located at the two opposite ends of a plasmodesma.

determinate growth Growth of limited duration such as that of leaves, flowers, and other lateral appendages.

development The change in form and complexity of an organism or part of an organism from its inception to maturity; combined with growth.

diacytic stoma A stomatal complex in which one pair of subsidiary cells, with their common walls at right angles to the long axis of the guard cell, surrounds the stoma.

diarch Arrangement of primary xylem in the root; having two protoxylem strands, or two protoxylem poles.

dichotomous venation A venation pattern in which the veins repeatedly branch into equal parts.

dictyosome A membranous cell organelle composed of stacked cisternae each producing secretory vesicles at the periphery; a *Golgi body.*

dictyostele A stele in which large overlapping leaf gaps dissect the primary vascular system into anastomosing strands, each with the phloem surrounding the xylem.

differentiation The physiological and morphological changes that occur in a cell, tissue, organ, or organism during development from a meristematic, or juvenile, stage to a mature, or adult, stage. Usually associated with an increase in specialization.

diffuse apotracheal parenchyma Axial parenchyma in secondary xylem occurring as single cells or as strands distributed irregularly within the tissue.

diffuse porous wood Secondary xylem in which vessels of one predominant size are distributed fairly uniformly throughout a growth layer; or with the vessels decreasing in size very slightly from early to late wood.

distal Farthest from the point of origin or attachment. Opposite of *proximal.* Often used in plant anatomy to mean in the direction of the apical meristem.

distichous Arrangement of leaves in two vertical rows; any two-ranked arrangement.

dorsiventral leaf Type of leaf possessing distinct upper and lower sides. Term derived from the reference to the abaxial and adaxial sides of a leaf as dorsal and ventral, respectively.

double fertilization The fusion of egg and sperm (resulting in a $2n$ fertilized egg, the zygote) and the fusion of the second male gamete with the polar nuclei (resulting in a $3n$ primary endosperm nucleus). A unique characteristic of angiosperms.

druse A globular, compound structure composed of calcium oxalate crystals.

duct An elongated space formed by separation of cells from one another (schizogenous origin), by dissolution of cells (lysigenous origin), or by a combination of the two processes; usually concerned with secretion.

early wood The wood (secondary xylem) formed in the first part of a growth layer and characterized by a lower density and larger cells than the late wood. Also called *spring wood.*

eccrinous secretion A secretion that leaves the cell as individual molecules passing through the plasmalemma and cell wall. Compare with *granulocrinous secretion*.

ectophloic siphonostele A stele composed of primary xylem and external primary phloem enclosing a pith.

ectoplast See **plasmalemma**.

egg apparatus The egg cell and two synergids located at the micropylar end of the female gametophyte (or embryo sac) in angiosperms.

elaioplast A leucoplast (a type of plastid) which functions as an oil-storage organelle.

elaiosome An outgrowth on a seed or a fruit which stores oil and serves as food for ants.

embryo sac The female gametophyte of angiosperms.

embryogeny (or **embryogenesis**) The process whereby the embryo develops.

enation A term applied to outgrowths of the stem that are considered to be primitive leaves. See also *microphyll*.

enation theory A theory that regards microphylls as simple enations in contrast to megaphylls which are considered to have evolved from branch systems.

endarch order of maturation Xylem which during development is characterized by centrifugal maturation of cells; that is, the oldest, mature elements (protoxylem) are closest to the center of the axis.

endocarp The innermost layer or layers of the pericarp.

endodermis A specialized, single layer of cells enclosing the vascular regions of roots and some stems, the cells of which are characterized by the presence of Casparian bands in the transverse and radial anticlinal walls.

endogenous Arising from a deep-seated tissue, as a lateral root.

endomembrane system Collective term for the membrane continuum of a cell consisting of plasmalemma, tonoplast, endoplasmic reticulum, Golgi bodies, and the nuclear envelope.

endoplasmic reticulum (often abbreviated as **ER**) A complex, three-dimensional system of membranes forming tubular or flattened compartments (cisternae) that permeate the cytoplasm. The cisternae appear like paired membranes in sectional profiles. The membranes may be coated with ribosomes (rough ER) or be free, or nearly free, of ribosomes (smooth ER).

endosperm A tissue of stored food in the embryo sac of angiosperms formed following fusion of a sperm cell and two polar nuclei. The endosperm provides nutrition for the developing embryo.

endothecium A wall layer in an anther, usually with secondary wall thickenings.

endothelium See *integumentary tapetum*.

enucleate Lacking a nucleus.

epiblast A small structure opposite the scutellum present in the embryo of some grasses.

epicotyl The young shoot of the embryo or seedling above the cotyledonary node, consisting of an axis and leaf primordia. See also *plumule*.

epidermis The outer layer of cells of the primary body of a plant.

epigeal Type of seed germination in which the cotyledon or cotyledons are raised above the surface of the substrate. Opposite of *hypogeal*.

epigynous Borne on or arising from the ovary; used of floral parts (petals, sepals, and stamens) when the ovary is inferior and the flower is not perigynous.

epipetalous stamen A stamen adnate to the corolla.

epithelium A layer of cells, often secretory in function, covering a free surface or lining a cavity.

epithem Mesophyll of a hydathode, a water-secreting structure of leaves.

ergastic substances Passive products of a protoplast; storage or waste products which may be synthesized within a protoplast or transported from other cells.

eukaryote A cell, or an organism containing cells, characterized by a membrane-bound nucleus, genetic material organized into chromosomes, and membrane-bound cytoplasmic organelles.

eustele A stele, lacking leaf gaps, in which the primary vascular tissue comprises axial vascular bundles and leaf traces arranged around a pith. Characteristic of gymnosperms and angiosperms.

exalbuminous seed A seed without endosperm in the mature state.

exarch primary xylem Xylem which during development is characterized by centripetal maturation of cells; that is, the oldest, mature cells are farthest from the center of the axis. Typical of roots of most vascular plants as well as the stems of many pteridophytes.

exine The outer wall of a spore or a pollen grain.

exocarp The outermost layer or layers of the pericarp.

exodermis The outer layer, one or more cells in depth, of the cortex in some roots. The cells often are characterized by Casparian bands in radial and transverse walls and suberin lamellae covered by a cellulosic layer forming the inner part of the tangential walls.

exogenous Originating in a superficial tissue.

external phloem Primary phloem located externally to, and in contact with, the primary xylem.

extrafloral nectary Nectary occurring on a plant part other than a flower.

extraxylary fibers Fibers in various tissue regions other than the xylem.

false annual ring One of two or more growth layers formed during a growing season; usually the result of severe drought, forest fire, or extreme temperature fluctuations.

fascicle A bundle.

fascicular cambium Vascular cambium that develops within a vascular bundle.

fertilization The fusion of two gametes, especially of their nuclei, resulting in the formation of a diploid ($2n$) zygote.

fiber An elongated, usually tapering, thick-walled cell of the primary and/or secondary xylem. The cell may, but does not always, have a living protoplast at maturity; the cell wall is often lignified.

fiber-tracheid A fiber-like tracheid in the secondary xylem; commonly thick-walled, with pointed ends, and bordered pits that have lenticular to slit-like apertures that extend beyond the pit cavities.

Fibonacci series A mathematical series, 1, 2, 3, 5, 8, 13, 21, etc., in which each value is the sum of the two values that precede it. These fractions are used to characterize phyllotaxy (leaf arrangement). The series is named after Leonardo Fibonacci of Pisa (*c*.1170–*c*.1250) who formulated the relationship.

fibril Submicroscopic thread-like structure.

fibrous root system A root system composed of many (often adventitious) roots of similar length and thickness that emanate from the base of a stem. Characteristic of grasses and other monocotyledons.

filament A fine, thread-like structure; also the stalk supporting the anther in a stamen.

file meristem See *rib meristem*.

filiform Thread-like.

filiform apparatus A complex of slender cell wall invaginations in a *synergid* similar to those in *transfer cells*.

filling tissue Loose tissue formed to the outside by the lenticel phellogen; may or may not have cells with suberized cell walls. Also called *complementary tissue*.

floral tube A tube or cup formed by the united bases of sepals, petals, and stamens, often in perigynous and epigynous flowers.

florigen A hormone that promotes flowering.

flower The reproductive structure of angiosperms.

follicle A dry, dehiscent, many-seeded fruit derived from one carpel and splitting along one suture.

fossil The remains or traces of an organism preserved in sediments of Earth's crust.

free nuclear division Nuclear division that occurs without cell wall formation as in the early stages of development of endosperm in certain taxa.

frond The compound leaf of a fern or cycad; any large, highly divided leaf.

fruit In angiosperms, a mature, ripened ovary or fused cluster of ovaries, and any associated floral parts.

fundamental system See *ground tissue system*.

fundamental tissue The ground tissue in which other tissue systems are embedded.

funiculus The stalk of an ovule.

fusiform initials Cells in the vascular cambium from which, by cell division, the cells of the axial system of the secondary vascular tissues are derived.

gametangium (pl. **gametangia**) A structure in which gametes develop.

gamete A haploid reproductive cell.

gametophyte The haploid, gamete-producing phase in plants that have an alternation of generations.

gelatinous fiber A fiber with little or no lignification and in which part of the secondary wall has a gelatinous appearance.

generative cell The cell of the male gametophyte in gymnosperms which divides to form the stalk and body cells; in angiosperms, the cell of the male gametophyte (pollen grain) which divides to form two sperms (male gametes).

genotype The genetic constitution of an organism; contrasted with *phenotype*.

geotropism The growth response of a shoot or root to gravity; also called *gravitropism*.

germination The resumption of growth by the embryo in a seed; also the beginning of growth of a spore, pollen grain, bud, or other structure.

gibberellins Growth hormones that influence, among other aspects of plant development, the elongation of stems.

gland A multicellular secretory structure.

glandular hair A trichome having a unicellular or multicellular head composed of secretory cells; usually borne on a stalk of non-glandular cells.

gluten Amorphous protein occurring in the starchy endosperm of cereals.

glyoxysome A *microbody* containing enzymes necessary for converting fats into carbohydrates.

Golgi body See *dictyosome*.

graft A union of two plants, a part of one of which, the scion, is inserted into the root or stem of the other, the stock.

grana (sing. **granum**) Stacks of disk-shaped cisternae, the *thylakoids*, which contain chlorophylls and carotenoids; sites of the light reactions in photosynthesis.

granulocrinous secretion Release from the protoplast of a secretion carried in a vesicle by its fusion with the plasmalemma. Compare with *eccrinous secretion*.

gravitropism See *geotropism*.

ground meristem A transitional meristematic tissue, derived from the apical meristem, which gives rise to the ground tissues.

ground tissue system Primary tissues derived from the ground meristem; comprised primarily of parenchyma and collenchyma. Also called the *fundamental system*.

growth layer A layer of secondary xylem or secondary phloem produced by the vascular cambium. See also *annual ring*.

guard cells A pair of cells flanking the stomatal pore and causing, by changes in turgor pressure, the opening or closing of the pore.

guttation Exudation from leaves of water derived from the xylem.

gymnosperm A seed plant, the seeds of which are unenclosed; well-known living taxa include conifers, cycads, and *Ginkgo*.

gynoecium Collective term for the carpels in an angiosperm flower.

half-bordered pit-pair A pit-pair consisting of a bordered and a simple pit.

haplocheilic stomatal complex Structural arrangement in which development of subsidiary cells is perigene, that is, they arise independently of the guard cells. Characteristic of some gymnsperms.

hardwood Term applied to the wood of dicotyledons.

haustorium A specialized projection from a cell or tissue that functions as a penetrating and absorbing structure.

heartwood Non-living, inner layers of secondary xylem that no longer function in transport. The vessels in such wood are typically infiltrated by *tyloses*, and contain waste metabolites, often giving the wood a dark color.

helical cell wall thickening In tracheary elements, secondary wall deposited on the primary or secondary wall as a continuous helix.

hemicellulose A polysaccharide more soluble and less ordered than cellulose; a common component of cell walls.

heterocellular ray A ray in secondary vascular tissues composed of cells of more than one form: in dicotyledons, of procumbent and upright cells; in conifers, of parenchyma cells and ray tracheids.

heterogeneous ray tissue system System of rays in secondary vascular tissues consisting of only heterocellular rays or both homocellular and heterocellular rays.

heterosporous Producing two kinds of spores, *microspores* and *megaspores*.

hilum (1) The central part of a starch grain around which the layers of starch are deposited concentrically. (2) The scar left on a seed by the detached *funiculus*.

histogen Hanstein's term for a meristem in the shoot or root tip that forms a definite tissue system in the plant body. Three histogens were recognized: *dermatogen, periblem,* and *plerome.*

histogen concept Hanstein's concept stating that the three primary tissue systems in the plant body (the epidermis, the cortex, and the vascular system with associated ground tissue) originate from distinct meristems, the histogens.

homocellular ray A ray in secondary vascular tissues composed of cells of one form only: in dicotyledons, of procumbent, or upright cells; in conifers, of parenchyma cells only.

homogeneous ray tissue system System of rays in secondary vascular tissues consisting of only homocellular rays.

homologous Having the same phylogenetic, or evolutionary, origin but not necessarily the same structure and/or function.

hormone A chemical substance produced in one part of an organism and transported to another part in which is has a specific effect.

hyaloplasm See *cytoplasm.*

hydathode A secretory structure, often in the margin of a leaf, by which water is released through an epidermal pore.

hydrolysis The disassembly of large molecules by the addition of water.

hydromorphic Having the structural features of hydrophytes, plants that live in moist or aquatic environments.

hydrophyte A plant that requires a large supply of water and may grow partly or entirely submerged in water.

hypha (pl. *hyphae*) A single tubular filament of a fungus which with others comprise the *mycelium.*

hypocotyl Axial region of an embryo or seedling located between the cotyledon or cotyledons and the radicle.

hypodermis A layer or layers of cells beneath the epidermis distinct from the underlying ground tissue cells.

hypogeal Seed germination in which the cotyledon or cotyledons remain beneath the surface of the ground. Opposite of *epigeal.*

hypogyny Floral condition in which the sepals, petals, and stamens are attached to the receptacle below the ovary.

hypophysis The uppermost cell of the suspensor from which part of the root and root cap in the embryo of angiosperms are derived.

hypsophylls Floral bracts.

idioblast A cell in a tissue that markedly differs in form, size, or contents from other cells in the same tissue.

imperfect flower A flower lacking either stamens or carpels.

indehiscent Of a fruit, remaining closed at maturity, e.g., a *samara.*

indeterminate growth Unrestricted growth in which an apical meristem will remain active during growing seasons over many years.

inferior ovary Condition in which the calyx is completely or partly fused to the ovary, and other floral parts appear to arise from above or upon the ovary, that is, they are *epigynous*.

initial A cell in a meristem that by division gives rise to two cells, one of which remains in the meristem, the other of which is added to the plant body.

integument Outer cell layer enveloping the nucellus of the angiosperm ovule and from which the seed coat differentiates.

integumentary tapetum The deeply staining innermost integumentary epidermis lining the embryo sac in some taxa and apparently assisting in the nutrition of the embryo. Also called *endothelium*.

intercalary growth Growth by cell division some distance from the meristem in which the dividing cells originated as, for example, in a leaf petiole or in the internodes of a stem.

intercalary meristem Meristematic tissue derived from the apical meristem located some distance from it and often intercalated between tissues that are no longer meristematic.

intercellular space A space between two or more cells in a tissue.

interfascicular cambium Vascular cambium that differentiates between vascular bundles in the interfascicular parenchyma.

internal phloem Primary phloem located to the inside of, and in contact with, primary xylem.

internode The part of the stem between two successive nodes.

intervascular pitting The pitting between tracheids or vessel members.

intine The inner wall layer of a pollen grain or spore.

intrusive growth A type of growth in which a growing cell intrudes between other cells that separate from each other along the middle lamella.

intussusception Growth of a cell wall by interpolation of new wall material within previously formed wall.

irregular flower A flower in which one or more members of at least one whorl of the perianth differ in form from other members of the same whorl. An irregular flower cannot be divided in two equal halves in more than one plane.

isobilateral mesophyll A mesophyll in which palisade parenchyma occurs beneath the epidermis on both lower and upper sides of the leaf.

isobilateral leaf A leaf in which the palisade parenchyma occurs beneath both the upper and lower epidermis.

isodiametric Regular in form, with all diameters of the same length.

isogamy Sexual reproduction in which the gametes are identical (or very similar) in size.

isomorphic Identical (or very similar) in form.

isotropic Having the same properties along all axes. Optically isotropic material does not affect the light.

karyokinesis Division of a nucleus as distinguished from the division of the cell, or *cytokinesis*.

kranz anatomy The arrangement of mesophyll cells around a conspicuous bundle sheath forming a wreath-like structure in leaves of C_4 plants.

lacuna (pl. **lacunae**) The air space between cells; also the parenchymatous region in an increment of secondary xylem above and often partially enclosing a leaf trace. Lacuna, in the latter sense, is *not* synonymous with leaf gap. See also *multilacunar node, trilacunar node, two-trace unilacunar node,* and *unilacunar node*.

lacunar collenchyma Collenchyma tissue characterized by intercellular spaces and cell wall thickenings facing the spaces.

lamella A thin plate or layer.

lamellar collenchyma Collenchyma tissue comprised of cells with wall thickenings deposited between the corners of the cells.

lamina The leaf blade; also used to refer to a thin cell wall layer.

late wood The secondary xylem formed in the outer part of a growth layer; denser and composed of smaller cells than the early wood. Also called *summer wood*.

lateral meristem A meristem located parallel to the surface of an axis such as the *vascular cambium* and the *phellogen* or *cork cambium*.

latex (pl. **latices**) A fluid contained in laticifers; consists of a variety of organic and inorganic substances often including rubber.

laticifer A cell or a system of cells containing *latex*.

leaf buttress A lateral protrusion below the apical meristem constituting the initial stage in the development of a leaf primordium.

leaf gap A region of parenchyma in the primary vascular system (*siphonostele* or *dictyostele*) of many ferns opposite a diverging leaf trace.

leaf sheath The lower part of a leaf that invests the stem more or less completely.

leaf trace A vascular bundle which extends into a leaf from its connection with another vascular bundle in the primary vascular system of the stem.

lenticel A specialized region in the periderm containing intercellular spaces which allow an interchange of O_2 and CO_2 between the inner tissues of the plant and the external atmosphere.

leucoplast A colorless plastid; often a site of starch formation.

libriform fiber A xylem fiber commonly with thick walls and highly reduced bordered pits which often appear simple; a major component of secondary xylem in dicotyledons.

lignification Impregnation with lignin.

lignin A component of many cell walls which increases their rigidity and resistance to compression; a polymer of high carbon content derived from phenylpropane.

lithocyst A cell containing a *cystolith*.

locule The cavity within a sporangium containing the spores or pollen grains, or within an ovule containing the ovules.

lumen In a non-living cell, the space enclosed by the cell wall.

lutoids Vesicles, also called vacuoles, in laticifers bound by a single membrane and containing hydrolytic enzymes capable of degrading most of the organic compounds in the laticifer.

lycophytes (Lycophyta) A group of primitive vascular pteridophytes which includes extant taxa such as *Lycopodium* and *Selaginella* and their extinct relatives.

lysigenous Of an intercellular space, originating by a dissolution of cells.

lysis A process of disintegration or degradation.

lysosome An organelle bound by a single membrane and containing acid hydrolytic enzymes capable of breaking down proteins and other organic macromolecules.

maceration The artificial separation of cells of a tissue by causing a disintegration of the middle lamella.

macrofibril An aggregation of cellulose microfibrils, usually visible with the light microscope.

macrosclereid Elongated, rod-like sclereid with unevenly distributed secondary wall thickenings.

marginal meristem A meristem on the margin of a leaf primordium which gives rise to the blade.

matrix A medium in which something is embedded, e.g., the cell wall matrix in which cellulose microfibrils are embedded.

medulla An archaic synonym for *pith*.

medullary bundles Vascular bundles located in the pith.

medullary rays A rarely used synonym for *vascular rays*, referring to the broad regions of secondary parenchyma between elongate masses of secondary tracheary tissues in the stems of some vines and other plants with anomalous structure.

megagametophyte Female gametophyte in heterosporous plants; the embryo sac within the ovule in angiosperms.

megaphyll Usually a large leaf with a complex venation; the leaf trace is associated with a leaf gap. These characters contrast with those of a *microphyll*.

megasporangium A sporangium in which megaspores are produced; the *nucellus* in the ovule of angiosperms.

megaspore A haploid spore which develops into a female gametophyte in heterosporous plants.

megaspore mother cell See *megasporocyte*.

megasporocyte A diploid cell that undergoes meiosis and produces four haploid megaspores. Also called *megaspore mother cell*.

megasporophyll A leaf-like organ which bears a megasporangium.

meristem A region of undifferentiated tissue from which, by cell division, new cells are produced.

mesarch order of maturation Development of the primary xylem in a vascular bundle in which metaxylem differentiates in all directions from centrally located protoxylem.

mesocarp The middle layer of the ovary wall, or pericarp, of a mature fruit.

mesocotyl The internode between the scutellar node and the coleoptile in the embryo and seedling of members of the Poaceae.

mesogene Stomatal development in which neighboring or subsidiary cells and guard cells have a common origin. May also be referred to as *mesogenous*.

mesomorphic Having the structural features of *mesophytes*.

mesoperigene Stomatal development in which at least one subsidiary cell is related by origin to the guard cells, the others are not. May also be referred to as *mesoperigenous*.

mesophyll The photosynthetic parenchyma of a leaf blade located between the two epidermal layers.

mesophytes Plants that require an environment containing moderate levels of soil moisture and a moist atmosphere, in contrast to *xerophytes* which thrive in dry conditions and *hydrophytes* which often live in water or soil that remains very wet.

mestome sheath The inner, thick-walled, endodermis-like sheath of two sheaths that enclose the vascular bundles in the leaves of some grasses.

metaphloem Part of the primary phloem which differentiates after the protophloem and before the secondary phloem, if any of the latter is formed in a given taxon.

metaxylem Part of the primary xylem which differentiates after the protoxylem, and usually after cessation of elongation in associated tissues.

micelle A region in a cellulose microfibril in which the cellulose molecules are arranged in a crystalline lattice structure.

microbody Organelle bound by a single membrane and containing various enzymes. *Peroxisomes* and *glyoxysomes* are microbodies.

microfibril A slender strand of cellulose molecules which, with the matrix, are the major components of the cell wall.

microgametophyte The male gametophyte of a heterosporous plant; the pollen grain in a seed plant.

micrometer A unit of measurement commonly used to measure cellular and subcellular structures. 1/1000 millimeter; 10^{-6} m. Also called a *micron*. The symbol for micrometer is μm.

micron See *micrometer*.

microphyll Small leaf, usually containing a single vascular bundle; not associated with a leaf gap. Characteristic of plants containing protosteles. See also *enation*.

micropyle The opening in the integuments of an ovule of seed plants through which the pollen grains or pollen tubes usually enter.

microsporangium The sporangium in which microspores are formed; the anther locule and its walls in angiosperms.

microspore A haploid spore that develops into a male gametophyte in heterosporous plants.

microspore mother cell See *microsporocyte*.

microsporocyte A diploid cell that undergoes meiosis and forms four haploid microspores. Also called *microspore mother cell*.

microsporophyll Leaf-like organ that bears microsporangia; in angiosperms, modified into the stamen.

microtubules Nonmembranous tubules about 25 nm in diameter and of indefinite length which occur in the cytoplasm of eukaryotes; comprising dimers of alpha tubulin and beta tubulin; associated with the movement of cytoplasmic organelles. They also make up the meiotic and mitotic spindles and the phragmoplast.

middle lamella A layer of intercellular material, chiefly pectic substances, which cements together the primary walls of contiguous cells.

mitochondrion (pl. **mitochondria**) A double membrane-bound cell organelle concerned with respiration in eukaryotic cells; the major source of ATP in nonphotosynthetic cells.

monocotyledon One of the two major groups of angiosperms, differing from dicotyledons in having embryos with only one cotyledon.

monosaccharide A simple sugar, e.g., a 5-carbon or 6-carbon sugar.

morphogenesis The development of form; the sum of the various processes of development and differentiation of tissues and organs.

morphology The study of form and its development.

mucilage cell A cell containing mucilages or gums or similar carbohydrate material characterized by the property of swelling in water.

mucilage duct A duct containing mucilage or gum, or similar carbohydrate material.

multilacunar node A node characterized by five or more leaf traces that supply a single leaf. Typically, associated with each leaf trace, and through which each leaf trace passes, is a parenchymatous region in the first-formed increment of secondary xylem, called a *lacuna*.

multiperforate perforation plate In a vessel member of the xylem, a perforation plate that has more than one perforation.

multiple epidermis A tissue two or more cell layers deep derived from the protoderm; only the outermost layer differentiates as a typical epidermis.

multiple fruit A fruit composed of several matured ovaries each produced in a separate flower as, for example, the pineapple.

multiseriate ray A ray in secondary vascular tissues that is several to many cells wide.

mycelium All of the *hyphae* which, collectively, comprise the body of a fungus.

mycorrhiza (pl. **mycorrhizae**) The symbiotic association of a fungus and the roots of a vascular plant; may be ectotrophic wherein hyphae invest the root of the host or endotrophic wherein hyphae are located within the root cells; mycorrhizae enhance the water and mineral absorption capability of plants.

nacré wall See *nacreous wall*.

nacreous wall A non-lignified wall that is often found in sieve elements and resembles a secondary wall when it attains considerable thickness; the designation is based on the glistening appearance of the wall. Also called *nacré wall*.

nanometer One millionth of a millimeter; 10^{-9} m. A nanometer is equal to 10 angstroms (Å or A). The symbol for nanometer is nm.

nectary A multicellular glandular structure which secretes nectar, a liquid containing organic substances including sugar. Nectaries occur in flowers (*floral nectary*) and on vegetative plant parts (*extrafloral nectary*).

netted venation See *reticulate venation*.

nodal diaphragm A septum of tissue at the node of a stem extending across the otherwise hollow pith region.

node That part of the stem at which one or more leaves are attached; not sharply delimited anatomically.

nodules Englargements on roots of plants, particularly in the Fabaceae, inhabited by nitrogen-fixing bacteria.

non-articulated laticifer A simple laticifer consisting of a single, commonly multinucleate, usually branched, tube which originated from a single cell through cell growth; contains *latex*.

nucellus The inner part of an ovule in which the embryo sac develops. Considered to be equivalent to the *megasporangium*.

nuclear envelope The double membrane enclosing the cell nucleus.

nucleolus A spherical body, composed mainly of RNA and protein, present in the nucleus of eukaryotic cells, one or more to a nucleus; site of synthesis of ribosomes.

nucleus An organelle in a eukaryotic cell bound by a double membrane and containing the chromosomes, nucleoli, and nucleoplasm.

ontogeny The development of an organism, organ, tissue, or cell from inception to maturity.

open venation Leaf venation in which large veins end freely in the mesophyll instead of being connected by anastomoses with other veins.

opposite pitting Pits in tracheary elements disposed in horizontal pairs or in short horizontal rows.

organ A distinct and visibly differentiated part of a plant such as a root, stem, leaf, or part of a flower.

organelle A distinct body within the cytoplasm of a cell, specialized in function.

orthostichy A vertical or nearly vertical line along which is attached a series of leaves or scales on an axis of a shoot or shoot-like organ; may be identical to the steepest parastichy in a phyllotactic system.

osteosclereid A bone-shaped sclereid having a columnar middle part and enlargements at both ends.

ovary The lower part of a carpel (simple pistil) or of a gynoecium composed of united carpels (compound pistil) containing the ovules, and which will develop into a fruit or part of a fruit.

ovule Structure in a seed plant enclosing the female gametophyte and composed of the nucellus, one or two integuments, and a funiculus (ovular stalk); differentiates into the seed.

ovuliferous scale In conifers, the appendage of a cone to which the ovule is attached.

paedomorphosis Delay in evolutionary advance in some characteristics as compared with others resulting in a combination of juvenile and advanced characteristics in the same cell, tissue, or organ.

palisade mesophyll Leaf mesophyll characterized by elongated cells with their long axes perpendicular to the surface of the leaf and in contact with the epidermis; may occur only in the adaxial part of a leaf blade or in both adaxial and abaxial regions (*isobilateral mesophyll*). May function in part in the transmission of light into the inner part of the leaf.

palisade parenchyma The mesophyll immediately beneath the upper epidermis of leaves, usually tubular in form. In leaves with *isobilateral mesophyll*, palisade parenchyma occurs in both upper and lower regions.

paracytic stoma A stomatal complex in which one or more subsidiary cells flank the stoma parallel to the long axes of the guard cells.

paradermal Parallel to the epidermis. Refers specifically to a section made parallel to the surface of a flat organ such as a leaf.

parallel evolution Evolutionary process that results in similar structures and functions in two or more evolutionary lines. See also *convergent evolution*.

parallel venation Arrangement in which main veins in a leaf blade are approximately parallel to one another, although converging at the base and apex of the leaf. Also called *striate venation*.

parastichy A helix along which is attached a series of leaf primordia in a shoot apex, or leaves or scales along the axis of a mature shoot. Compare *orthostichy*.

paratracheal parenchyma Axial parenchyma in secondary xylem associated with vessels. Includes *aliform, confluent*, and *vasicentric* patterns.

paraveinal mesophyll Mesophyll in a leaf blade that forms a sheet of tissue parallel to the venation system, often in the same plane as the veins.

parenchyma A tissue composed of parenchyma cells.

parenchyma cell A living cell in which various physiological and biochemical processes occur, usually thin-walled and of variable size and form. Parenchyma cells comprise the ground tissue of plant organs.

parthenocarpy The development of fruit without fertilization; the fruits are usually seedless.

passage cell A cell in an endodermis that remains thin-walled when other cells in the same tissue region develop thick secondary walls. *Casparian bands* are deposited within the transverse and radial walls of the passage cell and contiguous cells.

pectic substances A group of complex carbohydrates, derivatives of polygalacturonic acid, which occur in plant cell walls; major constituents of the *middle lamella*.

pedicel The stalk of an individual flower in an inflorescence.

peduncle The stalk of a solitary flower or of an inflorescence.

peltate trichome A trichome consisting of a discoid plate of cells borne on a stalk or attached directly to the basal foot cell.

perennial A plant that continues to produce reproductive structures for three or more years.

perfect flower A flower having both carpels and stamens.

perforation plate Part of a wall of a vessel member that is perforated.

perianth Petals and sepals of a flower considered together.

periblem The meristem which gives rise to the cortex. One of the three histogens, *plerome*, *periblem*, and *dermatogen*, according to Hanstein.

pericarp The wall of a fruit which develops from the ovary wall.

periclinal Having the orientation of cell wall or the plane of cell division parallel with the nearest surface of an organ. Opposite of *anticlinal*.

pericycle The tissue region located between the primary vascular tissues and the endodermis.

periderm Secondary protective tissue that replaces the epidermis in stems and roots, rarely in other organs; consists of *phellem* (cork), *phellogen* (cork cambium), and *phelloderm*.

perigene Stomatal development in which neighboring or subsidiary cells and guard cells have separate origins. May also be referred to as *perigenous*.

perigyny The arrangement of floral parts in which the sepals, petals, and stamens are attached to a cup-shaped extension of the receptacle, but appear to be attached to the ovary.

perisperm Storage tissue in a seed similar to endosperm but derived from the nucellus.

peroxisome A cell organelle of the *microbody* type that is involved in glycolic acid metabolism associated with photosynthesis.

petal A flower part, often colored, which with other petals comprises the corolla.

petiole The stalk of a leaf.

petiolule The stalk of a leaflet.

phellem (cork) An outer protective tissue of roots and stems composed of nonliving cells with suberized walls and formed centrifugally by the phellogen (cork cambium) as part of the periderm.

phelloderm A tissue resembling cortical parenchyma produced centripetally by the phellogen (cork cambium) as part of the periderm of stems and roots in seed plants.

phellogen (cork cambium) A lateral meristem which gives rise to the periderm which consists of phellem and phelloderm.

phelloid cell A cell within the phellem but distinct from other cells in this tissue in lacking suberin in its walls.

phenotype The physical appearance of an organism which results from interaction between its *genotype* (genetic constitution) and the environment.

phloem The food-conducting tissue of vascular plants which is composed of sieve elements, various kinds of parenchyma cells, fibers, and sclereids.

phloem loading The process by which photosynthate is transferred into sieve cells or sieve tube members.

photoperiodism Response to the duration and timing of day and night expressed in certain aspects of growth, development, and flowering in plants.

photorespiration The light-dependent production of glycolic acid in chloroplasts and its subsequent oxidation in peroxisomes.

photosynthates The carbohydrates produced during the process of photosynthesis.

photosynthesis The process by which carbohydrates are formed from carbon dioxide and water in the presence of chlorophyll, using light energy.

phragmoplast A subcellular structure composed of microtubules that arises between daughter nuclei at telophase and within which the cell plate forms during cell division; appears initially as spindle-shaped, but later spreads laterally in the form of a ring.

phragmosome A layer of cytoplasm formed across the cell where the nucleus becomes located and divides. The equatorial plane of the phragmoplast coincides with the plane of the cytoplasmic layer (phragmosome).

phyllode A flat, expanded petiole or stem replacing the leaf blade in photosynthetic function.

phyllotaxy (or **phyllotaxis**) The arrangement of leaves on a stem; the mathematical principles of such arrangement. See also *Fibonacci series*.

phylogeny Evolutionary history of a species or large taxon.

phytochrome A pigment occurring in the cytoplasm and serving as a photoreceptor for red and far-red light; involved in timing certain processes such as dormancy, leaf formation, flowering, and seed germination.

pinocytosis A process of uptake of a substance by invagination of the plasmalemma.

pistil The ovary, style, and stigma of a flower. May be simple, consisting of a single carpel, or compound, consisting of several fused carpels.

pistillate Of a flower, having one or more carpels but no functional stamens.

pit A depression in the cell wall where the primary wall is not covered by secondary wall.

pit aperture An opening into the pit from the interior of the cell. If a *pit canal* is present in a bordered pit, two apertures are recognized, the inner, from the cell lumen into the canal, and the outer, from the canal into the pit cavity.

pit canal The passage from the cell lumen to the chamber or cavity of a bordered pit. See also *pit aperture*.

pit cavity The space within a pit from pit membrane to the cell lumen or to the outer pit aperture if a *pit canal* is present.

pit membrane The part of the intercellular layer (middle lamella) and primary walls that separate the pit cavities of a pit-pair.

pit-pair Two opposite pits in the walls of contiguous cells plus the pit membrane.

pith Ground tissue in the center of a stem or root.

placenta The part of the ovary walls to which ovules or seeds are attached.

placentation The arrangement of ovules in an ovary.

plasma membrane See *plasmalemma*.

plasmalemma Single membrane enclosing the protoplast. A type of unit membrane. Also called *plasma membrane* or *ectoplast*.

plasmodesma (pl. **plasmodesmata**) Highly specialized regions of endoplasmic reticulum that extend through cell walls and connect the protoplasts of adjacent cells.

plastid An organelle with a double membrane in the cytoplasm of many eukaryotes. May be the site of photosynthesis (chloroplast) or starch storage (amyloplast), or contain yellow or orange pigments (chromoplast).

plastochron The time interval between the inception of two successive repetitive events, as the origin of leaf primordia.

plate meristem A meristmatic tissue consisting of parallel layers of cells which divide only anticlinally with reference to the wide surface of the tissue. Characteristic of ground meristem of plant parts that assume a flat form, as a leaf.

plerome The meristem forming the core of the axis composed of the primary vascular tissues and associated ground tissue, such as pith and interfascicular regions. One of the three histogens, *plerome*, *periblem*, and *dermatogen*, according to Hanstein.

plumule Embryonic shoot above the cotyledon or cotyledons in an embryo. See also *epicotyl*.

polar nuclei The two centrally located nuclei in an embryo sac which fuse with a male nucleus forming the triploid endosperm nucleus.

pollen A collective term for pollen grains.

pollen grain In seed plants a microspore included in an elaborately structured wall and which contains an immature or mature *microgametophyte*.

pollen mother cell See *microsporocyte*.

pollen sac A locule in an anther containing the pollen grains.

pollen tube A tubular cell extension formed by the germinating pollen grain; carries the male gametes into the ovule.

pollination In angiosperms, the transfer of pollen from the anther to the stigma; in gymnosperms, the transfer of pollen from the male cones to the immature ovules.

polyarch Primary xylem column of the root characterized by many protoxylem strands.

polyderm A type of protective tissue in which layers of parenchyma cells with suberized walls alternate with layers of parenchyma cells having non-suberized walls.

polyembryony Development of more than one embryo in a single seed.

polymer A large molecule composed of many similar molecular subunits.

polymerization Chemical union of monomers, such as glucose or nucleotides, resulting in the formation of polymers such as starch, cellulose, or nucleic acid.

polyribosomes Aggregates of ribosomes involved in protein synthesis.

polysaccharide A carbohydrate composed of many monosaccharide units joined in a chain, such as starch or cellulose.

pore A term of convenience for a vessel member as seen in transverse section of secondary xylem.

pore multiple (or **cluster**) A group of two or more pores (vessels in cross-sections of secondary xylem) crowded together and usually flattened along the surfaces of contact.

porous wood Secondary xylem containing vessels.

P-protein Phloem protein; found in cells of the phloem of seed plants, most commonly in sieve elements. Formerly called slime.

preprophase band A ring-like band of microtubules delimiting the equatorial plane of the future mitotic spindle in cells preparing to divide.

primary body The part of the plant, or the entire plant if no secondary growth occurs, that arises from the embryo and the apical meristems and their derivative meristematic tissues.

primary growth Growth resulting from activity of apical meristems of roots and shoots and their derivative meristems.

primary peripheral thickening meristem The meristem in many large monocotyledons, proximal to the apical meristem, responsible for increase in thickness of the stem.

primary phloem Phloem tissue derived from the provascular tissue during primary growth and differentiation of a vascular plant. *Protophloem* differentiates first, followed by *metaphloem*.

primary pit field A thin region in a primary wall traversed by plasmodesmata. Usually the site of development of pits in the secondary wall.

primary root The root that develops from the radicle in the embryo. If development of this root continues it becomes a tap root.

primary tissues Tissues derived from the apical meristem and the transitional tissue regions, protoderm, ground meristem, and provascular tissue; contrasting with secondary tissues derived from lateral meristems such as the vascular cambium and the phellogen.

primary vascular tissues The xylem and phloem that differentiate from provascular tissue during primary growth and differentiation of a vascular plant.

primary wall The wall layer that develops during cell growth as opposed to the secondary wall that develops after cell expansion has ceased.

primary xylem Xylem tissue that differentiates from provascular tissue during primary growth and differentiation of a vascular plant. *Protoxylem* differentiates first, followed by *metaxylem*.

primordium (pl. **primordia**) A cell or organ in an early stage of differentiation, e.g., a sclereid primordium, or a leaf primordium.

procambium See *provascular tissue*.

procumbent ray cell In secondary vascular tissues, a ray cell having its longest axis oriented radially.

proembryo An early development stage of an embryo before the main body of the embryo and the suspensor become distinct.

prokaryotes Single cells, or organisms consisting of cells lacking membrane-bound nuclei and membrane-bound organelles, e.g., bacteria, blue-green algae, mycoplasmas. See *eukaryote*.

promeristem The initiating cells and their most recent derivatives in an apical meristem.

prop roots Adventitious roots that develop on the stem above the soil level and which serve as additional support for the plant axis.

prophyll The first leaf, or one of two first leaves, on a lateral shoot.

proplastid A plastid in its earliest stages of development.

protective layer In the abscission zone, a layer of cells that, because of substances impregnating its walls, has a protective function in the scar formed by abscission of a leaf or other plant part.

protein bodies See *aleurone grains*.

prothallial cell Sterile (vegetative) cell or cells in the microspores of gymnosperms and the microgametophytes of other non-angiospermous vascular plants.

protoderm A single-layered meristematic tissue region that gives rise to the epidermis.

protophloem The first-formed cells of the primary phloem in a plant organ.

protoplasm Living substance of cells, excluding their organelles.

protoplast The organized living unit of a single cell excluding the cell wall.

protostele The simplest type of stele, containing a solid column of vascular tissue, with the phloem peripheral to the xylem.

protoxylem The first formed cells of the primary xylem in a plant organ.

protoxylem lacuna A space surrounded by parenchyma cells in the protoxylem of a vascular bundle; often develops when protoxylem elements are stretched and torn resulting from elongation growth.

protoxylem poles Term of convenience for loci of protoxylem strands as seen in transverse section.

provascular tissue The transitional tissue region, partly meristematic, from which the primary vascular tissue differentiates.

proximal Situated near the point of origin or attachment. Opposite of *distal*. Often used in plant anatomy to mean in a direction away from the apical meristem, i.e., toward the base of the plant.

pteridophyte Any plant that lacks seeds and that is characterized by free-sporing reproduction, such as ferns and other relatively primitive taxa.

pulvinus An enlargement at the base of the petiole of a leaf, or petiolule of a leaflet; has a role in the movements of a leaf or leaflet.

quantasomes Granules located on the inner surface of chloroplast lamellae; thought to be involved in the reactions in photosynthesis requiring light.

quiescent center A region in the apical meristem of roots which is relatively inactive in cell division; considered to be a region that has some control in root development.

rachis The petiole of a fern frond, or main axis of a compound leaf.

radial section A longitudinal section cut along a radius of a stem or root; in secondary xylem or phloem, a section cut parallel to a vascular ray.

radial seriation Arrangement of units, such as cells, in an orderly sequence in a radial direction; characteristic of cambial derivatives.

radial symmetry Of a flower, having parts that can be divided equally in more than one longitudinal plane passing through the floral axis. Contrasted with *bilateral symmetry*.

radicle The embryonic root.

ramified Branched.

ramiform pit A pit with a *pit canal* that branches.

raphe A ridge along the body of the seed formed by the part of the funiculus that is adnate to the ovule.

raphides Needle-shaped crystals which commonly occur in bundles.

ray A sheet of tissue, variable in height and width, formed by a ray initial in the vascular cambium and which extends radially in the secondary xylem and secondary phloem.

ray cell initials Meristematic cells in a ray initial from which ray cells are derived.

ray initial A cluster of meristematic cells in the vascular cambium that gives rise to a ray.

ray tracheid Non-living, radially oriented cells with walls containing circular-bordered pits that occur in the rays of the secondary xylem of several conifer and angiosperm families as well as in the extinct progymnosperm *Archaeopteris*.

reaction wood Wood with abnormal anatomical characteristics formed in parts of leaning or crooked stems and on the lower (conifers) or upper (dicotyledons) sides of branches. See also *compression wood* and *tension wood*.

receptacle The part of the flower stalk that bears the floral organs.

regular flower See *actinomorphic*; *radial symmetry*.

residual meristem A meristematic region in the shoot apex, below the apical meristem in some taxa, in which provascular strands develop.

resin duct A duct of schizogenous origin lined with resin-secreting cells (epithelial cells) and containing resin.

reticulate cell wall thickening In tracheary elements of the xylem, secondary cell wall deposits on the primary wall in an anastomosing or net-like pattern.

reticulate perforation plate In vessel members of the xylem, a multiperforate perforation plate in which the bars delimiting the perforations form a reticulum.

reticulate sieve plate A compound sieve plate with sieve areas arranged so that the wall between forms a net-like pattern.

reticulate venation A pattern of venation in a leaf blade in which the veins form an anastomosing system, the whole resembling a net. Also called *netted venation*.

retting Freeing fiber bundles from other tissues by utilizing the action of microorganisms causing, in a suitable moist environment, the disintegration of the thin-walled cells surrounding the fibers.

rhexigenous As applied to an intercellular space, originating by rupture of cells.

rhizodermis The epidermis of roots.

rhytidome The part of the bark consisting of intersecting regions of internal periderm between which are regions of cortex and/or non-functional secondary phloem.

rib meristem A meristem in which the cells divide perpendicular to the longitudinal axis of an organ and produce a complex of parallel, vertical files of cells; particularly common in ground meristem of organs that are of cylindrical form. Also called *file meristem*.

ribonucleic acid Nucleic acid formed on chromosomal DNA and involved in protein synthesis. Commonly abbreviated as RNA.

ribosome A cell component composed of protein and RNA; the site of protein synthesis.

ring porous wood Secondary xylem in which the vessels (pores) of the early wood are distinctly larger than those of the late wood, forming a well-defined zone or ring as seen in transverse section.

RNA See *ribonucleic acid.*

root cap A sheath of cells covering the apical meristem of the root.

root hair A tubular trichome on the root epidermis that is an extension of a single epidermal cell; occurs in the zone of maturation and facilitates absorption.

samara Simple, dry, one- or two-seeded indehiscent fruit; characterized by wing-like outgrowths of the pericarp.

sapwood Outer part of the wood of the stem (or trunk) in which active conduction takes place; usually lighter in color than the heartwood.

scalariform cell wall thickening In tracheary elements, secondary wall depositions on the primary wall in a ladder-like pattern. Similar, in some taxa, to a helix of low pitch with the coils interconnected at intervals.

scalariform perforation plate In a vessel member, a type of multiperforate end wall in which elongated perforations are arranged more or less parallel to one another so that the cell wall bars between them form a ladder-like pattern.

scalariform pitting In tracheary elements, elongated pits arranged parallel to one another so as to form a ladder-like pattern.

scalariform–reticulate cell wall thickening In tracheary elements, secondary wall thickening intermediate between scalariform and reticulate wall thickening.

scalariform sieve plate A compound sieve plate with elongated sieve areas arranged parallel to one another in a ladder-like pattern.

schizogenous Of an intercellular space, originating by separation of cell walls along the middle lamella.

schizo-lysigenous Of an intercellular space, originating by a combination of two processes, separation and degradation of cell walls.

scion See *graft.*

sclereids Sclerenchyma cells, varied in form, but (with a few exceptions) not very elongate, and having thick, lignified secondary walls with many simple pits.

sclerenchyma A tissue composed of sclerenchyma cells, also a collective term for sclerenchyma cells in the primary plant body; includes fibers, fiber-sclereids, and sclereids.

sclerenchyma cell A cell of variable form and size and having more or less thick, often lignified, secondary walls; belongs to the category of supporting cells and may or may not be devoid of a protoplast at maturity.

sclerification The process of becoming changed into sclerenchyma, i.e., the development of secondary walls, with or without subsequent lignification.

scutellum (pl. **scutella**) The single cotyledon in a grass embryo, specialized for the absorption of the endosperm.

secondary body The part of the plant body that is added to the primary body by the activity of the lateral meristems, vascular cambium, and phellogen; consists of secondary vascular tissues and periderm.

secondary cell wall The inner layer of the wall deposited upon the primary wall after cell growth (increase in size) has ceased.

secondary growth In gymnosperms, most dicotyledons, and some monocotyledons, a type of growth characterized by an increase in thickness of the stem and root, and resulting from formation of secondary tissues by the vascular cambium. Commonly supplemented by activity of the cork cambium (*phellogen*) forming periderm.

secondary phloem Phloem tissue formed by the vascular cambium during secondary growth in a vascular plant.

secondary plant body See *secondary body*.

secondary tissues Tissues produced by the vascular cambium and phellogen during secondary growth.

secondary vascular tissues Vascular tissues (both xylem and phloem) formed by the vascular cambium during secondary growth in a vascular plant.

secondary wall See *secondary cell wall*.

secondary xylem Xylem tissue formed by the vascular cambium during secondary growth in a vascular plant.

secretory cavity A space lysigenous in origin and containing a secretion derived from the cells that broke down in the formation of the cavity.

secretory cell A living cell specialized with regard to secretion of one or more, often organic, substances.

secretory duct A duct schizogenous in origin and containing a secretion derived from the cells (epithelial cells) lining the duct. See *epithelium*.

secretory hair See *glandular hair*.

secretory structure Any of a great variety of structures, simple or complex, external or internal, that produce a secretion.

seed coat The outer coat of the seed derived from the integument or integuments of the ovule. Also called *testa*.

sepal A unit of the calyx.

separation layer See *abscission layer*.

septate fiber A fiber with thin transverse walls (septa) that are formed after the cell develops a secondary wall.

septum (pl. **septa**) A partition.

sessile Of a leaf, lacking a petiole; of a flower or a fruit, lacking a pedicel.

sexine The outer layer of the exine of a pollen grain; the sculptured part of the exine.

sheath A sheet-like structure enclosing or encircling another; applied to the tubular or enrolled part of an organ, such as a leaf sheath, and to a tissue layer surrounding a complex of other tissues, as a bundle sheath enclosing a vascular bundle.

sheathing base A leaf base that encircles the stem.

sieve area A specialized region of a sieve element containing pores commonly lined with callose and through which traverse protoplasmic strands that connect the protoplasts of contiguous sieve elements.

sieve cell A sieve element that has sieve areas, usually with sieve pores of small diameter, on all walls; there are no end wall sieve plates. Typical of gymnosperms and seedless vascular plants.

sieve elements Cells in the phloem tissue concerned primarily with conduction of photosynthate and hormones; *sieve cells* and *sieve tube members*.

sieve plate The part of the cell wall (usually the end walls) of a sieve element bearing one or more highly differentiated sieve areas. Typical of angiosperms.

sieve tube A series of sieve tube members arranged end to end and interconnected through sieve plates.

sieve tube element See *sieve tube member*.

sieve tube member One of a series of cell components of a sieve tube. Characterized by sieve plates on the end walls and less highly differentiated lateral sieve areas. Also called *sieve tube element*.

silica cell Cell filled with silica, as in the epidermis of many grasses.

simple perforation plate A perforation plate of a vessel member with a single perforation.

simple pit A pit in a secondary wall which lacks an overhanging border.

simple pit-pair A pair of opposing simple pits in the walls of contiguous cells.

simple sieve plate A sieve plate composed of one sieve area.

siphonostele A stele in which the vascular system is in the form of a cylinder enclosing the pith.

slime Archaic term for *P-protein*.

slime body An aggregation of *P-protein*.

slime plug An accumulation of *P-protein* on a sieve area, usually with extensions into the sieve pores.

softwood The wood of conifers.

solitary pore A vessel, as seen in transverse section, in the secondary xylem surrounded by cells other than vessel members.

specialized Of organisms, having special adaptations to a particular habitat or mode of life; of cells or tissues, having distinctive functions.

spenophytes (Spenophyta) A group of primitive vascular pteridophytes which includes the extant taxon *Equisetum* and its extinct relatives.

sperm The male gamete.

spherosomes Single membrane-bound bodies in the cytoplasm containing lipids.

spindle fibers Microtubules aggregated in a spindle-shaped complex extending from pole to pole in cells with a dividing nucleus.

spongy parenchyma Leaf mesophyll parenchyma with conspicuous intercellular spaces and containing chloroplasts.

sporangium (pl. **sporangia**) A hollow structure (unicellular or multicellular) in which spores develop.

spore A reproductive cell, resulting from meiosis, from which, through mitoses, a haploid (1*n*) *gametophyte* develops.

sporophyll A leaf or leaf-like organ that bears sporangia.

sporophyte The diploid (2*n*) phase which produces spores in a life cycle characterized by an *alternation of generations*.

sporopollenin The substance composing the outer wall, or exine, of pollen grains and spores; a cyclic alcohol highly resistant to decay.

spring wood See *early wood*.

stamen Floral organ which produces the pollen and is usually composed of anther and filament. Collectively the stamens constitute the *androecium*.

staminate Of a flower, having stamens but lacking functional carpels.

starch An insoluble carbohydrate, the chief food storage substance of plants, composed of numerous glucose units.

stele A morphologic unit of the plant axis (stems and roots) comprising the primary vascular system and associated ground tissue (pericyle, interfascicular regions, and, in some concepts, the pith).

stellate Star-shaped.

stigma The region of the carpel, usually at the apex of the style, that serves as a surface upon which the pollen germinates.

stock See *graft*.

stoma (pl. **stomata**) An opening in the epidermis of leaves and stems bordered by two guard cells and functioning in gas exchange. Also called *stomate*.

stomatal apparatus See *stomatal complex*.

stomatal complex Stoma and associated epidermal cells (*subsidiary cells*) that may be ontogenetically and/or physiologically related to the guard cells. Also called *stomatal apparatus*.

stomatal crypt A depression in a leaf, the epidermis of which bears one or more stomata.

stomate See *stoma*.

stomium A fissure or pore in an anther lobe through which pollen is released. Its formation is a type of dehiscence.

stone cell See *brachysclereid*.

storied cambium Vascular cambium in which the fusiform and ray initials are arranged in horizontal rows as seen in tangential sections. Also called *stratified cambium*.

storied wood Wood in which the axial cells and rays are arranged in horizontal rows as seen in tangential section. In some taxa, the rays alone may be storied. Also called *stratified wood*.

Strasburger cells Marginal ray cells in the secondary phloem of conifers that have a function similar to that of companion cells in angiosperms, but which typically are not related developmentally to the sieve cells.

stratified cambium See *storied cambium*.

stratified wood See *storied wood*.

striate venation See *parallel venation*.

strobilus An axis bearing numerous sporophylls or ovule-bearing structures (scale-like in conifers); characteristic of gymnosperms, lycophytes, and sphenophytes.

stroma The ground substance of plastids.

style An extension of the ovary, usually columnar, through which the pollen tube grows.

suberin Fatty substance in the cell walls of cork (phellem) cells and in the Casparian band of endodermal and exodermal cells.

suberin lamella A thin layer of suberin deposited on the primary wall in cells of an endodermis or exodermis; may subsequently be covered by a cellulosic layer (secondary wall).

suberization Impregnation of the cell wall with suberin or deposition of a suberin lamella on a cell wall.

submarginal initials Meristematic cells beneath the protoderm along the margins of a developing leaf lamina that contribute cells to the interior tissue of the leaf.

subsidiary cell An epidermal cell associated with a stoma, sometimes developmentally related to the guard cells, and usually morphologically distinguishable from epidermal cells composing the groundmass of the tissue.

summer wood See *late wood*.

superior ovary See *hypogyny*.

suspensor An extension at the base of the embryo that anchors the embryo in the embryo sac and pushes it into the endosperm.

symbiosis A living in close association of two (or more) dissimilar organisms; included are parasitism in which the relationship is harmful to one of the organisms, and mutualism in which the relationship is beneficial to both.

symplast The living protoplasts of all cells in an organism, or a region of an organism, and the plasmodesmata by which they are connected.

symplastic growth Growth in which cells in a developing tissue grow in a coordinated manner and in which there is no intrusion of some cells between others or slippage between contiguous cells.

sympodium A vascular bundle of the stem and the associated leaf traces.

syncarpy A condition in a flower characterized by a fusion of carpels.

syndetocheilic Stomatal type of gymnosperms in which the subsidiary cells or their precursors are derived from the same protodermal cell as the guard cell mother cell.

synergids Two cells in the micropylar end of the embryo sac associated with the egg cell in angiosperms.

syngamy The process by which two haploid cells (gametes) fuse forming a zygote; fertilization.

tangential section A longitudinal section cut at right angles to a radius of a cylindrical structure such as a stem or root. A tangential section of secondary wood or secondary phloem is cut at right angles to the rays.

tannins A heterogeneous group of phenol derivatives; amorphous, strongly astringent substances widely distributed in plants, and used in tanning, dyeing, and preparation of ink.

tapetum A layer of cells in an anther lining the locule and absorbed as the pollen grains mature. In the ovule, an integumentary epidermis next to the embryo sac. Also called *endothelium*.

tap root The first or primary root of a plant; a continuation of the radicle of the embryo.

tap root system A root system based on the tap root which may have branches of various orders.

taxon (pl. **taxa**) Any one of the categories (species, genus, family, etc.) into which living organisms are classified.

telome One of the distal branches of a dichotomously branched axis in a primitive vascular plant.

telome theory A theory that regards the telomes as basic units from which the diverse types of leaves and sporophylls of a vascular plant have evolved.

template A pattern or mold guiding the formation of a negative or a complement. A term applied in biology to DNA duplication.

tension wood Reaction wood in dicotyledons, formed on the upper sides of branches and leaning or crooked stems; characterized by lack of lignification and often by high content of gelatinous fibers. See also *compression wood* and *reaction wood*.

tepal A member of a floral perianth that is not differentiated into calyx and corolla.

terminal apotracheal parenchyma See *boundary apotracheal parenchyma*.

testa The seed coat.

tetrarch The primary xylem of the root with four protoxylem strands (or poles).

thylakoids Sac-like membranous structures (cisternae) in a chloroplast combined into stacks (grana) and present singly in the stroma as interconnections between grana.

tissue A group of cells organized into a structural and functional unit. Component cells may be alike (simple tissue) or diverse (complex tissue).

tissue system A tissue or tissues in a plant or plant organ structurally and functionally organized into a unit. Commonly three tissue systems are recognized: *dermal, vascular,* and *fundamental* (ground tissue system).

tonoplast The cytoplasmic membrane (a unit membrane) bounding the vacuole. Also called the *vacuolar membrane.*

torus (pl. **tori**) The central, thickened part of the pit membrane in a bordered pit; consists of middle lamella and two primary walls. Typical of bordered pits in conifers and some other gymnosperms.

trabecula (pl. **trabeculae**) A rod–like extension of cell wall material across the lumen of a cell.

tracheary element A water-conducting cell, tracheid, or vessel member.

tracheid An elongate tracheary element of the xylem with tapered or rounded ends, and having no perforations, as contrasted with a vessel member; may occur in primary and in secondary xylem.

transection See *transverse section.*

transfer cell A parenchyma cell with wall ingrowths that increase the surface of the plasmalemma which lines the wall surface. Specialized for short-distance, apoplastic transfer of solutes.

transfusion tissue In gymnosperm leaves, a tissue surrounding or otherwise associated with the vascular bundle (or bundles), and composed of short tracheid-like cells and parenchyma.

transition region A region in the plant axis where root and shoot are united and which shows primary structural characteristics transitional between those of stem and root.

transitional tissue regions Term applied to regions of tissue between the apical meristem and mature tissues such as protoderm, ground meristem, and provascular tissues, tissue regions in which cell division continues in the more distal regions and growth and differentiation take place more proximally.

transmitting tissue The tissue in the style of a flower through which the pollen tube grows.

transverse section A section cut perpendicular to the longitudinal axis of a cell or plant part. Also called a cross-section or *transection.*

traumatic resin duct A resin duct that develops in response to injury.

triarch Primary xylem of a root with three protoxylem strands (or poles).

trichoblast A cell in the root epidermis that gives rise to a root hair.

trichome An outgrowth from the epidermis. Trichomes vary in size and complexity and include hairs, scales, and other structures; may be glandular.

trichosclereid A type of branched sclereid, usually with hair-like branches extending into intercellular spaces.

trilacunar node A node characterized by three leaf traces that supply a single leaf. Typically associated with each leaf trace, and through which each leaf trace passes, is a parenchymatous region in the first-formed increment of secondary xylem, called a *lacuna.*

triple fusion In most angiosperms, the fusion of one of the two sperm nuclei with the two polar nuclei forming the triploid ($3n$) primary endosperm nucleus.

tropism Movement or growth in response to an external stimulus; the direction of the movement or growth is determined by the direction from which the stimulus comes.

tube cell The cell in a pollen grain of some gymnosperms from which the pollen tube develops.

tunica Peripheral layer or layers of an apical meristem of a shoot, the cells of which divide anticlinally and thus contribute to the growth in surface of the meristem. Forms a mantle over the *corpus*.

tunica–corpus concept A concept of the organization of the apical meristem of the shoot under which it is differentiated into two regions distinguished by their methods of growth: the peripheral tunica, one or more layers of cells showing surface growth (anticlinal divisions); the interior, corpus, a mass of cells showing growth in volume (divisions in various planes).

two-trace unilacunar node A node characterized by two traces that supply a single leaf. Typically the pair of traces traverses a single *lacuna*.

tylose (pl. **tyloses**) An outgrowth from a parenchyma cell of the xylem (axial parenchyma or a ray cell) which extends through a pit-pair into the cavity of a tracheid or vessel member partially blocking, or with other tyloses, completely blocking, the lumen of the tracheary element.

undifferentiated In ontogeny, still in a meristematic state or resembling meristematic tissues or structures; In a mature state, relatively unspecialized.

unifacial leaf A leaf having a similar structure on both sides. Conceived ontogenetically, a leaf that develops from a growth center abaxial or adaxial to the initial leaf primordium apex and which thus includes tissues only from the abaxial or adaxial side of the primordium. The validity of the ontogenetic concept is questionable. Compare with *bifacial leaf*.

unilacunar node A node characterized by one leaf trace that supplies a leaf. Typically associated with the leaf trace, and through which it passes, is a parenchymatous region in the first-formed increment of secondary vascular tissues called a *lacuna*.

uniseriate ray In secondary vascular tissues, a ray one cell wide.

unisexual Of a flower, lacking either stamens or carpels; a perianth may be present or absent.

unit membrane A three-layered membrane consisting of a light layer of lipid between two dark layers of protein.

upright ray cell In secondary vascular tissues, a ray cell, the long axis of which is parallel to the axis of the stem or root.

vacuolar membrane See *tonoplast*.

vacuolation The development of *vacuoles* in a cell.

vacuole Cavity within the cytoplasm filled with a watery fluid, the cell sap, and bound by a unit membrane, the tonoplast; a component of the lysosomal compartment of the cell.

vacuome Collective term for all of the vacuoles in a cell, tissue, or plant.

vascular Consisting of or giving rise to conducting tissues, e.g., xylem, phloem, vascular cambium.

vascular bundle A strand of vascular tissue (usually primary xylem and phloem).

vascular cambium A lateral meristem from which secondary xylem and secondary phloem are produced in the stem and root. Periclinal divisions in cambial initials produce cells, some of which differentiate into phloem cells, others of which differentiate into xylem cells.

vascular rays Radially arranged sheets of (usually) parenchyma cells that extend through the secondary xylem and the secondary phloem, and are produced by the vascular cambium.

vascular system All of the vascular tissues in an organ or a plant.

vascular tissue A general term referring to either or both vascular tissues, xylem and phloem.

vasicentric paratracheal parenchyma Axial parenchyma in the secondary xylem which forms sheaths around vessels. See also *paratracheal parenchyma*.

vein A vascular bundle that comprises part of the vascular system of a leaf.

velamen A multiple epidermis covering the aerial roots of some tropical epiphytic orchids and aroids; also occurs in some terrestrial roots.

venation The arrangement of veins in the leaf blade.

vessel A superposed series of vessel members forming a tube.

vessel element See *vessel member*.

vessel member A non-living, conducting cell of the xylem characterized by perforations in the contiguous end walls of superposed cells that form a vessel. Vessel members function in the transport of water and minerals through the primary and secondary xylem of angiosperms.

vestured pit A bordered pit with wall projections from the inner surfaces of the overhanging borders.

wall See *cell wall*.

water vesicle An enlarged, highly vacuolate epidermal cell; a *trichome* in which water is stored.

whorl Leaves or flower parts arranged in a circle on an axis.

wood Secondary xylem.

wound cork See *wound periderm*.

wound periderm Periderm formed in response to injury. Also called *wound cork*.

xeromorphic Having the structural features typical of xerophytes.

xerophyte A plant adapted to a dry habitat.

xylem A complex tissue of parenchyma and tracheary elements that functions in the longitudinal transport of water and minerals. In secondary xylem, rays function in radial transport. The xylem is also a supporting tissue.

xylem elements Cells of the xylem tissue.

xylem initial A cambial cell on the xylem side of the cambial zone that is the source of one or more cells arising by periclinal divisions and differentiating into xylem elements either with or without additional divisions in various planes.

xylem ray The part of a vascular ray that is located in the secondary xylem.

zygomorphic Having an irregular flower form. Opposite of *actinomorphic*. See *bilateral symmetry*.

zygote The diploid ($2n$) cell that results from the fusion of male and female gametes.

Index

Page numbers in italics, indicate references to figures. Page numbers in bold, denote entries in the glossary.

abaxial leaf surfaces 318, 320, **387**
Abies balsamea 172
Abies concolor 125, *125*, 132
abscisic acid 171
abscission 341, *342, 343*, **387**
    layer **387**
    zone 341, *342, 343*, **387**
accessory cell *see* subsidiary cell
accessory parts in fruit **387**
accessory transfusion tissue **387**
Acer pseudoplatanus 342
acetylcholine 267
achene 368
acicular crystal **387**
acidic conditions, plant cell growth
    66
acropetal development 13, **387**
actin
    cell elongation 98
    see also F-actin
    microfilaments 48, *50*, 57, 201, 378
actinomorphic **387**
actinostele **387**
actomysin 201
adaptation 1–3
    structural 3–5
adaxial leaf surfaces 318, 320, **387**
adenosine triphosphate (ATP) 45
Adiantum capillis-veneris 100, 331
adnation **387**
adventitious roots 275, 297, **387**
aerenchyma 286, 322, **387**
Aesculus hippocastanum 201, 224
Agathis robusta 165
aggregate rays 205, **387**
Aglaophyton major 2, *2*
albedo **387**
albuminous cells (Strasburger cells)
    220, **388**
aleurone **388**
    grains 53
    layer **388**
algae 1
    green 1, 8
alkaloids 258
allelopathy 266
allium 330
Allium cepa 291

allometry 327
alternation of generations 350
amoeboid tapetum **388**
amyloplasts 38, 53, 301, *303*, **388**
anastomosis **388**
anatomy **388**
androecium **388**
angiosperms **388**
    apical meristems 84, *85, 86*
    axial system 181
    cytokinesis 86
    eusteles 125, *127*
    lateral root initiation 296, *296, 298*
    primary phloem 109
    reproduction *364*, 355–366
    self-incompatibility 377
    sieve tube members 109
angle of divergence 134, 136
anneau initiale (ring initial) 86
annual elements in protoxylem 107
annual rings (growth layers) 181, *182*,
    388
anthers *356, 357, 358*, **389**
anthesis **389**
anthocyanin pigments 47, **389**
anticlinal divisions 166, *167, 169*,
    389
anticlinal planes 84
antipodal cells 360, **389**
apex (*pl.* apices) **389**
apical cell **389**
apical dome 82
apical initials 82, *83, 84*, 277, *277*
    closed 278
    differentiation 14, *15*
    open 278
    organogenic region 86
    roots 277, 276–280
    single 276, *277*
    tunica 86
apocarpy **389**
apomixis **389**
apoplast 20, 58
apposition 64, **389**
aquaporins 289, 375
Arabidopsis thaliana 97, 280, 294, 295,
    307
arbuscules 305

archegonia 353
areole **389**
aril **389**
Artemisia annua 266
artemisinin 266
aspartate 339
Asplenium nidus 150, *151*
astrosclereids 21, 22, *22*
astroscleroid **389**
atactosteles 128, **389**
auxin 295
    cambial activity 171
    cell wall loosening 97, 133
    efflux transporters 97, 295
    leaf abscission 341, 342
    leaf primordia 88, 89, 90, 326
    non-directional transport 96
    polar transport 96, 197, 295, 329
    primary vascular system 116–118,
        134
    root primordia 296
    root tissue patterning 295
    synthesis 96
    systematic name 96
Avena sativa 150
axial system 28, 180, *181*, **390**
    angiosperms 181
    gymnosperms 181

band plasmolysis 284
bark 240, **390**
    see also rhytidome
bars of Sanio *see* crassulae
basal cells 363
basifugal development *see* acropetal
    development
basipetal differentiation 16, **390**
bast fiber **390**
Beta vulgaris 254, *254*
bijugate plants 134
blastozone 326
body cells 354
Boerhaavia diffusa 255, *255*
brachysclereids 22, *22*, **390**
branch gap **390**
branch traces **390**
Brassica oleracea 375
bryophytes 1

buds 12, *13*
  accessory **387**
  axillary **390**
  primordia 13
bulliform cells 139, *142*, 325, *326*, **391**
bundle caps 23, *24*, **391**
bundle sheaths 23, *24*, 320, 338, *338*, 339, **391**
  extensions 320, **391**
bundles
  amphicambral 24, *25*
  amphicribal 105, **388**
  amphivasal 24, *25*, 105, **388**
  axial 122, 129, *130*
  bicollateral 23, *25*, 105, **390**
  collateral 23, *25*, 105, *106, 107, 108*, 319, **392**
  concentric 24
  major axial 128, 129, *129, 130*
  medullary 250, **403**
  minor axial 128, *129*, 130, *130*
  provascular 92, *93*
  vascular 23

calcium carbonate 54
calcium oxalate crystals 53, 54, 92, 258, 259, *260*
callose 34, *34*, 218, 223, *224*, **391**
  definitive 218, *219*, 227
  hydrolysis 227
  platelets 222
  sieve areas 223
callus **391**
Calvin cycle 339
calyptrogen 277, 280, **391**
calyx 356, **391**
cambia, accessory 251
cambial initials 163, 164, *164, 165, 167, 168*, **391**
  submicroscopic structure 171–172
cambial zone 163, *164, 165*
cambium (*pl.* cambia) 163, **391**
  activity 166–170
  armpit 160, *160*
  cytokinesis in fusiform initials 174–175, *175*
  differential cell growth and immature derivatives 175–176
  dormancy and reactivation *173*, 172–174
  hormones and activity 170–171
  interfascicular 154, 155, *157*, **401**
  storied **416**
  structure *164*, 163–166

submicroscopic structure of initials 171–172
  see also vascular cambium
Cananga odorata 337
Cannabis sativa 266
capillaries, interfibrillar 60
capsules 368
carbon dioxide 149, 339
Carica papaya 334
carotenoid pigments 46, 47
carpels *356, 357, 358*, 367, 368, 371, *372*, **391**
  closure 373
  legumes 373
  open *373*
caruncle **391**
caryopsis 368
Casparian band 284, *284*, 285, 286, **391**
cataphylls 316, **391**
cauline vascular bundles 132–133, **391**
caulis **391**
cell **391**
  central 360
cell plate 67, *67, 68*, **391**
cell theory of multicellularity 8, *9*
cell walls 57–58, **391**
  development 66, *67, 68, 69*, 67–71
  growth 64–66
  plasmodesmata *72, 73*, 71–75
  secondary **413**
  structure and composition *59, 62, 63*, 58–64
  thickening
    annular **389**
    scalariform **413**
    scalariform–reticulate **413**
cellulase 342
cellulose 20, 53, **391**
  microfibrils 57, 59, *59, 62, 63, 69*, 199, *200*, 201, 331, *332*
    deposition angle 65
    phloem parenchyma cells 225
  synthase 69, 70
    synthase complexes (rosettes) 69, *70*, 198, 201
central column 72, 281, *282, 283*
central core 376
centrifugal development 115, **392**
centripetal development 115, **392**
Cercidiphyllum japonicum 192
chalaza **392**
Chamelaucium uncinatum 269

channels, intercellular 20, *21*, 286
Charophyceae 1
Chenopodium album 251, *253*
chimera **392**
chlorenchyma **392**
Chlorogalum pomeridianum 276
chlorophyll 1, 46, 47
chloroplasts 46, *46*, 47, 336, **392**
Cholodny–Went hypothesis 302
chromoplasts 46, 47, **392**
cicatrice **392**
1, 8-cineole 266
cinnamic acid 54
cisterna (*pl.* cisternae) **392**
citrus fruits 367
Citrus limon 260, *261*
cladophyll **392**
cleavage polyembryony 354, *355*
closing cells 246
closing layer **392**
coenocytes 263, **392**
cohesion **392**
Cohesion Theory 209
colchicine 201
Coleochaete 1
coleoptile 365, 370, **392**
coleorhiza 365, 370, **392**
Coleus blumei 111, *112*
collenchyma 15, *16, 21*, **392**
  angular 22, **388**
  lacunar 22, **402**
  lamellar 22, **402**
colleters 266, **392**
columella 278, *279*, **392**
  root cap 280, 281–308
commissural vascular bundle **393**
communication systems 9
  trafficking 74
companion cells 34, 35, 220, *220, 231*, 230–234, **393**
  apoplastic movement 231
  collection phloem 230
  derivation 230
  sieve tube–companion cell complexes 230, 233
Compensating Pressure Theory 209, 210
compitum **393**
complementary cells 246
complementary tissue **393**
  see also filling tissue
complex tissue 17
compression wood **393**
conducting tissue *see* vascular tissue

conifers
  axial parenchyma cells 229
  leaf traces 159
  resin ducts 181, *181, 189*, 189–190,
    260, *261*
  tracheids 183
  transfusion tissue 340, *341*
  venation 320
conjunctive tissue 250, **393**
connate **393**
  see also cohesion
contact cell **393**
contact parastichy 135, *135*
contractile roots 275, **393**
  contraction mechanism 276
  development 276
convergent evolution *see* parallel
    evolution
copal **393**
cork cambium *see* phellogen
cork cells 139, **394**
corolla 356, 372, **394**
  tube **394**
corpus 84, **394**
  initials 86
cortex 16, 286, **394**
cotyledons 316, 369, **394**
  primordia 365
p-coumarylic acid 64
crassulae (*sing.* crassula) 185,
    394
cristae (*sing.* crista) 45, **394**
cross-fields 188, **394**
crystallite 60
crystalloids of P-protein 228,
    394
  dispersal 228
cuticle 141, 143, 144, **394**
  epicuticular wax 143, *143*, 144
cuticular pores 143
cutin 98, 141, 143, **394**
cutinization **394**
cycads 1
  root nodules 306
  venation 320, *321*
cyclosis 172, **394**
cystoliths 54, 259, **394**
cytochalasin D 98
cytohistological zonation 86
cytokinesis 86, **394**
  fusiform initials *175*, 174–175
cytokinins 171, 197
cytological zonation **394**
cytoplasm 40, **394**

cytoplasmic annulus 72
cytoskeleton 98–99, 295
  in leaf development *332*, 331–332
  in pollen tube growth 378–379
  in wall development 224–225

decussate phyllotaxy 127, 136, **394**
dedifferentiation 241, **394**
definitive callose 218, *219*, 227
dehiscence 367, 368, 369, **394**
  loculicidal 368
  septicidal 368
deoxyribonucleic acid (DNA) 41
dermal tissue system **394**
dermatogen 278, **395**
desmotubules 72, *72*, 73, *73*, **395**
determinate aspects of plant
    development 12, 81
development **395**
diarch **395**
dichotomous system 320
dicotyledons
  epidermis of roots 146
  non-deciduous 159, *161*
  secondary vascular tissue 131
  secondary xylem *191, 192*, 190–194
  vascular cambium 159
dictyosome **395**
dictyostele 121, *122*, **395**
differentiation **395**
diffuse apotracheal system **395**
diffuse parenchyma 203, *205*
diffuse porous secondary xylem 202,
    203
diffuse porous wood **395**
diploid zygote 350
DNA *see* deoxyribonucleic acid
dormancy and reactivation 172–174
Drimys winteri 132
druses 54, 259, *260*, **395**
duct **395**
  lysigenous 260, *261*
  see also resin ducts, secretory ducts

early wood 181, **395**
eccrinous (eccrine) secretions 258,
    396
ectendomycorrhizae 305
ectomycorrhizae 304
egg apparatus **396**
elaioplasts 53, **396**
elaiosome **396**
electroosmosis 235
elementary fibrils 60

embolisms 187
embryo 350, 354, *355*, 365
embryo sac 359, 360, **396**
  chalazal end 360
  micropylar end *352*, 352, 359
  monosporic 360
  nutrient transport 362
  secretions 376
  sperm entry 361
  tetrasporic 360
embryogeny (embryogenesis) **396**
enation 316, **396**
  theory **396**
  see also microphyll
endarch order of maturation 115,
    396
endocarp 367, **396**
endodermis 16, **396**
  roots 284, *284*, 285
endomembrane system **396**
endomycorrhizae *304*, 305
endoplastic reticulum (ER) *39, 43, 57*,
    396
  cisternae 73
  rough *42*, 44
  smooth *43*, 44
endosperm 361, **396**
endothecium **396**
endothelium (integumentary
    tapetum) **396**
epiblast **396**
epicotyl 365, 369, 370, **396**
  see also plumule
epicuticular wax 143, *143*, 144
epidermal cell patterns 144
epidermis 16, 138, 287, 294, **396**
epipetalous stamen **397**
epistomatous 323
  guard cells 149
  hypostomatous 323
  leaves 323, *325*
  multiple 140, *142*, 288, **405**
  roots 146
  shoots *139, 140*, 141, *141*, 146
  stomata *147*, 147–150
epithelium 189, 259, *261*, **397**
epithem 263, **397**
Equisetum 93, *94*
ER *see* endoplasmic reticulum
ergastic substances 53–54, **397**
essential oils 53, 258, 260, *261*
ethylene 341, 342
etioplasts 47
eukaryote **397**

eusteles 105, *106*, 122, 123, *123*, 124, *124*, *125*, *126*, **397**
  angiosperms 125, *127*
  cauline vs. foliar vascular bundles 132–133
  evolution 132
evergreens 159
exalbuminous seed **397**
exarch 115, **397**
excision layers 245
exine **397**
exocarp 367, **397**
exodermis 286, **397**
expansin 64, 66
  cell wall loosening 66, 97, 332
  control of wall loosening 197
  formation of leaf primordia 88
  leaf primordia 326

F-actin 201, 361
face view 31
false annual ring **397**
fascicle **397**
fascicular cambium 154, 155, *157*, **397**
fats and oils (lipids) 53, 259, *260*
fenestriform pits 188
fertilization **397**
  double 361, **395**
fiber primordia 113
fibers 23, *24*, **397**
  extraxylary **397**
  gelatinous 193, **398**
  libriform 192, *193*, 194, 197, **402**
  septate **414**
  substitute 193
fiber-tracheids 33, 191, **397**
Fibonacci series 136, **397**
fibril **398**
fibrous root systems 275, **398**
Ficus elastica 54, 259
filament *356*, *357*, **398**
filiform apparatus 360, *362*, **398**
filling tissue **398**
  see also complementary tissue
flavonoid pigments 207, 258
floral tube **398**
florigen **398**
flowers 355, *356*, 358, **398**
  complete **393**
  epigynous 370, **397**
  hypogynous 367, **400**
  imperfect **400**
  induction 371, *372*
  irregular **401**

morphogenesis *370*, 370–374
  perfect **407**
follicles 368, **398**
fossil **398**
Frankia 306
free nuclear division **398**
frond **398**
frost hardiness 173
fruits 367, **398**
  accessory 367
    multiple 367
    simple 367
  aggregate accessory 367
  aggregate 367, **387**
  dehiscent 367, 368, 369
  development and distribution 367–369
  indehiscent 367, 368, **400**
  multiple 367
  seed dispersal 369
  simple 367
fundamental tissue **398**
funiculus 361, **398**

gametangium **398**
gametes 350, **398**
gametophytes 350, **398**
generative cells 354
generative spiral 123, 134, 136
generative wall **398**
genetics and cell growth 97–98
genotype **398**
geophytes 275, 276
geotropism **398**
germination 369, 370, **398**
gibberellic acid *see* gibberellins
gibberellins 96, 291, **399**
  cambial activity 171
  cortical microtubule rearrangement 134
  formation of leaf primordia 88
Ginkgo biloba 1, 125, *126*, 132, 351
Gladiolus grandiflorus 276
gland **399**
glandular hairs 266, **399**
glucan chains 70
gluten **399**
glycosides 258
glyoxysome **399**
Golgi bodies *39*, *40*, *42*, 44, *44*, 50, *51*, *52*
  accumulation of precursor compounds 51

dormancy 172
movement 51
Golgi cisternae 44
Golgi vesicles 41, 44, 51, 67, *67*, 68, *68*
graft **399**
grana (*sing.* granum) *46*, 47, **399**
grasses
  guard cells 148, *148*
  leaf development 330, *331*
  pollen 376
  xerophytes *333*, 334
gravitropism 301–302
ground tissue 90, 91, *91*, **399**
growth
  allometric 327
  coordinated **393**
  determinate **395**
  indeterminate **400**
  intercalary **401**
  intrusive 197, **401**
  isometric 327
  primary **410**
  secondary **413**
    anomalous **389**
growth layers 26, 181, *182*, **399**
  see also annual rings
guard cells 147, *147*, **399**
  anisocytic 150, *151*
  anomocytic 150, *151*
  diacytic type 150, *151*
  movement mechanism 149
  paracytic 150, *151*
guttation 263, **399**
gymnosperms **399**
  apical meristems 83, 85
  axial system 181, *181*
  distinctive features of phloem 229–230
  lateral root initiation 296, *296*, *298*
  nitrogen fixation 305
  primary phloem 109
  reproduction 351–355
  secondary vascular tissue 131
  secondary xylem 183–188
  Strasburger cells 233
gynoecium **399**

haplocheilic development 150
hardwood **399**
Hartig net 305
Hatch–Slack pathway 339
haustorium 354, **399**

heartwood 207, **399**
Hedera helix 134
hemicellulose **399**
heteroblastic development 336
heteroblastic series 335
heterocellular rays 204, *206*, **399**
heterogeneous ray tissue system **400**
heterosporous plants 351, **400**
hilum 53, **400**
histamine 267
histogen 278, **400**
holocrine secretions 259
homogeneous ray tissue system **400**
homology **400**
homoplasy 122
homosporous plants 350
hormone **400**
Hyacinthus orientalis 276
hyaloplasm 40
    see also cytoplasm
hydathodes 263, *263*, 264, **400**
Hydrangea paniculata 259
hydroids 3
hydrolysis **400**
hydrophytes 334, **400**
hypocotyl 365, 369, *370*, **400**
hypodermis 301, **400**
hypogeous germination 370
hypogynous flowers 370
hypophysis **400**
hypsophylls 316, **400**

idioblasts 53, 139, 259, *260*, **400**
indeterminate aspects of plant
    development 12, 81
indole-3-acetic acid *see* auxin
initials **401**
    fusiform 164, *165*, 168, 169, *169*,
        170, **398**
        cytokinesis 174–175, *175*
        non-storied 166
        storied 166
    in monocotyledons 277, *277*
    submarginal **416**
integument **401**
integumentary tapetum
    (endothelium) **401**
intercellular space **401**
internodes 15, **401**
interspersed rows 188
intine **401**
intussusception 64, **401**
involution 140
isobilateral leaf **401**

isobilateral mesophyll 322, *324*
isodiametric **401**
isogamy **401**
isotropic **401**

Kappe 279, *279*
karyokinesis **401**
Körper 279, *279*
Körper–Kappe (body–cap) concept
    279, *279*
Kranz 339
    anatomy **401**
    syndrome 339

lacuna (*pl.* lacunae) **401**
lamella **402**
    middle **393**, **404**
lamina **402**
Larix decidua 229
late wood 181, **402**
lateral roots 295–297
    endogenous development 295
    gravitropism 301, *303*
lateral trace 125, *126*
latex 261, *262*, **402**
laticifers 92, 260, *262*, 263, **402**
    articulated 261, **389**
    compound **393**
    non-articulated 261, *262*, **405**
leaf
    buttress **402**
    dorsiventral **395**
    gaps 121, 131, **402**
    primordia 326
        initiation 326, 327
        leaf primordium buttress 326
    sheaths 330, **402**
    traces 120, 123, **402**
        lacunae *131*, 131–132
leaves
    abscission *342*, *343*, 341–343
    autumnal coloration 47
    axils 12, **390**
    categories
        anisophyllous 334
        bifacial **390**
        cataphylls 316
        cotyledons 316
        foliage 316
        hypsophylls 316
        prophylls 316
        shade 334
    development 326–331
        frequency of cell division 330

initiation 326, 327
    primary morphogenesis 326, 327
    secondary morphogenesis 327
evolution 316
midvein 320
photosynthesis and assimilate
    loading 336–338
primordia 13, *14*, 87
    formation *87*, *89*, 88–90
    role of the cytoskeleton 331–332
structure *317*, *318*, *319*, 317–325
    epidermis 323, *325*
structure in relation to function
    336
structure of C_3 and C_4 plants *338*,
    338–339
supporting structures *340*, 339–340
transfusion tissue in conifers *341*,
    340–341
variations in form, structure and
    arrangement *333*, 333–336
see also nodes; phyllotaxy;
        sympodial systems
legumes 368
Lemma minor 223
lensing of light
lenticels 240, *244*, 245–246, **402**
leptoids 3
leucoplasts 46, **402**
life cycle of plants 350–351
light
    lensing
    scattering
lignans 54
lignification **402**
lignin 5, 54, 64, **402**
Lilium longiflorum 378
linear complexes 69
linear tetrad of megaspores 353
lipids *see* fats and oils
lithocysts 259, **402**
locule **402**
Lonicera japonica 268
lumen (*pl.* lumina) 20, **402**
lutoids 263, **402**
Lycopersicon esculentum 223, 226, *226*
lycophytes 272, 316
lysis **402**
lysosome **402**

maceration **403**
macrofibrils 59, *59*, **403**
macrosclereids 22, *22*, **403**
malate 339

marginal rows 188
margo 186, *186*
matrix 60, **403**
Matteuccia struthiopteris 82, 116
Medicago sativa
medulla **403**
megagametogenesis 360
megagametophytes 353, 355, *355*, 359, *361*, **403**
megaphylls 316, **403**
megasporangia 351, **403**
    multicellular 353, *353*
megasporangiate (female) cones 351, *351, 352*
megaspores 351, 354, **403**
megasporocyte (megaspore mother cell) 353, *353*, 359, *360*, **403**
megasporophyll **403**
membrane system, internal 42
meristem 12, 81, **403**
    adaxial **387**
    apical 13–16, 81–88, **389**
    axillary **390**
    cell growth and development 96
    cell shaping by microtubules 99–100
    cytoskeleton and cell growth 98–99
    effect of hormones on growth and development 96–97
    formation of leaf primordia *87, 89*, 88–90
    genetics and cell growth 97–98
    ground **399**
    intercalary *94*, 93–94, **401**
    lateral **402**
    marginal 326, 327, **403**
    permanent 81
    plate **409**
    primary peripheral thickening *95*, 94–95, 248, **410**
    promeristem 82
    residual 92, *93*, **412**
    rib 82, 84, **412**
    role in plant growth and development 81
    self-perpetuating 81
    transitional tissue regions *91*, 90–92
mesarch 115
mesarch order of maturation **403**
mesocarp 367, **403**
mesocotyl **403**
mesogenous development 150
mesomorphic **403**
mesoperigenous development 150

mesophyll 320, *332*, **403**
    amphistomatous 322
    centric **392**
    isobilateral 322, *324*
    palisade 320, *324*
    paraveinal 322, *323*
    spongy 320, *322, 324, 340*
mesophytes 334, **403**
mestome sheath **403**
metabolic compounds, secretion of 257
metaphloem 114, 216, **404**
metaxylem 108, 114, 293, **404**
micelles 60, *62*, **404**
microbodies 48, *48*, **404**
microchannels 144
microfibrils 29, *29*, 65, **404**
microgametophytes 358, **404**
microphyll 316, **404**
    see also enation
micropyle **404**
microsporangia 351, *356, 357, 358, 359*, **404**
microsporangiate (male) cones 351, *351, 352*
microspores 351, 358, **404**
microsporocytes (microspore mother cells) 358, **404**
microsporogenesis 354
microsporophylls *352, 353*, **404**
microtubules 48, *49, 71, 72, 73, 73*, 99, *198, 199, 199, 200*, 331, 332, *332*, 361, **404**
    cell shaping 99–100
    phloem parenchyma cells 225
    pollen tubes 378, 379
mitochondria *39, 40*, 45, *45, 51*, **404**
monocotyledons **404**
    epidermis of roots 146
    initials 277, *277*
    leaf vascular system 329
    outer protective layer 246
    primary peripheral thickening meristems *95*, 94–95
    primary vascular systems 128
    secondary growth *249*, 249–250
    seed germination 370
    zonation pattern 130
monojugate plants 134
morphogenesis **404**
morphology 17, *18*, **404**
    protoplasts *40*, 39–50
    stem primary vascular system 105, *106, 107*

mosses 3
mucilage cells 259, 301, **404**
multicellularity, origins of 8–10
    cell theory 8, *9*
    organismal theory 8, *9*
multiseriate ray cells 205, *206*, **405**
mycorrhizae 302–305, **405**
    vesicular-arbuscular mycorrhizae 305
myosin 201, 361

nacreous wall (nacré wall) 222, **405**
nectar 358
nectaries 268, 358, *358*, **405**
    extrafloral **397**
Nepeta racemosa 266
Nicotiana alata 52
Nicotiana tabacum 359, *360*
Nigella damascens 331
Nitella 48, *50*
nitrogen fixation in root nodules 305–306
nodal diaphragm **405**
nodal structure
    multilacunar 132, **405**
    trilacunar 132
    two trace unilacunar 132
    unilacunar 132
nodes 15, 120, **405**
    anatomy 120
    cauline vs. foliar vascular bundles in eusteles 132–133
    leaf trace lacunae *131*, 131–132
    phyllotaxy 133–136
    structure in pteridophytes *121, 122*, 120–122
nodules **405**
nucellus 353, *353*, 359, **405**
nuclear domains 362, *364*
nuclear envelope 42, **405**
nuclear spindle 48
nucleolus **405**
nucleus 41, **405**
    polar **409**

oils *see* fats and oils
ontogenetic spiral 123
ontogeny **405**
organ **406**
organelles 14, 50–52, **406**
orthostichies 135, 136, **406**
osmiophylic granules 46
Osmunda cinnammomea 82
osteosclereids 22, *22*, **406**

ovaries 357, *358, 360*, **406**
  inferior **401**
ovules *356, 357, 360*, 373, *373*, **406**
ovuliferous scales *352, 353*, **406**
oxaloacetic acid 339

paedomorphosis **406**
palisade mesophyll 320, *324*, **406**
Pandorina 8, *9*
Papaver rhoeos 377
Papaver somniferum 263
paradermal section 339, **406**
parallel evolution (convergent
  evolution) **406**
parastichy 134, *135*, 136, **406**
  contact 135, *135*
paraveinal mesophyll 322, *323*, **406**
parenchyma 15, *16, 18, 19, 20*, **406**
  apotracheal 203, 230–234, **389**
  axial 35, *204, 205*, 221, **389**
    wood 187, *187*
  boundary 203, **390**
  confluent paratracheal 203, **393**
  cells 113, **407**
  interfascicular 92
    paratracheal 203, **388**
  paratracheal 203, **406**
  spongy **415**
  storage 255
  terminal 203, *205*
  vasicentric 203, *205*, **420**
parthenocarpy **407**
passage cells 285, **407**
pathways, radial 289, *290*
pectic substances **407**
pectin 143
pedicel **407**
peduncles 351, **407**
pellicle 287, *287*, 288
peltate hairs 145
perennial **407**
perforation plate 405, **407**
  compound 194, *195, 196*
  reticulate **412**
  scalariform **413**
  simple 194, **415**
perianth 356, **407**
periblem 278
pericarp 367, 368, **407**
periclinal divisions 166, 167
periclinal planes 84, **407**
pericycle 17, *282, 283*, **407**
periderm 5, 27, 138, **407**
  internal 243

structure and development *241*,
  240–243
perigenous development 150
perigyny **407**
peripheral zone 84
periplast 67
perisperm **407**
petals 356, *356*, 371, 372, **407**
petiole **407**
phellem 27, *241*, 242, *242*, 301, **407**
  cell types 242
    thick-walled 242, *242*
    thin-walled *242*, 243
phelloderm 27, 241, *241*, 243, 301, **407**
phellogen *26, 27*, 159, 240, *241*, 301,
  407
  initials 240
phelloid cells 243, **408**
phenotype **408**
phenylpropane 64
phloem 163, **408**
  cells 33–35
  companion cells and Strasburger
    cells 230–234
  distinctive features in
    gymnosperms 229–230
  evolution 215–216
  external **397**
  fibers 35, 221
  internal **401**
  loading 230, **408**
  metaphloem 114
  primary phloem 23, 113, 216, **410**
    cellular composition and
      development patterns *110*,
      109–110
    differentiation 110, *111, 114*
    metaphloem 114
    protophloem 110, 113, 114, *115*
  P-proteins 227–229
  protophloem 110, 113, 114, *115*
  role of cytoskeleton in wall
    development 224–225
  secondary 157, 158, 216, *217*, **414**
    intraxylary 252
    sieve element cell walls *222, 223,
      224*, 221–224
    protoplast 225–227
  structure and development 216–221
  transport mechanism 234–235
3-phosphoglyceric acid 339
photoperiodism **408**
photorespiration **408**
photosynthesis 20, 38, 336–338, **408**

phragmoplast 174, 175, **408**
phragmosome 174, **408**
phyllode **408**
phyllopodium 326
phyllotaxy 123, 133–136, **408**
  distichous 127, 136, **395**
  helical 134
  patterns 88
phylogeny **408**
phytochrome **408**
phytotoxins 266
Picea abies 354, 379
PIN efflux transporters 96, 295
PIN genes 97
pinocytosis **408**
Pinus strobus 188
Pinus virginiana 353
Piper betle 250, *252*
Piper excelsum 250
pistillate **408**
pistils *356, 357, 358*, 372, **408**
  compound 358
  interactions with pollen *374*,
    374–377
  simple 358
Pisum sativum 200, 307, 329
pit 30, *30*, **408**
  aperture 31, *31*, 185, *186*, **408**
  aspirated **389**
  blind **390**
  border 31, *31*, 188, **390**
  boundary 297
  canal 185, *186*, **408**
  cavities 31, *31*, 186, **408**
  circular-bordered pits 183, *183*, **392**
  fields, primary 30, *30*, 166, **410**
  membrane 30, *31, 32*, 186, *186*, **408**
  ramiform **414**
  simple 18, *20*, 30, **415**
  vestured **420**
pith **409**
pit-pairs 18, *20*, **409**
  bordered 31, 183, *184, 184*, 185, *185,
    186, 188*, **390**
  half-bordered 32
  simple 23, **415**
pitted elements 108
pitting
  alternate **388**
  intervascular **401**
  opposite **406**
  scalariform **413**
placenta **409**
plagiogravitropism 275

plagiotropic shoots 334
plasmalemma (plasma membrane) 40, *40*, 41, **409**
 orifice 73
plasmodesmata 8, *10*, 30, 38, 39, 57, 72, *73*, 71–75, **409**
 central cylinder 72, *72*, 73, *73*, *392*
 differentiation 86
 primary 73, *74*
 secondary 73, *74*
 size exclusion limit (SEL) 74, 75, 86
plasmodesmata–pore connections 231
plasmodesmograms 233, *234*
Plasmodium falciparum 266
plastids 46, 47, 172, 229, **409**
plastochron 88, 135, **409**
plerome 278, **409**
plumule **409**
 see also epicotyl
polar auxin transport 295
pollen 357, **409**
 adhesive foot 375
 dehydration 374
 grains 354, *354*, 358, 361, *363*, **409**
 calcium synthesis 377
 grass 376
 interactions with pistils *374*, 374–377
 sac **409**
 tubes 354, *354*, 361, *363*, 376, *376*, **409**
 role of cytoskeleton in growth 378–379
pollination **409**
polyarch **409**
polyderm **409**
polyembryony 354, **409**
polygalacturonase 342
polymer **409**
polymerization **409**
polyribosomes 42, **409**
polysaccharides **409**
polystelic structure 254
pore multiple **410**
pores 217, 223, **410**
 differentiation 229
 plasmodesmata–pore connections 231
 solitary **415**
potassium ions 149
Potentilla 126
Potentilla fruticosa 132
P-protein (phloem protein) *224*, 226, *228*, **410**

absence in gymnosperms 229
 bodies 225, 227
 crystalloids 228
 nature and function 227–229
 sieve tube conductivity 227, 229
preprophase **410**
pressure flow hypothesis 234
primary body 25, **410**
primary tissues 16–23, *282*, *283*, 281–289, **389**
 development *292*, 291–295
 differentiation of mature tissues *293*, *293*
 differentiation of pith 293
 endodermis 284, *284*, 285
 pericycle *282*, 283
 primary xylem 281, *282*, *283*
primordium (*pl.* primordia) **410**
procambium *see* provascular tissue
proembryo 354, *355*, 365, **410**
prokaryotes **410**
prolamella bodies 47
promeristem 82, 278, **410**
promesophyll 328, *329*
prop roots **410**
prophylls 316, **411**
proplastid **411**
prostele *282*
protective layer 342, **411**
protein synthesis 41, 47
prothallial cell **411**
protoderm 90, *91*, 287, *287*, 294, **411**
protophloem 110, 113, 114, *115*, 216, **411**
protoplasm **411**
protoplast **411**
protoplasts of eukaryotic cells *39*, 38–39
 ergastic substances 53–54
 morphology *40*, 39–50
 movement of organelles 50–52
protosteles 105, *106*, 120, *121*, **411**
protostelic roots 293
protoxylem 107, 108, 110, 111, *111*, *112*, *113*, 114, *115*, 293, **411**
 helical elements 107
 lacunae 109, **411**
 poles **411**
provascular strands 92, *93*
provascular tissue 90, *91*, 105, *106*, *107*, 328, *329*, **411**
Pteridium aquilinum 82
pteridophytes
 adventitious roots 297

lateral root initiation 295
 nodal structure *121*, *122*, 120–122
 primary phloem 109
 reproduction 3
 scalariform elements 108
 spore sizes 350
 sporophytes 350
pulvinus **411**

quantasomes **411**
Quercus suber 285
quiescent center 278, *278*, 280, 281, *283*, **411**
 role in root development 280–281
 root cap 280, 281

rachis **411**
radial section 17, *19*, **411**
radial system 28, 180, *181*
radicle 365, 369, 370, **411**
raffinose
raphe **411**
raphides 53, 259, **412**
ray cell
 biseriate 205, *206*, **390**
 derivatives 167
 initials 166, 167, *169*, **412**
 procumbent 204, *206*, **410**
 uniseriate 205, *206*, **419**
 upright 204, *206*, **419**
ray initials 164, *165*, 166, 169, *169*, 170, **412**
 multiseriate 170
ray parenchyma 182
ray parenchyma cells 35, 187, 188
ray tracheids 183, 188, **412**
rays 28, 182, **412**
 homocellular 204, **400**
 medullary 251, **403**
reaction wood 193, **412**
receptacle **412**
recognition locus 377
release phloem companion cells 230
reproduction in plants 350–351
 angiosperms *364*, 355–366
 development and distribution of fruits 367–369
 floral morphogenesis *370*, 370–374
 gymnosperms 351–355
 pollen–pistil interactions *374*, 374–377
 role of cytoskeleton in pollen tube growth 378–379
 seed germination 369–370

seedling development 369–370
self-incompatibility 377–378
resin ducts 181, *181*, 187, *189*, 189–190, **412**
respiration 20
reticulate cell wall thickening **412**
reticulate elements 108
retting **412**
Rhizobium 306
rhizoderms 138, **412**
rhizoids 2
rhizosheaths 286
rhytidome 138, 159, 240, 301, **412**
    formation *245*, 243–245, *246*
ribonucleic acid (RNA) 41
ribosomes 41, 47, **412**
ring initial (anneau initiale) 86
ring porous wood **412**
RNA *see* ribonucleic acid
Robinia pseudoacacia 165
root cap 13, 274, 365, **413**
    columella 280, 281–308
    function and role in gravitropism
        301–302
    quiescent center 280, 281
root contraction 276
root hair 146, 288, 294
    cells 294, **413**
    zone 288, *288*
roots *273*, 272–274
    adventitious 297
    auxin and tissue patterning 295
    cluster 275
    epidermis 146
    gravitropism 301
    highly specialized systems 275–276
        cluster roots 275
        contractile roots 275
    lateral root development 295–297
        angiosperms 296, *296*, 298
        gymnosperms 296, *296*, 298
        pteridophytes 295
    lateral transport of water and
        minerals 289–291
    morphology 274
        adventitious roots 275
        fibrous root systems 275
        tap roots 274
    mycorrhizae 302–305
    nitrogen fixation in root nodules
        305–306
    quiescent center in development
        280–281
    root–stem transition 306, *307*

secondary growth *299*, *300*, 299–301
    cambium 299
    hypodermis 301
    unusual features 248, *251*, *252*, *253*,
        250–256
    see also primary tissues; radicle
rosettes (cellulose synthase
        complexes) 41, 69, *70*, 198, 201
rotated-lamina syndrome 334, *335*
rows
    marginal 188
Rudbeckia laciniata 337

S1 layer 58, *59*, *60*
S2 layer 58, *59*, *60*
S3 layer 58, *59*, *60*
salt glands 267, *267*, 268
samaras 368, **413**
Sambucus nigra 342
sap wood 207, **413**
scalariform elements 108
schizogenous ducts (cavities) 259, **413**
sclereids 22, *22*, **413**
    astrosclereids 22, *22*
    brachysclereids 22, *22*
    macrosclereids 22, *22*
    osteosclereids 22, *22*
    trichosclereids 22, 23
sclerenchyma 15, *16*, 22, 23, 24, **413**
sclerification **413**
scutellum 365, **413**
secondary body **413**
secondary tissues 17, 26, **414**
    effect of growth on leaf and
        branch traces *160*, 159–162
    effect of growth on primary body
        158, 157–159
    vascular cambium *155*, *156*, 154–156
secretions 257
    external secretory structures *263*,
        263–269
    granulocrine 258, **399**
    internal secretory structures *260*,
        259–263
    mechanisms of secretion 258–259
    non-metabolic compounds 258
    substances secreted 257–258
secretory cavity **414**
secretory cells 259, *261*, 269, **414**
secretory ducts 181, **414**
secretory structure **414**
seed coat **414**
seed dispersal 369
seedlings 369–370

seeds 357, 369–370
    albuminous **388**
self-incompatibility 377–378
sepals 356, *356*, 371, **414**
septate 192
septum (*pl.* septa) **414**
Sequoia sempervirens 188
seriation, radial **411**
sheath **414**
sheathing base **414**
shoot system of vascular plants 10, *11*,
    12
    epidermis *139*, *140*, 141, *141*, 146
    shoot apex 81, 82, *83*, *84*
short day length stress 171
sieve areas 33, *34*, 217, *218*, *219*, **414**
    gymnosperms 229
    sieve element cell walls *222*, *223*,
        224, 221–224
    sieve element protoplast 225–227
sieve cells 33, *34*, 216, *218*, **414**
sieve elements 216, *218*, **414**
sieve plates *218*, *219*, 226, 227, **414**
    compound 34, *34*, 220, **393**
    reticulate **412**
    scalariform **413**
    simple 34, *34*, 218, *219*, **415**
sieve pores *34*, *34*, *223*, 229
sieve tube–companion cell complexes
    230, 233
sieve tubes 34, *34*, **414**
    control of conductivity 227, 229
    members 34, *34*, 109, 217, *218*, *219*,
        220, 226, **415**
    thick-walled 232
signal transduction 4
silica 54
    bodies 54
    cells 138–139, **415**
siphonosteles 105, *106*, 121, *121*, **388**,
    396, **415**
size exclusion limit (SEL) 74, 75, 86
slime **415**
    body **415**
    plug **415**
softwood **415**
Sorghum bicolor 144
spermatozoids 1
sperms *354*, 358, **415**
    isomorphic 359, *360*, **401**
    non-motile 361
    nuclei 354
    twin 359, *360*
spherosomes 47, 53, 171, **415**

spindle fibers **415**
sporangium **415**
spore 2, **415**
  haploid 350
sporocytes 350
sporophyll **415**
sporophytes 350, **415**
  diploid phase 350
  pteridophytes 350
sporopollenin 1, 4, **415**
spring wood *see* early wood
stamens *356, 357, 358,* 371, **415**
starch 53, 172, **415**
  grains 53, 171
statocytes 301, 302
statoliths 301, *303*
stele 117, *117,* **415**
stellate **415**
stems
  differentiation of primary vascular tissues *111,* 110–115
  morphology and development of primary vascular system 105, *106, 107*
  primary phloem 109–110
  primary tissues 16–23
  primary xylem 106–109
  role of auxin in primary vascular system development 116–118
  root–stem transition 306, *307*
  unusual features 248
    anomalous structures *251, 252, 253,* 250–256
    primary peripheral thickening meristem 248
    secondary growth in monocotyledons *249,* 249–250
stigma 357, *357,* 361, *363,* **415**
stigmatic surface 374, *374, 375*
  dry 374
  wet 374
stoma (*pl.* stomata) 4, *147,* 147–149, **389, 416**
  anisocytic **388**
  development 149–150
  diacytic **395**
  paracytic **406**
stomatal aperture 147, *147*
stomatal complex **399, 416**
stomatal crypts 325, **416**
stomium **416**
stone cell *see* brachysclereid
"stop" signal 52

Strasburger cells 35, 215, 221, *221,* 230–234, **416**
strobilus **416**
stroma *46, 47,* **416**
structures of plants
  adaptations 3–5
  apical meristem 13–16
  determinate aspects of plant development 12, 81
  indeterminate aspects of plant development 12, 81
  origin of multicellularity 8–10
    cell theory 8, *9*
    organismal theory 8, *9*
  phloem cells 33–35
  primary tissues of stems and roots 16–23
  secondary growth 25–28
  shoot system of vascular plants *11, 12,* 10–12
  vascular bundle types 23–24, *25*
  xylem cells 28–33
stylar canal 376
style 357, *357,* **416**
suberin 5, 208, **416**
  lamellae 284, **416**
suberization 246, **416**
subsidiary cell **416**
substomatal chamber 148, *148*
sucrose 171
summer wood *see* late wood
sun leaves 334
superior ovary *see* hypogyny
suspensor 365, **416**
symbiosis 274, 305, **416**
symmetry
  bilateral **390**
  radial 370, **411**
  zygomorphic 371, **420**
symplast 9, 57, **417**
symplastic domains 85, 206
symplastic growth 176, **417**
sympodia 122, **417**
sympodial systems 120, 122–131
synapylic acid 64
syncarpy **417**
syndetocheily 150, **417**
synergids 360, 361, *361, 362,* **417**
syngamy **417**

Tamarix aphylla 267, 268
tangential section 17, *19,* **417**
tannins and tanniniferous substances 54, 258, 259, *260,* **417**

tap roots 274, 301, **417**
tapetal wall 354
tapetum *359,* **417**
taxon (*pl.* taxa) **417**
telome 133, 316, **417**
template **417**
tension wood **417**
tepal **417**
Teratophyllum rotundifoliatum 325, 334
terpenes 258
terpenoids 265
testa 365, **417**
tetrahydrocannabinol 267
tetrarch **417**
thylakoids 47, **417**
tissue **418**
tissue regions 16
tissue system **418**
tonoplast 49, 225, **418**
torus 186, *186,* 187, **418**
trabecula (*pl.* trabeculae) **418**
trace
  median 125, *126*
tracheary elements 28, 195–202, **418**
tracheids 29, *29,* 190, **390, 418**
trafficking 74
transection **418**
transfer cells 35, 41, 120, 258, **418**
transfusion tissue **418**
transition region 84, 90, *91,* **418**
transmitting tissue 376, **418**
transpiration 209
transport phloem companion cells 230
transverse section 17, *19,* **418**
traumatic resin duct **418**
triarch **418**
trichoblasts 146, 294, **418**
trichomes 144, 145, 146, **418**
  glandular 264, *264,* 265, 266
  peltate 264, *265,* **407**
trichosclereids *22, 23,* **418**
triglycerides 376
trilacunar nodal structure 132, **418**
triple fusion **418**
Triticum aestivum 331
tropism **418**
Tsuga canadensis 229
tube cell **419**
tubulin 48
tunica 84, 86, **419**
  tunica-corpus concept 84, **419**
turgor pressure 14
tyloses 180, *208,* 207–209, **419**

unbranched hairs 145, *145*
unit membrane 41, **419**
Urtica dioica 266, 267

vacuolar membrane *see* tonoplast
vacuolation 49, **419**
vacuome **419**
vascular bundles 16, 23, *24*, **419**
 amphicambral 24, 105
 amphivasal 24, 105
 bicollateral 23, 105
 collateral 23, 105, *106, 107, 108*
 concentric 24, **393**
 in eusteles 132–133
vascular cambium 26, *26, 27, 28, 155, 156,* 154–156, **419**
 see also cambium
vascular parenchyma cells 231, *232*
vascular plants 1
 shoot system *11, 12,* 10–12
vascular rays 187, **420**
vascular systems 127, 128, **420**
 closed 127, 128
vascular tissues 16, **420**
 primary **410**
 secondary **414**
velamen 140, 288, **420**
venation **392**, 406, **420**
 dichotomous **395**
 morphology 327, *328*
 parallel 320, **406**
 reticulate 320, *321,* **412**
verticulate phyllotaxy 127
vesiculate hairs 146

vessel **420**
vessel members 32, *32,* 193, 194, *194,* **420**
Vicia faba 223, 226
Vinca 372
Volvox 8

wall
 primary 29, *29,* 58, *59, 61,* 64, 66, **410**
 secondary 29, *29,* 58, *59, 61,* 66
wall labyrinth 259
wall thickening
 helical **399**
 tertiary 185
waste metabolites 53
water stress 171
water transport mechanism in secondary xylem 209–210
water vesicle **420**
waxes 98, 143
 epicuticular 143, *143,* 144
 soluble 141
whiplash flagella 1
whorl **420**
wood 163, **420**
 fibers 33
 parenchyma 33
 porous **410**
 storied **416**
wound periderm **420**

xeromorphic plants 322, **420**
xerophytes 333, *333,* 338, **420**
xylem **420**
 cells 28–33

elements **420**
initial **420**
metaxylem 108, 114
primary 23, 110, 112, *112,* 114, *115,* 216, **410**
 cellular composition and development patterns 106–109
 differentiation 110, *111,* 114
 metaxylem 108, 114
 protoxylem 107, 110, 111, *111, 113,* 114, *115*
 roots 281, *282, 283*
protoxylem 107, 108, 109, 110, 111, *111, 113,* 114, *115*
secondary 157, 158, 180, **414**
 dicotyledons *191, 192,* 190–194
 differentiation of tracheary elements 195–202
 diffuse–porous 202
 distribution patterns of xylary elements and rays 202–207
 gymnosperms 183–188
 ring-porous 202, *203*
 structure *181,* 180–183
 tyloses *208,* 207–209
 water transport mechanism 209–210
xylem pressure probe 209
xylem ray **420**
xyloglucan chains 61, *62*

Zea mays 281, 287, 289, 291, 337, 370
 embryo *366*
zosterophyllophytes 272
zygote 350, 354, 361, 363, **420**